D1823874

India Studies in Business and Economics

The Indian economy is considered to be one of the fastest growing economies of the world with India amongst the most important G-20 economies. Ever since the Indian economy made its presence felt on the global platform, the research community is now even more interested in studying and analyzing what India has to offer. This series aims to bring forth the latest studies and research about India from the areas of economics, business, and management science. The titles featured in this series will present rigorous empirical research, often accompanied by policy recommendations, evoke and evaluate various aspects of the economy and the business and management landscape in India, with a special focus on India's relationship with the world in terms of business and trade.

More information about this series at http://www.springer.com/series/11234

S. P. Mukherjee

Quality

Domains and Dimensions

 Springer

S. P. Mukherjee
Department of Statistics
University of Calcutta
Howrah, West Bengal, India

ISSN 2198-0012   ISSN 2198-0020   (electronic)
India Studies in Business and Economics
ISBN 978-981-13-1270-0   ISBN 978-981-13-1271-7   (eBook)
https://doi.org/10.1007/978-981-13-1271-7

Library of Congress Control Number: 2018946581

Printed on acid-free paper

This Springer imprint is published by the registered company Springer Nature Singapore Pte Ltd.
The registered company address is: 152 Beach Road, #21-01/04 Gateway East, Singapore 189721,
Singapore

Dedicated to the memories of
My father
Late Tara Prasad Mukherjee
who moulded my academic life
And My Teacher
Late Purnendu Kumar Bose
who initiated me to the field of Quality

Preface

Quality is not just a buzzword that generally begets a lip service. It is today a serious concern to all individuals and institutions who realize the all-important role that quality plays in the economic development of a country. Quality today is no longer confined to the manufacturing arena alone or even just the field of public services. Quality considerations now embrace a wide array of entities, both concrete and abstract. Managing quality now does involve policies, plans and practices guided by a congenial quality philosophy.

Quality is transnational—quality of goods and services has a place and fetches a good price everywhere in the world, except when fettered by trade barriers and other extraneous hindrances.

A National Quality Movement needs guidance of policy-makers, leadership from competent preachers, professionals and practitioners and patronage from state and industry and, above all, support and co-operation from the community at large.

In fact, demand for better quality in all that we acquire for our current or eventual consumption, whoever may be the suppliers—large, medium or small manufacturing units in public or private sectors as also public and private service providers—is an essential prerequisite for quality improvement, particularly in situations where quality is not built in suo moto but is infused on demand or insistence.

Quantification of concepts and measures followed by quantitative analysis of situations leading to inferences with controlled uncertainty has permeated almost all spheres of scientific investigations . No wonder, lots of mathematical and statistical methods and tools have found useful applications in the context of quality improvement, starting from quality planning and analysis.

It must be remembered, however, that to push the frontiers of quality movement across people with different profiles, a clear and holistic comprehension of quality is more important than a detailed quantitative approach to analysis and control. In fact, implementation of a Quality System in any organization has to be dovetailed into the structure and even culture of the organization. And for this, a very important consideration is the role of quality in business. Depending on the business in which the organization is engaged, we have to weave quality into the pores of that business. Detailing the requirements of a Quality System calls for an appreciation of the processes carried out and products and services turned out. Thus, a discussion on economics of quality should be an integral component of any discourse on quality.

We live in an era where we have come to realize the necessity to minimize the consumption of materials as also energy for producing goods and services to meet the growing human needs for better and more varied products. One crucial step in this direction is provided by innovation. We have to think systematically about development and deployment of new materials, new processes, new designs, new control mechanisms and new products, so that by consuming less energy and materials we can meet the ever-growing needs of humans. New product development to build in multi-functionality and flexibility may have a pulling-down effect on the diversity and multiplicity of products and services and may imply lesser emissions, wastes and losses.

The present volume embodies a modest attempt to convey such a holistic and comprehensive view of quality and its increasing ramifications in various spheres of human activity to suit respective special features and requirements.

It is universally accepted that people who are adequately motivated and committed to work for improvement in all they do are the backbone of all improvement efforts. And unless the Quality of Working Life in all its implications—physical and psychological—is upto the mark, quality-related activities will suffer and outputs of quality improvement exercises will fall short of the targets. In fact, Quality of Life for all of us—producers and consumers alike—should get due recognition in the ambit of Quality Management.

Demand for better quality in all that we need to live decent lives and corresponding aptitude to build quality in whatever we do will depend a lot on the quality of education received by all stakeholders. Quality education should not only equip the recipients with adequate knowledge and skills to perform relevant tasks, but should also create right attitudes and positive thinking as well as the spirit of appreciative enquiry. Equally important is the quality of research and development taking place in both publicly funded and private institutions. We have academic institutions in the field of higher education where teachers and scholars are engaged in research activities, besides dedicated laboratories to carry out both research and development activities to enhance the national income and improve the quality of life.

The role of information in making right decisions and taking appropriate actions in any organization needs no emphasis. And information industry has become a strategically important service industry. Thus, information defines an emerging domain for a discussion on quality. Information integrity brings a new concept of quality, and international standards on different aspects of information quality have entered the arena of Quality Management.

Quality is important in the IT and ITES sectors, and IT itself can provide a big support to the quality improvement process. IT is ushering in unheard—of changes in workplaces—in designs, in processes, in controls and support services. Use of relevant though sophisticated tools for analysis involving huge volumes of data captured through automation in various processes would have been impossible without the use of softwares which have the required capabilities. And software quality analysis takes due advantage of concepts, measures, methods and techniques which have emerged over the years in the arena of Quality Management.

Reliability is best understood as quality of performance that can be assessed only after the product has been deployed or put to use. And this quality depends on both the quality of design—the most important determinant of quality of an entity—and the quality of conformance (to design requirements) during manufacture. That way, 'good quality' at the end of production is necessary but not sufficient for the ultimate quality of a product as perceived by the user. Reliability analysis for different types of products depending on the pattern of their use and the possibility or otherwise of a failed item to be repaired has grown to be a full-fledged subject today, especially because of the strategic and economic consequences of unreliability. Any published material on quality is regarded as incomplete if it fails to make at least a brief reference to reliability.

Statistical concepts, methods and techniques along with sofwares to facilitate their use have become a major component of any quality analysis and improvement exercise. Here again, more important than knowing the detailed derivation of procedures and corresponding computations are the understanding of the purpose, the output and the limitation of any statistical procedure. A broad idea without mathematical formulae and computational details may appeal to those who think about the world of quality.

The present volume is not intended to be a textbook on Quality or Quality Management. It is meant to convey a broad understanding of Quality pervading different spheres in a manner that is mostly non-quantitative, except in places where quantification should not be avoided. Due emphasis has been placed on services and not merely on manufactured products. The materials presented are mostly reflections of the author's own realizations and writings, though some materials contained in some books or reports or documents have also been used with due reference. Not meant to be a handbook for quality professionals in manufacturing and service industries, the book does not contain case studies. However, examples have been provided in many domains and illustrative calculations have been included in a few cases.

The absence of a volume that, in the perception of the author, conveys a comprehensive concept of quality in any context and can provide some broad guidance for improving the present state of affairs in a wide diversity of human activities motivated the present author to take up the challenging task of filling this void.

Howrah, India
September 2017

S. P. Mukherjee

Acknowledgements

Exhortation, encouragement and entreaties from a host of well-wishers and students to write a compendium on Quality, based on the many articles I had written over the years, the lectures I had delivered, and the experiences I had gained during my involvement as the leader of a team from the Indian Association for Productivity, Quality and Reliability to render advisory services to some industries, have eventually resulted in this volume. I remain grateful to all of them and I hasten to add that in an attempt to prepare the material in a less-than-adequate period of time, I cannot claim to have fully met their expectations.

I like to record my sincere appreciation of the help received from Sri Subrata Ghosh, Sri Sudipta Chatterjee and Dr. Aniruddha Banerjee for their perusal of several chapters and for offering some suggestions for improvement in content and presentation.

To the authors of many books and articles published in a wide variety of journals which have enriched my thoughts and ideas on the topics covered in this volume, I remain sincerely indebted. I have freely quoted them in some parts of the text where their writings have become too well known, of course with due reference.

Thanks are due to Prof. Cathal Brugha of the School of Management, University of Dublin, for partnering with me in a collaborative study of Quality Management in some industries in India.

The manuscript was essentially typed by me, and Sri Dipankar Chatterjee helped me a lot in preparing some portions and in formatting the entire typescript. I thankfully acknowledge his help and support.

I should not hesitate to admit that only a persistent persuasion by my wife Reba and my sons Chandrajit and Indrajit could transform the idea of writing a book into a reality. My wife gave me complete relief from many household duties during the last one year and meticulously attended to my health problems, and my sons enquired about the progress through overseas calls quite frequently.

Contents

About the Author

S. P. Mukherjee retired as the centenary professor of statistics at the University of Calcutta, where he was involved in teaching, research and promotional work in the areas of statistics and operational research. He was the founder secretary of the Indian Association for Productivity, Quality and Reliability and is now its mentor. He received the Eminent Teacher Award from the Calcutta University, P.C. Mahalanobis Birth Centenary Award from the Indian Science Congress Association and Sukhatme Memorial Award for Senior Statisticians from the Government of India.

Chapter 1
Quality: A Big Canvas

1.1 Introduction

The history of Quality is as old as that of production. Whatever was produced—through a manufacturing process involving machines or simply through manual operations—had some level of Quality built into it, essentially in terms of skills in designing, producing and checking conformity with the design. This level in some cases was quite high and in some others pretty low. Taken somewhat simplistically as the collection of features of the product which could meet some use requirements or had some aesthetic or prestige value or had a long life or was user-friendly, a product and its quality were and are still now inseparable. Of course, with emphasis gradually growing on differentiation among products in terms of their quality, a 'quality product' has come to mean a product with a high level of quality.

The panorama of Quality has evolved through many paradigms (new ones not to be always branded as 'old wine put in a new bottle') encompassing the history of production, starting with those who produce (and obviously design) taking all care of quality in a phase referred to as 'operator Quality Control', through inspectors assessing quality subjectively though dispassionately (in a phase branded as 'inspector Quality Control'), Statistical Quality Control providing a more efficient anatomy of the production process, followed by Total Quality Control or Company-Wide Quality Control with a clarion call for involvement of all to boost quality, Total Quality Management throwing up a more humane but comprehensive and customer-oriented approach, to many other brands like right first time, zero-defect, Six Sigma, and what not.

The story of quality has been unfolded extremely rapidly in recent times in consonance with the amazingly fast developments in Science and Technology, society, economy and even polity. Now there are engineers, statisticians, management experts and similarly branded experts recognized as 'Quality Professionals'. Quantitative tools (essentially statistical)—often quite sophisticated —are being widely used for quality and reliability assessment as also improvement

© Springer Nature Singapore Pte Ltd. 2019
S. P. Mukherjee, *Quality*, India Studies in Business and Economics,
https://doi.org/10.1007/978-981-13-1271-7_1

exercises. At the same time, roles of design engineering, of customer voice analysis and customer satisfaction measurement, of methods and practices to enhance people's involvement in quality efforts are also being accorded good priority.

Everyone talks about quality—about quality of diverse entities like food and fodder, safety locks and entertainment electronics, communication devices, health care, administration, primary education, banking services, news reporting, environment, and what not. Even an article submitted for possible publication in a peer-reviewed journal is assessed by the referee(s) for the quality of its content as well as of its presentation.

All of us are concerned about quality—at least in regard to goods and services consumed by us and environment for life and work given to us. If not, in relation to goods and services produced/rendered by us and environment created by us for others.

Any discussion on quality has to reckon with the fact no one can say the last word on quality. Perceptions about quality, standards bearing on quality, quality practices, and even quality ethics reveal a lot of diversity over the expanding spectrum of stakeholders and are changing fast, keeping pace with developments in Science and Technology. However, to our relief stand out certain realizations about quality shared by professionals and philosophers alike. A few of these are mentioned below.

Quality is not a goal—it is a march forward. Even zero-defect is not the ultimate to be achieved. Since we can always have a more stringent definition of defect, scope for improvement always remains—in materials, processes, procedures and systems.

Quality is not a freak—it is the outcome of a deliberate, planned and systematic effort. Such an effort has to be deliberately taken up by the producer. This will or, rather, zeal to produce goods of quality has to be translated into a policy (preferably documented). To put policy into action, a quality planning exercise has to be undertaken. Level of quality to be achieved and the way to achieve these should be clearly spelt out in necessary.

In a somewhat controversial article entitled 'Improved Quality Tactic', Budyansky (2009) argued that the definitions and simplified methodologies for quality articulated over the years have created a 'vague' situation. He mentions five principal 'vagueness factors', viz. (1) absence of a common and generally accepted definition of quality (2) absence of effective and well-defined rules of measurement and quality evaluation (3) absence of a methodology to go from physical parameters to quality parameters (4) absence of a of a methodologies for fulfilling a quasi-objective analysis and synthesis of quality and (5) absence of a common technology for quality assurance. He feels that the vagueness around quality is analogous to vagueness about concepts like' freedom' or 'democracy' or 'illness'.

The point to be remembered is that quality corresponds to a big canvas that combines richness with complexity and deserves a multi-pronged discussion, right from the philosophical angle to the mundane consideration of business interests.

1.2   Ubiquity of Quality

Quality is all pervasive, residing in all tangible and concrete forms and figures, processes and products. Quality has been a reflection of human civilization of progress in Science and Technology, of an ever-growing demand by human society for more of better goods and services, and a consequent effort to satisfy the demand. The quest or craving or passion for quality had a rather recent entry in the arena of industry or more generally the arena of organized human activities, though it had been too long an engagement in the field of fine arts and also in the behaviours and actions of even lower organisms.

Quality is associated with everything around us, judged in terms of features or characteristics or parameters of the particular entity that we consider. Thus, quality is often discussed in respect of the following entities.

* *Raw Materials*—like coal raised from mines, water flowing along a river, air that we inhale, fruits that are plucked from trees and plants. We even talk about quality of soft materials like raw data arising from a given source. In the case of coal, quality is judged in terms of properties like dust content or useful heat value; water quality is assessed in terms of dissolved oxygen, total dissolved solids, pH value, coliform count, ratio between biological oxygen demand to chemical oxygen demand and a few other parameters. Adequacy, accuracy, timeliness, etc. could characterize quality of data.

* *Processes (including process technologies)*—like machining of a metal, calibration of measuring equipment, auditing, developing a quality plan, communication of a message, etc. Quality of any such process can be judged in terms of rotation per minute of a machine, feed rate of input materials, duration of a conversion operation, yield and the like—all compared against specified norms or standards.

* *Processed Materials*—like washed coal, treated water, boiled fruits, spun fibre, alloy steel, tabulated data, etc. For spun fibre, we may use tensile strength as an important quality parameter. In many cases, change in quality parameters of corresponding raw materials during the process provide good ideas about quality of processed materials.

* *Products (including augmented and extended products)*—like a lathe machine, an item of furniture, cooked food, a steel utensil, a data-based report, and the like. Quality parameters are given out in terms of fitness for use requirements, value derived by the user against the price paid, manufacturing cost, costs of usage, maintenance, disposal and similar other features.

* *Services (loaded with products to varying extents)*—like transportation, marketing, health care, recreation or entertainment, education, security, etc. Quality is judged in terms of both measurable properties as also subjective perceptions.

[Products and processed materials have to be carefully distinguished in terms of use, viz. a product for final consumption and a processed material for as an intermediate. Further, services may be heavily loaded with products, the product content may be moderate or the service may not involve much of a product content.]

* *Work Environment*—like the physical facilities for work, security, safety, hygiene, etc. where the most important quality parameter could be adequacy and availability.

* *Environment*—covering humans, flora and fauna, natural resources. Here diversity and abundance are quite important to discuss quality. Aspects like pollution, natural resource depletion or conservation become relevant in this context.

We also started talking about quality of life covering various aspects of individual and public life. We speak of quality of measurements, of information, of communication and many other entities. In an interesting article entitled 'The Golden Integral Quality Approach from Management of Quality to Quality of Management' Galetto (1999) wrote about quality of management in terms of intellectual honesty, holistic cooperation, quality integrity and scientific approach to decision-making.

Concerns for quality have been felt and expressed even in spheres far removed from the world of goods and services. Among the six plenary talks in the 2017 Stu Hunter Research Conference in Denmark attended by 60 participants from 18 countries in America, Europe and Asia, one was on *Quality and Statistical Thinking in a Parliament and Beyond by P. Saraiva*. The speaker and the three discussants considered such matters as choice of issues taken up by the legislators for discussion and debate, the use of facts and figures, the credibility of figures rolled out, the use of logical reasoning and the outcomes of discussions.

Except for the case of raw materials, where quality is being continuously altered by anthropogenic activities which affect ecology, quality is the outcome of a planned, deliberate and systematic effort put up by an individual or an organization.

1.3 Quality Variously Perceived

It is only expected that quality will be perceived differently by different persons from different angles or perspectives. Such perceptions vary widely from abstract, philosophical ones to concrete, quantifiable ones. We can possibly start from *Quality is beautility* to imply that quality combines beauty with utility—beauty that is eulogized and utility that is demanded. We are reminded of what a great poet wrote, viz. a thing of beauty is a joy forever. If we look at a pendulum clock once adorning the courtroom of a royal family and now kept in a museum or even a well-crafted jewelled wristwatch, we appreciate this statement.

By quality, we tend to identify and assess certain characteristics or properties which contribute to the value of a product. Value of a product or service is something that is created through some effort, is appreciated by the customers and can justify a higher than usual price. This concept of value includes (a) use or functional value (b) esteem or prestige value and (c) exchange or re-usability value. While all these are quite important as value contributions of a product or service, their relative importance would vary depending on the nature of the product. For a household decorative item, prestige or esteem value is more important, consideration of use value may not arise no one bothers about exchange value. On the other hand, for a household device like a ceiling fan use value dominates over esteem value and exchange value may be trivial.

It will be quite rational to look upon quality as a value addition and to argue that customers want value for money to acquire the product This value can be influenced by the quality of design as also by quality due to product support. In this connection, one may note that product augmentation and product extension can definitely enhance value addition and the nature and extent of these two should be incorporated right in the product design phase. Augmentation in terms of packing and packaging of the product and extension in terms user guide/manual, provision maintenance support and of spares and accessories, etc. do influence customer-perceived quality.

Quality is also perceived as the degree of excellence in

1. material or process or product;
2. service, procured or delivered;
3. work environment and;
4. performance, individual or collective.

This excellence is expected to yield a certain amount of satisfaction—to both the producer and the consumer—and results in a certain extent of profit to a business. In fact, excellence is one desideratum to judge the situation in a higher education system, besides access and equity.

Several other perceptions about quality may be considered to get an idea about the diversity in perceptions about quality.

The ultimate state of quality is not the 'absolute best', since quality is strongly related to market, to technology and to cost.

Quality is not static, goes on diminishing in relation to continuously increasing customer expectations.

Quality is not characterized by a single dimension. At least has two important dimensions to be taken into account for assessing the quality effort of an enterprise, viz. value addition and cost differential.

Quality has several different aspects to be taken care of by different groups of people within and outside the enterprise.

Quality Management is meant eventually to improve organizational performance. Improvement calls for measurement of quality—existing and desired—and has to be planned for.

1.4   Dimensions of Quality

In an article in Harvard Business Review, November–December, 1987, pages 101–109, David A. Garvin listed eight dimensions of quality and these have been clearly and concisely stated by Cathal Brugha (in a private communication to the author), with illustrations drawn from a Personal Computer.

Dimension 1. Performance—A product's primary operating characteristics. Clock speed, RAM, hard drive size, etc.
Dimension 2. Features—Characteristics that supplement basic functioning like wireless mouse, flat-screen monitor, DVD-RW, etc.
Dimension 3. Reliability—Probability that the product will not malfunction before a specified time period. Meantime between failures.
Dimension 4. Conformance—Degree to which the product's design and operating characteristics meet established standards. Underwriter laboratories labelled, mouse, monitor, keyboard included with CPU.
Dimension 5. Durability—Expected product life. Time to technical obsolescence, rated life of monitor.
Serviceability—Speed, courtesy, competence and ease of repair. Warranty conditions, availability of customer service and replacement parts.
Aesthetics—How does a product look, or feel, or sound, or taste or smell (depending on the nature of the product). Computer housing colour scheme. Keyboard touch, etc.
Perceived quality—Reputation and other indirect measures of quality like previous experience of self or others. Brand name, advertising.

While different dimensions of quality of a product may have different appeals to different persons or groups, we also talk about several Aspects of Quality which are of interest to different groups of people in different contexts. Some of these aspects are:

1. FundamentalAL: Quality of design (specifications/standards/grades)
2. Measurable: Quality of conformance
3. Consumer (marketable): fitness for use, life, performance, price, availability, delivery
4. Operational: Quality of management, system and operations, minimum waste and cost
5. Conservational: Optimum use of resources—materials and energy
6. Environmental: Safe waste, tolerable noise, etc. minimum air and water pollution and
7. Human: Quality of working life

Aspects 1, 4 and 5 should be great concern to the producer in particular; aspects 2 and 3 affect the customer, aspect 6 should be a common concern for all and aspect 7 affect the people engaged in the producer–provider organization.

From the point of industry or business, three broad dimensions of quality stand out prominently in the supplier–customer transactions. These are

Physical (technical) in terms of product features (including features of the augmented product) that can be quantitatively assessed or qualitatively judged at the time of product–service delivery.

Functional (interactive) in terms of features (including features of the extended product) which are revealed during product use/deployment.

Corporate in terms of features of the supplier–provider organization that influence customer perceptions about product quality over continued use as also their expectations about quality of a product to be acquired.

1.5 Approaches to Defining Quality

We all agree that quality is multi-dimensional and is comprehended in different ways by different individuals or groups in the light of different concerns and demands. Thus, we all agree that quality is not expected to have a unique or even a consensus definition among all interested individuals who are concerned with quality in a great diversity of situations. The question that arises in our minds is 'how badly do we need to have a formal definition of quality to appreciate good quality or deprecate poor quality, to pay for quality, to assess or compare quality in terms of some standard? Notwithstanding the validity of such a question, we may proceed with alternative definitions which have come up from time to time in a bid to find out some broad consensus. There is no need to consider these definitions chronologically, though some pattern could emerge if that exercise is taken up.

Anyone interested or involved in the quality profession will come across a plethora of definitions of Quality offered by eminent exponents. This led Garvin (1988) to suggest the following five approaches to defining quality.

* *Transcendent Approach*—This approach relates to 'a condition of excellence implying fine quality as distinct from poor quality and achieving or reaching for the highest standard against being satisfied with the sloppy or the fraudulent'.

Examples of fine quality that meet this approach are present in arts and literature. The sculptures in the caves of Ajanta and Ellora, Beethoven's symphonies and Leonardo Da Vinci's 'Mona Lisa' are all examples of 'achieving or reaching for the highest standard'. This approach, however, lacks objectivity in the form of an absolute standard of performance which is equally acceptable to all. It is not much relevant to business and industry. It does not enable a worker to claim with certainty that his product is of the highest standard.

* *Product-based Approach*—It focuses on certain features or attributes that indicate higher quality. Leather upholstery on a sofa set, for example, is deemed to have higher quality than cotton. The disadvantage of this approach is that it implies the

presence of leather as the criterion to define higher quality and does not pay attention to its colour or finish.

* *User-based Approach*—It is defined by Juran as 'fitness for use'. Here the user determines the level of quality. The product–service that fulfils his requirements is of higher quality. In this approach, the producer or service provider must know how the customer intends to use the product–service and for which purpose and act accordingly. Customer satisfaction levels reflect levels of quality. This approach is most suited to modern competitive market. This approach needs to identify the target market and its requirements, and then design, develop and deliver the appropriate product–service to the customers. Here cross-functional teams should be in place to contribute to the product–service quality. This approach emphasizes on better brand image and increased market share. This approach may require the producer or service provider to create quality consciousness among the customers about the wide variety of benefits that may arise from the use of the product–service. This will help the customer to know about the potentials of the product–service and thus help him develop a better idea the versatility and quality of the product–service.

* *Manufacturing-based Approach*—According to Crosby (1996), this approach focuses on 'conformance to requirements'. Engineers specify the characteristics of a product. The more closely manufacturing conforms to these requirements; the better is the quality of the product. Conformance within a smaller range is better than conformance within a larger range.

The advantages of this approach are

(1) objectively measurable quality standards and
(2) reduction in cost of quality.

Conformance to requirements brings about substantial reduction in costs such as rework, scrap, re-inspections and returns from the customer. A major disadvantage is its lack of concern for the requirements of the customer. Product–service characteristics specified by the manufacturer or service provider may not in alignment with what the customers need. The underlying assumption is that customer satisfaction has a direct relationship with the fulfilment of the given specifications of a product.

* *Value-based Approach*—This approach to quality introduces the elements of price and cost. According to Broh (1982) 'quality is the degree of excellence at an acceptable price and the control of variability (resulting in uniformity) at an acceptable cost.' The degree of excellence at an acceptable price is decided by the customer, and the manufacturer decides on control of variability at an acceptable cost. The purchase decision of the customer is based on a trade-off between quality and price. This approach is not effective in providing objective quality criteria since many of the quality attributes could be subjective in nature.

1.6   The Deluge of Definitions

Many definitions of quality have emerged over the years and some of these are worth a perusal. There is—as is expected—a large communality among these. At the same time, each definition claims some individuality. Some definitions are conformity-based, while some others are oriented towards excellence. Some consider the set or totality or composite of features and characteristics—of the product as such or of associated functions such as design, manufacture, use, maintenance and even reuse as constituting quality of the product. On the other hand, quality is looked upon as an ability to meet certain requirements. Of course, the former emphasizes on what determines the latter. Let us look at some of the definitions that may serve the purpose of revealing this diversity in definitions. Some definitions apply directly to individual units (products or work items) while some others emphasize on uniformity or consistency among units in a suitably defined population or aggregate. Some definitions facilitate development of measures of quality, while such measures are difficult to arise from some definitions which stress on abstract or latent features or properties of the units. Some definitions tend to highlight the role of the designer–producer of a product or service to state that 'Quality is a way of Life'. Such definitions need not be debated for their meaningfulness, but since we cannot measure quality defined in such a manner, we cannot adopt such a definition for the purpose of comparison, control and improvement.

Let us look at definitions proposed by some eminent exponents of quality who have guided the quality movement over the decades beyond the Second World War.

Quality is the 'Level of Usefulness to the Consumer' Or
Quality is Fitness for (intended) Use—Juran (1988)

This fitness is generally assessed in terms of quality attributes or characteristics used as surrogates or constituents or, better, determinants of 'fitness'. Requirements for the latter entities are often mutually agreed upon between the producer–supplier and the customer–user and sometimes given out by the producer only or even dictated by the customer–user and accepted by the producer–supplier, and thus is justified a second definition, viz.

Quality is conformance to requirements or specifications—Crosby (1979). Another important definition attributed to Deming, though not explicitly stated this way by him runs as

'Quality is predictable degree of uniformity and dependability at low cost and suited to the market.'

This definition refers to uniformity (among different units or items in respect of quality attributes) and dependability (during use—a concept that can apply to even an individual item or unit), rather than conformity to requirements or specifications directly. It also harps on 'low cost' and suitability to the market, keeping in mind needs, expectations and purchasing power of the people visiting the market. Thus, this definition differs from those suggested by Juran or Crosby.

Alternatively, quality is the totality of features and characteristics of a product or–service that bear on its ability to satisfy given needs—ANSI–ASQC Standard A3 (1978). This definition leads us to consider several functions that precede and follow manufacturing, as was indicated by A. V. Fiegenbaum in taking quality as 'the total composite of product and service characteristics of marketing, engineering, manufacturing and maintenance through which the product or service in use meet the expectations of the customer.' It is presumed that marketing and engineering together should be able to incorporate those features in the product which are expected (might not be so stated) by the customers.

Galetto (1999) defines quality in terms of 'the set of characteristics of a system that makes it able to satisfy the needs of the customer, of the user and of the Society.'

The user's point of view is also reflected in the definition of Quality as the 'Quotient between what the user receives and what he expected to receive' (green). Accepting this definition, we are reminded of the fact that quality of a product or service is not something static and that in the context of increasing customer expectations (motivated by knowledge about technological advances, development of new and better materials and entry of new producers), the numerator has also to increase proportionately in order that quality as is perceived by the customer does not decrease over time. This calls for product quality improvement through innovation and updation of production processes.

The International Organisation for Standardisation (ISO) had first offered the definition 'Quality is the totality of features and characteristics of a product that bear on its ability to satisfied stated or implied needs' (1994). Later, this definition has been modified to indicate the outcome as 'Quality is the degree to which a set of inherent characteristics meet requirements.' (2005). One wonders if one should read too much into the ISO definition that does command some respect in some quarters. This definition takes Quality as something relative and not absolute—relative to 'requirements' (explicitly stated or just tacitly implied). If one argues that requirements may be circumscribed by purchasing power or access or even by individual attitudes and lifestyles, we may then justify different degrees of the same product to meet differing requirements or assign different labels of quality to it. It may not be out of place to mention that even now—as also earlier—we get the same product in different variants some marked as 'export quality', implicitly better than those not so marked.

Budyansky (2009) recognizes five vagueness factors associated with quality, viz. absence of a generally accepted definition of Quality, absence of effective and well-defined rules of measurement and quality evaluation, absence of a methodology allowing a transition from physical parameters to quality parameters, absence of a methodology to carry out a quasi-objective analysis and synthesis of quality and lastly the absence of a common technology for quality assurance. He argues that the distinction between procedure-oriented and result-related definitions should not be obliterated and that the universal definition of Quality should be 'Quality is a degree of compatibility of an obtained result with an objective'. He goes on to say that 'compatibility' theory is applicable to situations characterized by vagueness like fuzziness as in systems engineering or fuzzy set theory. He advocates three possible definitions, viz.

- Quality is the degree of achievement of a formulated objective.
- Quality is the degree of achievement of progress under certain conditions.
- Quality is the degree of 'wishing' satisfaction into a realistic environment.

He provides a somewhat complicated procedure to get at the 'compatibility' measure.

An Unconventional definition

Quality is a situation in which a product once sold does not come back (with a complaint or a request for repair or replenishment or recall), but the customer does (Margaret Thatcher, former Prime Minister of UK.). This definition does make a lot of sense in business and needs an appropriate context-specific interpretation.

All these definitions apply directly to products, but are also relevant to services (which involve products to varying extents).

1.7   The Longest Definition

It is worth recalling the almost prophetic words of a great visionary P. V. Donkelaar, President, Outward Marines, Antwerp, Belgium (1978) uttered some 40 years back that can be accepted as an early pointer to various exercises taken later on to broaden the concept of Quality, taking due account of concerns expressed (or felt, though not always explicitly stated) by different stakeholders. This definition—probably the longest one cited in the literature—and a quite debatable one in the context of modern industrial practice, runs as

> A product is of good quality if and only if at a minimum life-cycle cost it provides a maximum contribution to the health and happiness of all people engaged in its manufacture, distribution, use, maintenance and even recycling with a minimum use of energy and other resources and with an acceptable impact on Society and on Environment.

This comprehensive understanding of quality goes much beyond the current concerns with 'customer satisfaction' or 'customer delight' in two ways. It talks about 'happiness' and 'happiness' in social sciences is something tat implies but is not necessarily implied by 'satisfaction or delight'. It goes further to take care of 'health' and, that way, of 'safety'. Secondly, it does not speak of customers only, but goes on to embrace all others engaged in the entire life-cycle of the product from design to disposal.

It will be apparent from this definition itself that Quality Management involves decision-making using multiple criteria by different persons (person groups) with mutually opposed interests in different quality-related activities to achieve several different objectives. To minimize consumption of energy and other resources and, at the same time, to ensure solidity of product life and to provide a maximum contribution to the health and happiness of all people concerned at various stages with the product is an extremely difficult optimization exercise. Consumption of energy and resources is a concern to the producer while life-cycle cost is a matter of concern for the customer or the user.

This definition puts an emphasis on the impact of the product and its use on the society and the environment. And it is worth noting that these latter aspects are now being considered as important concerns of Quality Management.

A much bigger conflict that this definition begets concerns the fact business particularly in consumer goods as also of capital goods grows in recent times under a veiled policy of 'planned obsolescence' so that items of consumption become obsolete or out of fashion or difficult to maintain in order that new brands or even new variants or new products are introduced into the market to take over the still-operating or functional items or product units.

Thus, Quality Management today involves multiple stakeholders not all directly related to products and services that are traded between the producer or provider and the customer or the consumer, Custodians of public interest including regulatory bodies as also public interest activists have a say and Management—taken broadly as Quality Management here—can no longer brush aside their objections and suggestions. Of course, some may argue that such issues are too big for 'Quality Management' in usual parlance.

This definition has not escaped criticisms from the consumer goods industry, where emphasis is placed on the product's appeal to the customers in terms of user-friendliness, conformity to changing customer tastes and likes or dislikes, esteem or prestige value and similar other features without a direct reference to the concerns about consumption of energy and materials and the impact on environment.

1.8 Quality Through the Life Cycle

Quality as an attribute of a product or a process or a service is not a static concept. Before proceeding further, we may note that whatever is produced can be generally taken as a product. This way we need not differentiate between a product and a service. The life history (or better, life cycle) of a product can be comprehended in four phases and each phase some processes are involved. In fact, processes are inseparably associated with a product. The following different phases encompass a product's life cycle, viz.

1. Concept or Mission: developed prior to the design of a product, meant to capture the intentions behind designing and manufacturing the product by stating the desired or intended functions. To this end, we usually bring in some performance parameters or functional characteristics, some of which can be conveniently quantified while some others could be just qualitative. An automobile's intended performance may be stated in terms of some performance parameters like average speed or maximum attainable speed, fuel consumption per mile,

safety features, comfort feeling, etc. While such a list could be pretty long, some of these could be critical in describing the expected functioning and requirements on these critical parameter are taken as critical events, like

Maximum speed attainable $(S) \geq 150$ mph., miles per litre of fuel (FE) ≥ 22.

For a household electric lamp of a given wattage, the mission requirements could be

$L \geq 1000$ and $I \geq i_0$ where L is the length of life in hours, and I is the intensity of light emitted.

In general, if $Y_1, Y_2, \ldots, Y_k$ stand for k critical parameters in the mission and for satisfactory performance of the product $Y_i \in S_i$ $i = 1, 2, \ldots, k$ where S_i denotes the one-sided or two-sided interval in which Y_i should lie to ensure satisfactory performance. These inequalities should be satisfied in a prescribed sequence by a product during use and define what is called a mission plan or mission profile. Requirements on critical parameters are recognized as critical events which may or may not take place in the life of a particular unit of the product, in the context of a population of units to be turned out.

2. Design: developed in terms of requirements or specifications on physical or chemical or even biological parameters, generally referred to as quality parameters of materials, processes, checks and tests, controls, etc. which should be met during manufacture. Knowledge about the functional requirements and of their dependence on these quality parameters as also domain knowledge about the process(es) involved will yield the Design as $\{X_j \in I_j j = 1, 2, \ldots, l\}$ where X's are the quality parameters that influence performance and I's are the corresponding intervals within which these quality parameters are allowed to vary. Thus, for the case of a lamp, there could be design requirements like $T \geq 1800$ °F where T stands for the melting point of the filament used. The design would also specify the firmness of soldering of the cap on the bulb. Take, further, the case of a missile meant to hit and harm a target. The mission plan can be stated in terms of several critical events taking place in a given sequence, viz. the missile takes off from its launching pad or base almost immediately after being fired, should hit a point very close to the target and release some destructive energy. The following would be the critical events: T (take-off time) ≤ 2 s, D (distance between the target and the point hit) ≤ 3 m and E (energy released on hitting a point) $\geq e_0$.

In fact, design should and does specify requirements about process parameters like the feed rate of an in-process material into a machine or the alignment between the jaws of a machine or time during which a certain treatment should take place, etc. Design also specifies checks to be carried out at various stages and control to be exercised on the basis of the results of such checks.

Design as a comprehensive set of requirements should be developed keeping in mind the mission plan or profile.

3. Manufacturing: p) in terms of operations or activities to turn the initial input along with inputs at intermediate stages into the final product. Most often this is a multi-state process where the output at the end of any stage depends on the output of some previous stage(s). It is expected that the process plan takes care of the design requirements along with equipment performance. That way, manufacture or production has a big interface with procurement, engineering, inspection and quality control and related functions.

4. Use or deployment of the final product in environments which may or may not have been foreseen and accounted for during the design phase and for purposes which might not had been envisaged in the mission phase. All this may lead to malfunction or failure. The actual performance of the product gets revealed as we use it—at a point or over an interval, depending on the mission. More often than not, performance is noted by way of its converse, viz. failure and we may note the time-to-failure or, in case of a repairable product, time between consecutive failure or time to repair, etc. Functional or field-performance parameters noted during use or deployment are included directly or otherwise in the mission plan.

From the last phase, having noted the actual performance of a product in relation to the performance expected in the mission and targeted in the design, we like to revisit even the mission to find out if we were too ambitious in the sense of an expected performance that could hardly be achieved or if we could do better and achieve something more what we intended modestly. Either would lead to some design modification that would have to be accommodated during manufacture. Finally we would get to know the changed performance. And thus the cycle goes on.

The term 'Quality' has different implications during different phases of this life cycle. In fact, all the implications take the mission as granted and then proceeds to identify and quantify quality for each of the remaining three phases. Thus, the three different determinants of quality of a manufactured product are relative to a given mission and will change from one mission to another. And this mission usually takes care of technology, market and cost (of production).

The primary determinant of product quality should be the ability of a design to meet mission requirements and this ability is recognized as quality of design (sometimes referred to as design reliability). Second comes the ability of a manufacturing process to meet design requirements and this is accepted as process capability, also referred to as quality of conformance. Quality of design coupled with quality of conformance will imply that a product during use will meet mission requirements. And this ability as directly concerns the customer and the user is recognized as quality of performance, also known as product reliability. We thus come to the fundamental quality equation to be stated as

$$\text{Quality of Performance} = \text{Quality of Design} * \text{Quality of Conformance} \quad (1.1)$$

where the symbol * denotes 'composite of.'

A fourth determinant of quality for a repeated-use product is provided by quality due to product support by way of customer education to enable proper use and maintenance, etc. Thus, quality due to product support enhances quality of performance.

For the purpose of quantification, each of the above abilities can be viewed as obtained as a 'probability' which can be estimated from relevant data. Thus, quality of design is the conditional probability that products (turned out by a process that meets requirements of the design fully) meet mission requirements as can be estimated from field-use data on performance parameters compared with the mission-required values. Similarly, quality of conformance is an unconditional probability that design requirements are met during the production–manufacturing process that can be estimated from inspection data—covering incoming materials, work-in-progress items and finished goods—compared with corresponding values prescribed in the design. And product reliability is now a joint probability that both design and mission requirements are met, as can be estimated by considering use data. These probabilities can now be stated in terms of parameters X and Y as

$$\text{Quality of Design} = \text{Prob.}\left[Y_i \in S_i, \quad i = 1, 2, \ldots, k/X_j \in I_j, \quad j = 1, 2, \ldots, L\right]$$
$$\text{Quality of Conformance} = \text{Prob.}\left[X_j \in I_j, \quad j = 1, 2, \ldots, L\right] \text{ and}$$
$$\text{Quality of Performance} = \text{Prob.}\left[Y_i \in S_i, \quad i = 1, 2, \ldots, k\right]$$

Defined as probabilities, these determinants of quality yields the modified quality equation

$$\text{Quality of Performance} = \text{Quality of Design} \times \text{Quality of Conformance} \quad (1.2)$$

1.9   Quality of a Process

Quality of a process to transform some input—hard or soft—into some output (which also could be hard or soft) as distinct from the quality of input is comprehended indirectly in terms of the quality of the output. For this purpose, a process is segmented into work items each of which may have a specified intermediate output or a specified start or finish time. This way, there are seven generic ways to measure quality of a process (taken as a collection of work items) besides the Cost of Quality, viz.

1. Defect, i.e. deviation from the relevant specification given in the process plan regarding any observable feature(s) of some items or elements in the process
2. Rework—when the deviation can be corrected by repeating some work item(s)
3. Reject or Scrap—when the deviation cannot be corrected
4. Late Delivery—in terms of the work output arriving late for the next work item or process

5. Work Behind Schedule—may be caused by the delayed completion of the preceding work item or otherwise
6. Lost Items—when the output of some work item(s) was not properly preserved or saved and could not be retrieved when required
7. Items Not Required—in terms of some work item (s) not relevant in the process or are redundant.
   [Items mean work items or elements]

We can apply these quality measures only when a comprehensive process plan has been in place, identifying the features of the process—may be in the different work items or elements—and specifying limits for these features. Features of a process may include feed rate of some in-process material into a machine, time during which an operation like heating or cooling should be carried out, temperature to be maintained during an operation like calcination of raw petroleum coke within a calcinator, etc.

The above measurements apply to office outputs or outputs of some service operation as well as to outputs of production–laboratory–warehouse processes.

Regarding the output, the process plan should specify the target and we can speak of the following levels or values of the output before or after the process is completed.

1. Target—the level of performance to be achieved (in terms of performance parameters related to the different work elements of the process).
2. Forecast—the level of performance which may be better or worse than the target depending on current business situation. The target also shows when the target will be reached, if at all it can be reached.
3. Actual—the level of performance achieved till date.

We can add two more entities here to be noted and acted upon by process management, viz.

4. Problem—the gap between the actual and the target levels, when the actual is worse than the target
5. Opportunity—for improving quality when the actual is better than the target, at no extra cost.

1.10   Measures of Quality

For any item—a product as the output of a manufacturing process or a processed material or a work-in-progress unit—it may be difficult to measure quality. One possibility is to relate this measure to different parameters or characteristics of the item that influence its quality defined in a particular manner and to compare accrual or realized values or levels of the parameters for the given item with intended or targeted or specified values or levels.

Sometimes it is easier to measure quality inversely, viz. in terms of deviations from the above values or levels from those specified (may be in the design). Such a deviation in one parameter may or may not imply that the product will fail to satisfy some 'use' requirements—for use in some subsequent production stage or by the customer or the user. A deviation that renders the product unfit for use or for satisfactory functioning is usually referred to as a 'defect'. The presence of at least one defect will render the product 'defective'. An inverse measure of item quality could then be a count of defects. One has to note that such a defect may arise whenever a parameter is assessed through inspection. Thus, any property or parameter that is inspected corresponds to an 'opportunity' for a defect to arise. Of course, there could remain properties or parameters which are not inspected, because of complacency leading to a conviction that there would not be a defect even if this property was inspected or because it is inconvenient or difficult or too expensive to inspect the property. Such a property is regarded as a passive opportunity for a defect to arise. Thus, considering an item or product unit with many active opportunities for defects, a count of defects should be related to the number of opportunities rather than the number of items inspected. Since product quality has been improving almost everywhere, defects per opportunity may turn out to be pretty small and may rather consider 'defects per million opportunities' of DPMO as a measure of quality, not for a single product unit but for a sample of units taken out of a process. An analogous measure could have been the total number of defects in the sample or in case the sample size varies and measures of quality across samples need to be compared, one could better take the proportion or fraction defective as the sample quality measure.

Measures like 'defects per unit' or DPU and 'defects per million opportunities', i.e. DPMO, are applicable to manufactured articles as also for office or administration work outputs. For example, a purchase order to be forwarded to a vendor for supply of a material may be inspected for its accuracy, adequacy as well timeliness. The communication could have an incorrect or incomplete address of the vendor, a wrong material or a wrong quantity of material to be delivered, a wrong delivery date or a wrong specification attached or an incorrect statement of financial terms, etc. Each one is an opportunity for a defect. If, say, 20 such orders are inspected and in each 25 elements are checked before the orders are passed on to vendors, the total number of opportunities would be $20 \times 25 = 500$ and if a total of, say, three defects are come across, DPU would be $3/20 = 0.15$ and DPMO would be $(3/500) \times 2000 = 12$.

As has been pointed out in an earlier section, we need to develop two types of measures, viz. (1) a measure for an individual item or unit and (2) a collective measure that applies to some aggregate of items or units defined to suit some context. It is just obvious that the second type of measures will be appropriate and effective summary of individual quality measures for the individual items or units constituting the aggregate of interest. Thus, dealing with on-line process control activities where the outputs of the process under consideration are distinct items or units, we may either examine selected items or units or consider sub-groups of several consecutively produced items or units. In the latter case we need a sub-group quality measure. Similarly, dealing with the problem of sampling inspection from lots of discrete units or items, we need to speak of a lot quality measure as also a sample quality measure.

Number of defects in an item and total number of defects in a sample or sub-group illustrate the two types of measures. The number of defectives in a sample or the fraction defective is collective measures of quality. To make such measures more effective for the purpose of control and improvement, we may classify defects and defectives into several categories, associate suitable weights for the different categories and consider weighted totals as the collective measures of quality. In cases of quantitative quality measures for individual items or units, sample statistics like the mean, the range or the standard deviation could be taken as collective measures. Defects and defectives can be defined conveniently for both qualitative and quantitative quality measures for individual items or units. More sophisticated measures of sample quality like the trimmed mean dropping outliers, progressive averages, moving averages, exponentially weighted moving averages or cumulative totals or cumulative means and comparable measures of variability or dispersion are also being used increasingly, since we now wield great computational power and it is of little concern which measure do we use.

1.11 Quality–Reliability Interrelationship

'Quality' or 'reliability' of an entity (a material or process or product or service) cannot be defined uniquely, and even definitions offered by standards organizations (including ISO) change over time.

In fact, the latest ISO definition (2005) of quality is quite broad, and in some sense, subsumes Reliability as one aspect of quality. This definition runs as:

Quality is the degree to which a set of inherent characteristics meets requirements.

There are three noticeable features of this definition. Quality has to be assessed only in relation to some requirements. Quality is built into the entity through a set of characteristics which are incorporated in the entity during the design and manufacturing phases, and quality is measurable, though the measure could be ordinal as also cardinal.

Amplifying this definition, it has been generally appreciated that there are four dimensions or even determinants of quality, viz. quality of design, quality of conformance, quality of performance and quality due to product support. In fact, quality in the usual sense (as accepted in SPC) is just quality of conformance to design requirements.

Assuming that the design requirements truly reflect all relevant aspects regarding materials, processes, checks and modifications which have a bearing on the quality of the output, quality of conformance does imply the desired quality, quality of conformance can be checked a pre-, during and post-production (at the finished goods stage).

Quality of performance is the most important dimension of quality that concerns the customer or user. And this dimension can be assessed only during use of the Output. This dimension along with quality due to product support (specially in cases of intermittent-duty and continuous-duty products) really determine the ability of the product to carry out the required functions during use of the product throughout its life cycle.

Product reliability is the composite of quality of design and quality of conformance (as also of quality due to product support, wherever necessary). With this explanation, reliability goes beyond quality of conformance. Sometimes, we argue that reliability is an extension of quality to 'to use of deployment' phase or reliability concerns begin where quality concerns end.

Whatever be the views adopted, it is well appreciated that quality and reliability are definitely related closely. In a bid to establish some differences between quality and reliability, we should go by the conventional definition of quality which can be assessed in terms of physical, chemical, biological and other relevant properties (known to influence the quality of output) which are either explicitly stated or just implied or even meant to be commonly in place, which are specified in the design and which can be checked at the incoming materials stage or during different stages of manufacture or during inspection of final output.

On the other hand, reliability can be assessed in terms of functional or performance parameters like time-to-failure or time between consecutive failures or time to repair or even values or levels of performance parameters (like luminosity or energy or load, etc.) that determine the ability of the product to function as desired. Quite of ten the requirements about performance parameters are not clearly stated and some of these parameters are not easily or conveniently measurable.

1.12   Concluding Remarks

Quality is enigmatic; it is elusive. No one can say the last word on any aspect of quality. Some critics of modern Quality Management with its emphasis on formal and rigid systems, regimented workers and alluring rigmaroles argue that products of days long gone by when none of the above aspects of Quality Management were present had generally better quality than the products offered in the markets today. This, of course, does not mean that we can afford to neglect all that have been developed over the recent years to comprehend quality in all relevant details and to improve quality in products and services, leading to a better quality of living. We can hardly ignore the fact the nature of production processes and the variety as well as volume of products and services have dramatically changed over the years at an exponentially increasing rate.

We should note that users in recent times insist on better and better quality in newer and newer products and more and more reliable services. The ever-increasing quantity and variety of goods and services with the growing demand for better quality is not felt only in the areas of manufactured items—as was the situation earlier—but also in all services that constitute a large segment of the consumption basket today. Thus, the need to start with a comprehensive understanding of quality in all its facets and in all spheres of human activity is just a bad necessity nowadays.

Quality considerations embrace both soft philosophical approaches as also hard scientific and technological approaches. To strike a balance is a difficult task. However, taking care of both the approaches along with their merits and limitations is highly desirable in any discourse on quality and Quality Management.

References

ANSI/ASQ Standard A3. (1978). Quality System Terminology.

Broh, R. A. (1982). *Managing quality for higher profits*. New York: McGraw Hill.

Budyansky, A. (2009). Improved quality tactic. *Total Quality Management and Business Excellence, 20*(9&10), 921–930.

Crosby, P. (1979). *Quality is free*. New York: McGraw Hill.

Crosby, P. (1996). Illusions about quality. *Across the Board, 33*(6), 37–41.

Dale, B. G. (1999). *Managing quality* (3rd ed.). Oxford: Blackwell Business.

Deming, W. E. (1986). *Out of the crisis* (2nd ed.). Cambridge: Cambridge University Press.

Donkelaar, P. V. (1978). *Quality: A valid alternative to growth* (Vol. 4). EOQC Quality.

Fiegenbaum, A. V. (1956). Total quality control. *Harvard Business Review, 34*(6), 93.

Fiegenbaum, A. V. (1983). *Total quality control* (3rd ed.). New York: Mcgraw Hill.

Fiegenbaum, A. V. (2004). *Total quality control* (4th ed.). New York: Mcgraw Hill.

Galetto, F. (1999) The golden integral quality approach: From management of quality to quality of management. *Total Quality Management & Business Excellence, 10*(1), 17–35.

Garvin, D. A. (1984). What does "Product Quality" mean? *Sloan Management Review*, Fall, 25–43.

Garvin, D. A. (1987). Competing on the eight dimensions of quality. *Harvard Business Review, 65*(6), 94–105.

Garvin, D. A. (1988). *Managing quality—The strategic and competitive edge*. New York: The Free Press.

ISO 9001. (1994). *Quality management systems—Requirements* (1st revision).

ISO 9000. (2005). Quality Management Systems: Vocabulary.

Juran, J. M. (Ed.). (1988). *Quality control handbook*. New York: Mcgraw Hill.

Juran, J. M., & Gryna, F. M. (1993). *Quality planning and analysis* (3rd ed.). New York: McGraw Hill.

Oakland, J. S. (2003). *Total quality management, text with cases*. Oxford: Elsevier.

Saraiva, P. M. (2018). Quality and statistical thinking in a Parliament and beyond. *Quality Engineering, 30*(1), 2–22.

Chapter 2
Quality in Services

2.1 Introduction

The pattern of human consumption has changed dramatically over the last few decades, services accounting for more than 40% of the total consumption expenditure against only less than 25% earlier. And quite expectedly, the service sector now contributes a much larger share in the total national income, quite so compared to 'manufacturing'. Recent estimates put this figure at 58%. Consequently, diversity and quality of services have attracted a lot of attention, and many initiatives have been being taken to improve these.

Services of various types rendered by the service sector of industry, as well as different product-related services executed by the manufacturing sector, have assumed increasingly important places in the competitive global economy today. Naturally, quality of such services—as is particularly perceived by the customers—has come to focus, and attention is now being paid to develop suitable definitions and measures of service quality as well as to appropriate models and methods for analysing service quality. Such models and methods help us in initiating measure to improve service quality.

Quality of service cannot ignore considerations of accessibility, affordability, dependability, consistency and similar other attributes. It must be remembered that services are loaded with engineered products—some to almost a negligible extent and some others to a significant level. That way quality of the hard product involved does affect quality of service. The perceived quality of a service also depends on the service delivery process quality, which has been improving substantially through use of technological advances.

The present chapter takes off with a widely accepted definition of service, gives out distinctive features of services and of service quality, briefly considers some important models for analysing quality in services and provides hints at possible improvement in service quality. Obviously, it takes into account the diverse dimensions of service quality and offers a new approach to assessing service

© Springer Nature Singapore Pte Ltd. 2019
S. P. Mukherjee, *Quality*, India Studies in Business and Economics,
https://doi.org/10.1007/978-981-13-1271-7_2

quality. A section deals with problem of assessing quality in e-services that are becoming more and more common.

2.2   Service—Definition and Distinctive Features

Formally defined, service is in terms of the results generated, by activities at the interface between the supplier (provider) and the customer (receiver) and by supplier's internal activities, to meet customer needs. The supplier and the customer are represented by personnel and/or equipment. To comprehend service quality, one has to explicitly recognize the service delivery process as 'those supplier activities (internal as well as at the interface) necessary to provide the service'.

It may be relevant to differentiate the service production process from the service delivery process. The latter involves activities (internal as ell at the interface with the customer) necessary to provide the service. Depending on the situation, service production and service delivery processes have different impacts on service performance or results. Going by the standard definition of a 'product' as the 'result of a process', services are also products.

One aspect that distinctly differentiates between 'products' and 'services' is that services necessarily involves at least one activity performed at the interface between the producer and the customer, while this may not be true in the case of many manufactured items.

Provision of a service can involve some of the following activities

- an activity performed using a service provider supplied tangible product like a screw or a bolt as part of a contract or on request (e.g. repair or fixing some operational problem with a furniture item at the customer's site);
- an activity performed on a customer-supplied tangible product (e.g. a sofa set or an automobile or an almirah to be repaired or a shirt to be prepared from some cloth piece provided by the customer);
- an activity performed on a customer-supplied intangible product (e.g. income statement provided to a person tasked to prepare the tax return);
- the delivery of an intangible product (e.g. delivery of information or instruction in the context of knowledge transmission);
- the creation of ambience for the customer (e.g. providing for some accessories or artefacts or decoration in hotels and restaurants).

Running through all these types of services, sometimes requiring some tangible or intangible product, is several features in terms of which quality of service can be fully comprehended. Important among such features are intangibility (of the service operations), simultaneity (of the service production and service delivery processes) and heterogeneity (variation from situation to situation).

First, most services—as distinct from the product content in them—are intangible, Whatever may be the level of product content in a service, assessing quality of service is in terms of certain operations or activities carried out at the back-end with no interface with the customer as also some operations or activities at the front-end involving the customer (or his agent or representative). Since service involves such operations, service quality is analysed in terms of the performance of these operations. Of course, service quality in many cases will depend on quality of input materials or semi-finished products or finished components for which precise specifications can be developed and implemented. But for performance involving many subjective elements, such specifications are quite often too broad or generic and can hardly ensure the same quality consistently over repeated occasions. Most services cannot be counted, measured, inventoried, tested and verified in advance of sale or offer. Because of intangibility, the service provider may find it difficult to understand how consumers perceive the services and evaluate service quality. And, we must recognize the fact that such perceptions are sometimes fed back with distortion by some agent or representative of the consumer or user.

Second, services, specially those with a high labour content, are heterogeneous; their realized performance often varies from producer to producer, and their perceived performance varies from customer to customer, and from one occasion to another. Consistency in behaviour and in performance of service personnel is difficult to assume. And service quality depends a lot on human behaviour, unlike quality of manufactured products. Hence, what the firm intends to deliver may be quite different from what the customer receives.

Third, production and consumption of certain services are inseparable. As a consequence, quality in services is not engineered at the manufacturing or production site, then delivered intact to the customer. In labour intensive services, for example, quality issue arises during service delivery, usually in an interaction between the client and the contact person representing the service provider. The service firm may also have less managerial control over quality in services where consumer participation is intense (e.g. haircut, doctor's visits), since the client affects the process. In such situations, customer's input (description of how the haircut should look, description of symptoms) becomes critical to the quality of service performance.

2.3 Service Quality

Service quality is generally understood as the degree to which a service as received by the customer meets customer's requirements. In case service as is delivered by the service provider and service as is received or even assessed on consumption by the customer or user are different, service quality has to account for both. Requirements take into account both 'needs'—felt and/or communicated to the provider—as also 'expectations' with which a customer approaches a service provider. In fact, 'expectations' often dominate over 'needs' (unlike in the case of a product) and service quality, as perceived by customers, stems from a comparison of what they feel service providers should offer (i.e. their expectations) with their perceptions of the performance of service as provided. Thus, perceived service quality is viewed, inversely, as the degree and direction of mismatch between expectations and perceptions of customers.

And, in the case of services as distinct from other types of products, perceived quality is what counts more, rather than quality of service performance as is planned for and claimed by the service providers.

Customers' expectations are determined primarily by (1) personal needs (2) past experience about service performance of service providers (3) word-of-mouth communications from others about their experiences and (4) external communications from the service provider and from others including regulatory bodies.

Looking back to 'service' as the result of a 'process', it has to be noted that two distinct processes yield the result viz. the service production process (primarily a provider activity) and the service delivery process (at the customer-provider interface) Quality of the first activity is determined by the provider in terms of various parameters—of the process as also those of the input materials and products procured from outside—as anticipated to fulfil customer expectations and as required to meet business objectives. Quality of the second is affected more by human communication and behaviour. As a consequence, service quality is more difficult to define and measure compared to quality of other products. It is true that in some cases, quality of service delivery begets an adverse reaction from the customer, despite the product delivered being quite satisfactory. For example, quality of service may be poor in a restaurant, though the food served is excellent. The customer, however, may not come back to such an organization.

Just as reliability has been regarded as an extension of quality for manufactured products to the time of use or deployment, service quality in relation to some service performed by a service provider on a customer-supplied product like a household gadget or an automobile does include reliability considerations. The service performance will be judged in terms of dependability and duration of failure-free operation of the product.

Customer assessment is the ultimate measure of the quality of a service. Customer reaction could be immediate—right at the interface with the provider during service delivery or immediately after consumption or use, often at the provider's place, e.g. adverse reaction of a customer after taking food in an eatery. It may be delayed or retrospective, e.g. when food from the eatery was taken home for consumption, found below expectation and reported sometime later as unsatisfactory. Often subjective evaluation will be the sole factor in a customer's assessment of the service provided. And customers seldom volunteer their assessment of service quality to the service organization. Dissatisfied customers often cease to use or purchase services without communication of their findings and suggestions that would permit corrective action to be taken. Thus, reliance on customer complaints as a measure (inversely) of satisfaction may lead to unwarranted complacency and misleading conclusions.

It is worth while to note that time has been recognized as a distinct element of quality in a service context. This is not so in the case of manufactured products and in some situations where service implies providing an off-the-shelf item with not many customers in queue. And time has to be broken down into different components to find out components where action should be taken to effect a reduction. Thus, time to access, time to process or register a service request, time taken to provide the service and time to complete the service process including billing and payment and, more importantly, time taken to respond to a complaint and to resolve the same—all these should be tracked separately. It is not an uncommon experience in a healthcare service provider that patients have to wait too long to register their requests or to receive the service needed and to complete formalities at the end.

Finally, one could distinguish—at least for the limited and somewhat didactic purpose of understanding service quality—between quality in the service production process and quality of the service delivery process. The first should be called 'quality *in* service' and is more amenable to objective analysis, while the latter should be taken as 'quality *of* service' and is more a perception by the customer and, that way, is more subjective in nature and less amenable to quantification.

2.4 Measuring Service Quality

There have been several approaches to measuring service quality. Some of these are by way of surveys, some by way of benchmarking and some others involve workshops. Thus, we have

(1) Customer satisfaction surveys—through personal interviews or mailed questionnaires. In the latter case, a carefully crafted questionnaire has to be developed and sent out by mail with a forwarding letter explaining the purpose and safeguarding respondents' confidentiality. In either case, a more or less complete list of customers has to be developed and a distinction has to be made between those who buy services and those who use or consume those. Usually, a sample survey is carried out, and the sample has to be selected from this list in an objective manner. The response rate may be lower or higher, depending on the nature, extent and frequency of use and importance of service.

(2) Focus group discussion—the choice and engagement of a focus group may sometimes pose problems. A focus group has to involve, besides users of service by a particular service provider, users of other service providers and experts in design and delivery of services and even representatives of concerned regulatory bodies Responses are likely to be more useful in assessing the merits and demerits of the service. At the same time, these may not be easy to absorb always.

(3) Employee quality surveys—involving employees as interviewers who are expected to provide clarifications needed by respondents. However, responses may be fed into the mouths of less-informed respondents, and responses may be eventually biased. Thus, this type of data collection exercise should better be avoided.

(4) Customer complaint analysis—examining each complaint without prejudice against the complainant for any possible lapses on the part of personnel or possible deficiencies in physical facilities or possible defects in the materials used in the service production or delivery process or in both provides inputs for quality improvement.

(5) Random checks by management—an internal activity to detect and remove any deviation from internal specifications or directions.

(6) External benchmarking—comparing the service processes and the service performance against performance of 'best in class' service providers may help a service provider to get clues about shortfalls in quality.

(7) User group meeting—to get a direct feedback about quality of service as perceived by them as also their suggestions about possible ways to improve quality.

(8) Generic industry forum meet—may serve the purpose of benchmarking to some extent.

(9) Customer value workshops—customers required to fill up a brief questionnaire on spot to identify aspects of aspects which irritate them. Three levels of irritation, viz. disappointment, annoyance and anger. Sometimes, scale values of 1, 3 and 6 are assigned to these 3 levels of irritation to indicate lack of quality by summarizing numbers of participants giving such responses.

(10) Rating performance against internal standards through tests and checks carried out internally by simulating customers and service delivery processes

Some of these approaches may yield quantitative measures of quality, e.g. service operation performed on a customer-supplied manufactured product. Here, we can use the post-service duration of failure-free operation of the product or even the ability of the serviced product to carry out some intended operation when needed (yielding a yes/no response as a measure). In case of a customer value workshop, the average irritation score provides an inverse measure of service quality. With internal comparisons of service with internal standards under simulated conditions, the proportion of cases wherein the internal standards could be met can be taken as a measure, though not necessarily reflecting customers' perceptions truly. Customer satisfaction surveys properly designed and conducted can also offer reasonably good measures of service quality.

2.5 Measures of Service Quality

Measures like defect rate and proportion of items not required (but provided) involve subjective perceptions of customers. A defect is an instance of non-conformity to requirements stated and/or implied by the customer or even assessed by the provider on the basis of customers' statements, others' statements, etc. The important point is the agreement of the provider to comply with the requirements.

Depending on the nature and extent of non-conformity—detected internally by the provider before delivery or detected by the customer on delivery—the item(s) delivered may have to be reworked or even rejected. These eventualities are quite damaging to the provider's reputation. Sometimes, for some problem at the provider's end, the service process starts a little behind schedule (taking account of the resources available as also the urgency in delivery possibly indicated or simply implied by the customer). The extent of this delay as also its causes can be analysed **by the producer**. In some cases, a delayed start coupled with a fast production process may result in timely delivery. Delayed delivery is definitely a reflection of poor quality of service, even if this is associated with delivery of some products of good quality, unless the nature of the likely delay in delivery is explained by the provider to the satisfaction of the customer.

Certain items and/or features initially agreed upon may be missing from what has been delivered, mostly through oversight, and can make for poor quality in the perception of the customer. Similarly, some items or features might have been indicated as not necessary by the customer but were still delivered, probably to conform to certain practices rigidly. This may also lead to customer dissatisfaction.

2.5.1   Defect Rate

The defect rate—the most important of quality measures in case of services—is primarily based on customer perceptions and determined from customer responses to questions that are designed to reflect customers' assessment of quality as delivered in relation to what was expected, beyond their needs. Of course, certain non-conformities or deficiencies during service production and/or during service delivery can be identified by the provider. These also have to be taken into account in determining the defect rate. Disposition of provider-identified non-conformities depends on the Quality Policy of the provider organization.

In most cases, customers are required to give a rating on a five-point or seven-point scale, or to assign some rank (among named competitors for the same service). In the first case, we get an 'absolute' assessment of service quality for a given provider, while in the second we get a 'relative' assessment of quality of one provider compared to that of several providers.

Quality of what has been received is usually multi-dimensional, and we can obtain a rating for each dimension from each responding customer. It is a problem to combine the different ratings, and deriving only one scaled score corresponding to the single combined rating or to combine the scaled scores for the several ratings as these are. To introduce weights based on preferences or priorities, we invite the problem of aggregation once again, unless we go by the most agreed upon preferences or priorities.

As can be easily imagined, rating given by a respondent to service provided by a given provider will depend on the respondent's (1) expectations (2) previous experiences involving this provider as also other providers of the same service and

(3) peer influences. And responses may not really reflect assessments since different customers have different intentions to put in realizations truthfully. This problem may invite the use of randomized response technique to derive reasonably good estimates.

It may be desirable to analyse responses from a homogeneous group of respondents like those who are visiting the present provider organization for the first time or the group that has visited it earlier also.

In situations where the quality of service provided by a given organization has to be determined from rankings of several organizations with respect to quality of the service under consideration, we have to define an analogue of 'defect rate' from such rankings.

2.5.2   Use of Scaling Techniques

Let us assume a five-point scale with ratings Poor, Fair, O.K. (undecided/neutral) Good and Excellent. One can associate the first two ratings with a 'defect' and can take the proportion of responses in these two categories as the defect rate.

Alternatively, we assume an underlying continuum in terms of a normally distributed latent variable, viz. perceived quality, take the five categories of responses to correspond to ranges $(-\infty, x_1)$, (x_1, x_2), (x_2, x_3), (x_3, x_4) and (x_4, ∞) of the underlying latent variable and assign the scaled score

$$y_i = [\Phi(x_i) - \Phi(x_{i-1})]/[P(x_i) - P(x_{i-1})]$$

to category i. Here, Φ stands for the ordinate and P for the left-tail area under the standard normal distribution.

If f_i is the frequency of responses in category i (1–5), we can consider

$$(f_1 y_1 + f_2 y_2)/\Sigma f_i \quad \text{as the defect rate, say } z.$$

Some criticism of this approach centres round the use of a normal distribution with the highest concentration of responses (middle area) in the category 'O.K.' or 'Undecided'.

One can think of pair-wise comparison of perceived quality (be the same respondents) between two comparable (in terms of cost, convenience and other pertinent considerations) providers of the same service. And make use of Thurstone's method of paired comparisons to derive scale values for each service provider. Thurstone's product scaling is based on the matrix $((p_{ij})) = P_{m \times m}$ where p_{ij} is the proportion of respondents who prefer provider i to provider j. The scale values are obtained as

Another possible defect rate for a given service provider could be the measure

1 − (scale value for a given provider)/(maximum value across providers)

2.5.3 Modelling Defect Rate

To control or reduce the defect rate, we can build a model like

$$Z = f(X, Y) + e$$

where

X   a set of tangibles, which again can be graded by the customers
Y   a set of features of the employees (in the service organization) like their responsiveness, empathy.

X may take care of accessibility, availability of the service in its different presentable variants, etc.

We can use categorical data analysis techniques like logistic regression to study f. Once the form of f has been worked out, values or levels of X and Y can be found to minimize Z. Optimization will have to take into account any constraints on X and Y.

2.6 Measuring Relative Quality

One may like to place different providers of the same service on a linear scale where equal distances imply equal differences in quality. And quality assessment has to give due importance to several facets or aspects of quality. We can start with a homogeneous group of respondents and seek their rankings of the provider organizations in respect of each dimension of quality. For the sake of simplicity in analysis, tied ranks may be avoided; but respondents may find it difficult to avoid ties in some cases.

Before the ranks can be averaged across respondents, it may be desirable to test for the homogeneity of rankings by using Kendall's coefficient of m-rankings and Friedman's chi-square test.

Once averaging has been justified, one can build up a symmetric distance matrix among the organizations and then proceed to use multi-dimensional scaling to represent the organizations as points in a two-dimensional plane that allows easy visualization of the relative positions occupied by the different organizations in respect of quality in service.

Quality as is perceived by the user is now recognized as more important than quality as is declared by the provider or certified by some authorized body. This is more so in the case of services. And because perceptions are expected to be more variable across users, quality of services becomes more difficult to comprehend.

2.7   Dimensions of Service Quality

As is pretty well established, any attempt to improve the present state of affairs must be preceded by a consideration of the various factors or dimensions of the problem along with an analysis based on some model that links up factors with responses. In the context of service quality, different dimensions or facets of quality merit distinct recognition in any improvement effort.

Service quality has been discussed in terms of many dimensions including those related to the human beings involved. Not all these that follow apply to every service and some particular service involving a special customer and a special supplier may call for some other quality dimensions. The following is an indicative list.

Accessibility/availability;
Timeliness;
Adequacy/completeness;
Consistency;
Dependability/reliability;
Security and confidentiality (whenever needed);
Flexibility;
Complaint resolution;
Credibility and reputation;
Competence;
Courtesy;
Communication;
Responsiveness;
Tangibles.

As can be easily seen, some of these are physical or technical, some are functional or interactive, and others are corporate. The corporate dimension includes parameters like tangibles implying equipment, space, materials and other physical resources available with the service provider, which has to offer reasonable access to potential customers and also provide the services on offer, without pleading inability on a valid ground. It must have engaged competent personnel to man the service process, essentially the internal activities within the supplier organization and people who are courteous and who can communicate properly with customers and potential customers, respond to their requests, suggestions and complaints at the interface with the customers or potential customers. This way the organization has to earn good reputation and credibility. Dependability of a service can be assessed by the customer after its use or consumption, and a feedback is important to initiate any corrective or improvement action.

Adequacy or completeness of a service is a crucial consideration in some cases, where a potential customer may not like or afford to get segments of a service from different service providers and will usually look for a provider that has an umbrella to provide complete service which is in terms of a bunch of related services. A typical example would be the composite service required by a car that needs washing, fuel, some repair work, painting, etc., at one place.

Cloud computing—introduced in 2006—has become commonplace to deploy IT resources available in the virtual space essential for day-to-day operations of an organization relying on information and data security. 'Cloud Computing' can be simply described as IT services sold and delivered over the Internet. Three services are provided by different clouds for different service models, viz. Software as a Service (SaaS), Infrastructure as a Service (IaaS) and Platform as a Service (PaaS). Cloud can be private, public, community and hybrid.

In 'cloud computing', customers will attach importance to the following factors as determinants of service quality

Flexibility in payment;
Reduced IT infrastructure cost;
Data backup and recovery;
Ubiquitous access;
Scalability on demand;
Vendor reliability;
User-friendliness;
Customization
Speed and quality of broadband connection;
Availability on demand on a 24 × 7 basis;
Platform independence; and
Data security.

In a study involving a convenience sample of 95 IT professionals using a questionnaire and Likert's scaling of responses, it came out that the key service quality dimension that motivates adoption of cloud services is price competitiveness, while flexibility, security and reliability are moderately important, but they need improvement in hardware.

2.8 SERVQUAL Model

The need for improving quality was identified as an important strategy for marketing of services (Zeithaml 1985). There have been two major approaches to improvement of service quality; the first is in terms of an attempt to capture the gaps between service quality as is expected by the customers and the quality actually received in the perception of the customers, while the second analyses gaps between quality perceived by customers for one particular service provider and that for competing service providers or for the service provider judged as the 'best' by them. The first is based on some models to capture the service quality gaps (SQG's), and the other is essentially referring to Quality Function Deployment. The latter is directly focused on system features, component features and process plans which should be taken care of by an organization trying to improve its service quality. Incidentally, QFD is equally, if not more frequently, applied in manufacturing organizations. And in either approach, we have to contend with variations in perceptions of different customers with the same service provided by the same service provider as also in perceptions of comparative quality over different service providers. Thus, both the models have to go by a collective analysis. A popular generic instrument for assessing service quality which has been implemented with reasonable success in a wide variety of situations is the one suggested by Parasuraman et al. (1985, 1988) and popularly known as SERVQUAL. This was reviewed by Asubonteng et al. (1996). Several other models for this purpose have been discussed by Gronroos (1984), Lewis & Mitchell (1990) and Ghobadian (1994). In the SERVQUAL model, five broad dimensions of service quality were finally taken as a result of extensive investigations covering different types of services, starting with ten major determinants of perceived service quality, viz. access, communication, competence, courtesy, credibility, reliability, responsiveness, security, tangibles and ability to understand customer requirements.

- Tangibles: appearance of physical facilities, equipment, personnel, printed and visual communication material;
- Reliability: ability to perform the promised service dependably and accurately;
- Responsiveness: willingness to provide appropriate service and, generally, to help customers and provide prompt service;
- Assurance: possession of required knowledge and skill as also courtesy of employees and their ability to inspire trust and confidence;
- Empathy: caring, individual attention that is provided to customers.

In some applications, a total of ten dimensions have been created by essentially preserving the first three and expanding the remaining two as competence, courtesy, credibility, security, access, communication and understanding.

The original SERVQUAL instrument consists of 22 statements, each of which is related to one of the five major dimensions. Each statement appears in two forms: one designed to measure Customer Expectations (E) about organizations in general in the service area being examined and the other to measure Customer Perceptions (P) about the particular organization whose service quality is being assessed. Statements in both sections are scored by the customer on a seven-point Likert scale ranging from Strongly Disagree (1) to Strongly Agree (7), with no labels attached to scores 2–6. It is possible to associate labels like Disagree and Agree with these two scores.

The perceived service quality (or gap) score, usually denoted by Q, is then calculated for each statement as P score $- E$ score, implying a Q score ranging between -6 and $+6$. A negative Q score implies a level of service quality below that expected by the customer of a service provider in this Industry. A high positive score would imply a level of perceived service quality higher than that expected by the customer. For example, a Q score of 6 to be obtained as a P-score of 7 against an E-score of 1 really means that the customer was in some sense excited to perceive service quality as much beyond what he had expected. The customer is delighted. A score of -6 can be similarly interpreted to imply extreme customer dissatisfaction or even customer disgust.

Such gap scores admit of analysis in several different directions providing together a comprehensive view of service quality trends over time and differentials over customer groups differentiated in terms of their expectations. To start with, these gaps should be analysed across the five dimensions to find out dimensions crying out for improvement and, if necessary, to prioritise the dimensions for improvement efforts. This will directly call for an analysis of changes in gap scores over time as improvement actions continue to be taken. Discrete temporal changes —in fact reduction—can be possibly associated with distinct quality improvement exercises. Further, if large positive gaps are noted, associated costs of improvement efforts should be examined to work out feasibility of retaining such scores. Gap scores should also be compared over different market segments or customer groups to identify what measures are needed to enhance levels of satisfaction among all customer groups or even to work out adjustments in marketing strategies if costly improvements are not on the cards. Improvement exercises are usually fortified by benchmarking against 'best in class' service providers in the perception of customers. This means a comparison of gap scores to identify gaps between P-score of any one service provider and that of another or that of the 'best in class' organization. And such gaps may be more prominent when separate dimensions of service quality are considered in any comparison. In fact, one may go further to examine different parts or component of a service and analyse gap scores to work out targeted improvements in those parts or components.

Such a gap analysis can also provide potential information on the relative importance attached to each of the five dimensions by the customers, given that each responding customer is requested to indicate the importance of each dimension in yielding the overall service quality on a scale from 1 to 1000 in his perception.

Commenting on the nature and range of gap scores, Dotchin and Oakland (1994) comment 'It is unlikely, for economic considerations, that a firm will attempt consistently to exceed their customer expectations. Even if they did, it is probable that expectations would actually rise and create a new datum. It is usual, therefore, for SERVQUAL quality scores to be negative values'.

Other gaps in the service process may contribute to customer dissatisfaction, like

- Understanding gap—between customer expectations and management perceptions of these expectations;
- Design gap—between management-perceived expectations and service quality specifications;
- Service delivery gap—between service quality specifications and actual delivery of service

- Communication gap—between service specification and delivery on the one hand and the service provider's external communications to customers (what has been promised in media, service provider's publicity, etc.).

One more gap, viz. the potential discrepancy between the expectation and the perception of the customer, has also been considered by some investigators.

Design gap and service delivery gap are matters of internal concern and measures to reduce these gaps should elicit adequate attention of executive management. The first and the last gaps, viz. understanding gap and communication gap, involve interactions with customers for which management has to initiate measures to analyse feedback from customers and regulate its promotional activities not to create unnecessary or confusing claims of high quality.

Criticisms of the SERVQUAL model include (a) neglect of the price factor; (b) confusion among outcome, process and expectation; (c) dominance of perceptions; etc. Expectations are not necessarily consistent or predictable and are influenced by management communications and advertising. Value seems more stable.

A synthesized model comprising 14 service quality gaps was suggested by Candido and Morris (2000). The literature speaks of static and dynamic models, and Candido and Morris have proposed a mixed model (2001).

Several major concerns should be addressed by research workers and practitioners engaged in measuring service quality. These are (1) Why do we measure? may be to provide guidance for improvement programmes and optimum use of available resources. (2) What should be measured—quality or value? It is accepted (Bennington and Cummane 1998) that value creation is the factor that differentiates between service providers and that value, rather than quality, as perceived by the customer, better reflects the customer's overall assessment of a service. The perceived value gap between what is experienced by the customer and what is claimed or designed by the provider should be captured. This, incidentally, calls for an interface between the customer and the provider (or its representative), and hence, mail questionnaire method should be avoided, whenever possible. Expectations and that way perceptions vary across customer groups and markets catering to their needs, and thus, we should examine each segment of such groups separately. Findings from a survey or on the different gaps should better be displayed graphically for easy comprehension of their implications by management of service providers.

2.9 Quality in E-services

Recent years have seen a spurt in e-services offered by web-based service providers and accepted by technology-friendly users. As the novelty in electronic services wears off, consumers become less willing to tolerate poor service quality. Electronic service delivery is different from traditional service delivery, owing to the lack of human interactions that are so fundamental in traditional service encounters, but which have been largely replaced by technology. Electronic service quality is an emerging concept. One of the most widely accepted definitions of e-service quality was conceptualized by Zeithaml et al. (2000) as 'the extent to which a website facilitates efficient and effective shopping, purchasing and delivery of products or services'. This, however, can be regarded as simply website quality and ignores other important facets of electronic service like order fulfilment and returns.

Several scales to measure e-service quality were offered by a host of authors (Yaya et al. 2012) considering different sets of parameters or dimensions of e-service quality. Some of these failed to capture service quality related to e-commerce, some centred around user satisfaction rather than service quality, some provide transaction specific assessment of quality and some even were focused on physical off-line. In most of the previous scales, dimensions of service quality were chosen arbitrarily and the scales lacked adequate validity and reliability.

To address deficiencies in the original definition of e-service quality that eventually boiled down to website quality or to Web interface design quality, Zeithaml and associates (2000) analysed a number of website features at the perceptual attribute level and categorized them in terms of an e-SERVQUAL scale. Subsequently, (2002) seven dimensions for evaluating e-service quality were grouped into two classes, and two scales were derived from them to measure e-SQ. The dimensions for the first scale are

Efficiency: the ease and speed of accessing and using the site
Fulfilment: the extents to which the site's promises about order delivery and item availability are fulfilled
System Availability: the correct technical functioning of the site
Privacy: the degree to which the site is safe and protects customer information.

The second scale E-RecS QUAL was designed to suit customers who had occasions to encounter problems in handling and return, compensation for problems faced and availability of assistance. This scale is composed of 11 items (as is the first scale) grouped into three quality dimensions.

Responsiveness: effective handling of problems and returns through the site
Compensation: the degree to which the site compensates customers for problems
Contact: the availability of assistance through telephone or on-line representatives.

While these two scales have been widely used in several countries, shortcomings have been pointed out by research workers; e.g., E-S-QUAL does not make a distinction between e-retailers who sell products and those who sell services.

Issues pertaining to the choice of service quality dimensions, the number of questions to be put across in respect of each dimension along with the conversion of responses to scores, the demonstration of content validity, of discriminative ability and of test-retest reliability continue to pose problems to researchers.

2.10   Improving Quality in Services

Initiatives for improvement in service quality have to take place in both the soft and hard aspects, in both the service (generation) process as also in the service delivery process. And for the latter, we have to focus attention on the human aspects, the interaction between the provider and the recipient. To quote from Harvard Business Review 'too many service companies have…concentrated on cost-cutting efficiencies they can quantify, rather than on adding to their products' values by listening carefully …providing the services their customers genuinely want …the cost of losing this battle is unacceptably high'.

Measures recommended to improve service quality apply equally well to efforts to improve manufactured product quality also. Some of these are

1. Carry out annual image surveys and more frequent customer satisfaction surveys.
2. Measure the satisfaction of all direct and indirect customers—users, agents, representatives, etc.
3. Be objective and quantitative by introducing categories, if not measurements.
4. Do not ignore complaints which appear to be 'not symptomatic' of a larger problem.
5. Technology and technology support system like computerized registration/reservation/billing system has a big role to influence customers.

Properly planned and effectively executed customer satisfaction surveys will yield information about factors that beget satisfaction or otherwise. Factors have been generally categorized as

- Hygiene factors—things expected by the customer and causing dissatisfaction when not delivered or provided, e.g. paper napkins in an eatery;
- Enhancing factors—things which may lead to (or add to) customer satisfaction, but when not delivered do not necessarily cause dissatisfaction, e.g. mouth fresheners in an eatery;
- Dual threshold factors—things which when delivered above a certain level of adequacy lead to satisfaction, but when delivered at a performance level perceived to be below that threshold level are dissatisfaction.

Initiatives to improve service quality must emphasize on training of people involved in service delivery or in interacting with customers before or after delivery to ensure courtesy, helpfulness, sympathy, and similar other aspects of corporate quality. Of course, to be equally emphasized will be physical quality in terms of tangible facilities offered to customers as also technical quality which corresponds to the hard core of service quality. And there again, competence of people in the service production process is a must.

A programme to improve quality in services delivered and effectiveness as well as efficiency of the complete service process has to include an effort to identify.

Characteristics, which if improved, would satisfy the customer and benefit the service organization most.

Any changing market needs or needs of concerned regulatory bodies, if any, that are likely to affect the grade of service provided.

Any new technology or any innovation that can be adopted to bring about a perceptible improvement in the service (production) process. A simple automation in either service production or service delivery or transactions with customers can always be introduced.

Any new entrant in the set of service providers, in the particular service(s) under consideration that could open up new avenues to customers to search and select a service provider.

Any deviation from the specified service quality (as detected internally or reported by customers and others) due to ineffective or insufficient quality checks or controls.

Opportunities for reducing cost, while maintaining and even improving the service quality as is being provided currently.

Service quality improvement activities should address the need for both short-term improvement as well as long-term improvement. Short-term improvement can be worked out by some more training and retraining of the people involved and by a stricter check and control schedule at different stages of the service production and service delivery processes. Materials used in either of these processes should be inspected adequately to take care of possible deficiencies. Sometimes changing the internal arrangement of certain existing physical facilities and of the manner in which services are being provided may yield some concrete improvement results. And to achieve this, we may not necessarily need to augment our physical resources. However, long-term improvement may require more investment in augmenting and improving physical facilities available to customers.

An improvement programme should include

Identifying relevant data needs, sources and mechanisms to collect such data (data could be opinions, reports, test results, certificates etc.);

Data analysis and interpretation, giving priority to those activities which exert the greatest adverse impact on perceived service quality;

Feedback of results of analysis to operational management so that appropriate corrective and preventive actions can be initiated at the earliest;

Periodic review by senior management of long-term quality improvement actions, their implementation and results derived.

Members from different areas of the organization working together may offer fruitful and novel ideas directed towards improving quality and reducing cost. In fact, management should encourage personnel at all levels to contribute to the quality improvement programme, with recognition for their participation and effort. Often, suggestions offered, may be pro-actively, by some junior staff members like those involved in service delivery, are just ignored by Management.

A popular approach to service quality improvement is the use of Quality Function Deployment (QFD) technique. The QFD is a system for designing a product or a service, based on customer demands and competitor evaluation, with the participation of members of all functions of the supplier organization. It is a cross-functional planning tool which is used to ensure that the choice of the customer is deployed through all the stages within an organization. The goal of QFD planning effort is to maximize customer satisfaction. QFD is a kind of conceptual map that provides the means for inter-functional planning and communications. QFD builds up the House of Quality in four stages. In the first stage, performance or other physical requirements of the customer about the service to be provided along with their respective priorities are considered and are related to different specifications about components of the service. Associations among the component features are also taken into considerations. In the second stage, component features are translated into component properties.

With rising expectations of customers or clients or citizens or any other stakeholders and rapidly advancing technologies, service quality improvement eventually calls for innovation—innovation in the nature and type of service, the mode of delivery, the mechanism to elicit feedback, the method for redressal of grievances, the maintenance and deployment of resources, and related processes. And what is required to introduce innovation is a drastic change in the mind-set of the people engaged in the service-providing organization. For the management at the highest level, this implies allowing people in the organization adequate freedom and flexibility to think out-of-box, to experiment with novel ideas and to try out novel solutions—even on a pilot scale—without affecting services being currently provided.

2.11 Quality Management in Public Service

The government in any country in terms of its departments and branches along with all agencies whose operations and interactions with members of the public at large are overseen and regulated by the government encompasses a whole range of service providers to the people who depend a lot on their services. Quality of such services has a significant impact on the Quality of Life enjoyed by the people and hence deserves appropriate consideration.

An organization which provides service(s) to public at large and/or whose activities influence public interest is usually recognized as a public service provider. Government departments, regulatory bodies, public utility service providers, like banks and financial institutions, health care providing organizations, educational institutions specially those administered by the government, transport service organizations, municipal corporations and other local self-government bodies illustrate such organizations. Over the years, recipients of such services have been demanding better quality (Gaster 1995). IS 15700 (2005) is the Indian Standard on Quality Management Systems—Requirements for Service Quality by Public Service Organizations to provide guidance on service quality for such organizations. Service providers in the private sector may also derive strength from this standard.

As in other Indian Standards, there are some clauses which are generic and some others which provide relevant definitions, In fact, the last two clauses are the more important ones. Clause 7 of this standard which is specific to such organizations refers to 'Citizen's Charter, Service Provision and Complaints Handling'.

The Citizen's Charter, according to this standard, shall contain

(a) vision and mission statements of the organization;
(b) list of key service(s) being offered by the organization; and
(c) measurable service standards for the service (s) provided and remedies available to the customers for non-compliance with the standards.

Further, the Charter shall

(a) represent a systematic effort of the organization to focus on its commitment towards its customers.
(b) be simple and easily understandable and also available in local language, as required.
(c) be non-discriminatory.
(d) describe or refer to complaints handling process.
(e) include the name, address, telephone number and other details of the public grievance officer.
(f) be periodically reviewed for updation and continual improvement.
(g) highlight organization's expectations from its customers, wherever required, and
(h) provide information the date of issue of the citizen's Charter and persons who were consulted during its preparation.

Clause 8 discusses 'Implementation, Monitoring, Measurement and Improvement'.

It is important to remember that the Citizen's Charter is analogous to a process plan which provides the yardstick against which a process including the service process or the service delivery process can be assessed.

2.12 Concluding Remarks

While perceptions of stakeholders particularly end-users. dominate the assessment of quality of any product or service and render it vulnerable to subjective variations across individuals and, even for the same individual, from one context to another, quality of service is really a challenging entity. Despite improvements that can be brought about in the service production by objective means and people at the provider-customer interface are geared to provide satisfaction to the customer in terms of their behaviour and communication, service quality as judged by the customer right during its delivery or later on during service consumption may still be assessed as poor. As pointed out earlier, expectations about the service requested dominate over the needs explicitly stated and expectations are difficult to be read by the service provider in the service organization. May be, experience in dealing with different customers can be of some help.

Service quality has been an issue of great importance in Market Research and has attracted application of quantitative tools and also principles and methods in human behaviour. And a complete removal of humans with their behaviour patterns from the arena of services is not a feasible proposition, even if we think of e-services or of introducing as much of automation as possible in service production and delivery.

References

Asubonteng, P., et al. (1996). SERVQUAL revisited: A critical review of service quality. *Journal of Services Marketing, 10*(6), 62–81.

Bennington, L., & Cummane, J. (1998). Measuring service quality: A hybrid methodology. *Total Quality Management, 9*(6), 395–406.

Candido, C. J. F., & Morris, D. S. (2000). Charting service quality gaps. *Total Quality Management, 11,* s463–S472.

Candido, C. J. F., & Morris, D. S. (2001). The implications of service quality gaps for strategy implementation. *Total Quality Management, 12,* 825–833.

Dotchin, J. A., & Oakland, J. S. (1994). TQM in services. Part I understanding and classifying services. *International Journal of Quality and Reliability Management, 11*(3), 9–26.

Gaster, L. (1995). *Quality in public services.* Buckingham: Open University Press.

Ghobadian, A., et al. (1994). Service quality: Concepts and models. *International Journal of Quality and Reliability Management, 11*(9), 43–66.

Gronroos, C. (1984). A service quality model and its marketing implications. *European Journal of Marketing, 18*(4), 36–44.

IS 15700. (2005). Requirements for service quality by public service organisations.

Lewis, B. R., & Mitchell, V. W. (1990). Defining and measuring the quality of customer service. *Marketing, Intelligence and Planning, 8*(6), 11–17.

Parasuraman, A., et al. (1985). A conceptual model of service quality and its implication for future research. *Journal of Marketing, 49*(4), 41–50.

Parasuraman, A., et al. (1988). Servqual: A multiple item scale for measuring customer perception of service quality. *Journal of Retailing, 64*(Spring), 12–40.

Parasuraman, A., et al. (1991). Refinement and re-assessment of the SERVQUAL scale. *Journal of Retailing, 67*(4), 420–450.

Parasuraman, A., et al. (1993). More on improving service quality measurement. *Journal of Retailing, 69*(1), 140–147.

Yaya, L. H. P., et al. (2012). Measuring e-service quality: Reviewing E-S-QUAL. *Economic Quality Control, 27*(1), 19–41.

Zeithaml, V. A., et al. (1985). Problems and strategies for services marketing. *Journal of Marketing, 49*(Spring), 33–46.

Zeithaml, V. A., et al. (2000). *E-service quality: Definition, dimensions and conceptual model.* Working paper, Marketing Science Institute, Cambridge, MA.

Zeithaml, V. A. (2002). Service quality delivery through web-sites: A critical review of extant knowledge. *Journal of the Academy of marketing Science, 30,* 362–375.

Chapter 3
Managing Quality Today

3.1 Introduction

While Quality Management has been a much-discussed and much-documented subject not merely in the Industrial World but also in all sorts of transactions between a provider and a recipient, managing quality today has become a significantly different task coming out of its pristine boundaries, responding to the remarkable changes that have characterized industry and business today. And a key feature of the changes is the exponentially increasing demand for better quality emanating from greater awareness about more and more advanced production processes.

Quality Management today cannot be regarded as a natural continuation of concepts, constructs and practices bearing on quality since the days of Industrial Revolution. Recent times have introduced some fundamental changes in Quality Management in response to certain socio-economic changes and the advances in Science and Technology. The following stand out clearly to indicate the paradigm shifts in Quality Management:

- Quality directly integrated with business.
- Focus shifted from products to processes.
- Proactive process adjustments dominate reactive product rectification.
- More emphasis is placed on planning (design as well as experimentation) than on execution.
- Technological, managerial and human aspects are reckoned as more important than purely technical (including statistical) considerations.
- Data-based decisions, data-dependent (sequential) plans and procedures are preferred.
- Greater use is made of computers, artificial intelligence and expert systems.
- Emphasis is placed on a systems' approach.

© Springer Nature Singapore Pte Ltd. 2019
S. P. Mukherjee, *Quality*, India Studies in Business and Economics,
https://doi.org/10.1007/978-981-13-1271-7_3

This is not an exhaustive list and some more shifts in Quality Management practices will meet the eyes in some forward-looking and excellence-seeking organizations.

The material that follows is meant to bring out some of the nuances of Quality Management as can be made out from the best practices and as can be used to build up a broad framework that may appear to be too abstract and overt but can explain some of the fine points involved in todays' Quality Management and some seemingly incredible combination of skills displayed by the modern Quality Manager.

3.2 Some Defining Developments

The function of managing quality in manufacturing industries and, later on, in service-providing organizations—at the corporate or at the operational level—has been greatly influenced by directions and guidance provided by some eminent and committed exponents of quality (often referred to as Quality Gurus) over the last seven decades or so. Subsequently, the emergence some new concepts, methods and techniques and the appearance of some standards on Quality Management and related subjects developed by National Standards Bodies as well as the International Organisation for Standardisation (ISO) gave a comprehensive coverage of Quality Management in generic terms, followed up by some standards meant for some specific manufacturing situations.

While Quality Systems and System Standards have been discussed in a later chapter, we now provide a glimpse into the Quality Philosophies and Practices recommended by some Quality Gurus. Although their approaches differ in areas of emphasis, priority considerations for the various task related to quality, techniques to be used and a few other issues like the role of management, the common thrust has been on the concept of continuous improvement and its pursuit. They all have emphasized removal of unwanted variation and planning and improving work processes. In fact, the whole gamut of wise counsels can be summarized in terms of the following:

1. Top management must lead the change process.
2. The change process requires a cultural transformation.
3. Quality is integrated into all functions and looked upon as a separate function.
4. People and not machines are the driving force behind quality.
5. Quality requires participation of everyone in the organisation.
6. Motivation alone cannot bring about change, though it is important.
7. Company-wide education and training is essential for long-term improvement.
8. Continuous improvement demands commitment and single-minded attention of top management.

Among the names recalled almost always in a discussion on quality, we may start with William Edwards Deming (1986). His fourteen points of management which, according to him, aimed at making it possible for people to work with joy. He stressed the role of management to lead the comprehensive and continues improvement of the system and the continuous development of people as individuals and teammates. Deming is also remembered as one who introduced the Plan-Do-Check-Act cycle—also called the Deming Cycle—for a systematic approach to managing quality.

Next comes Juran (1988) who held the view that quality improvement takes place project by project and in no other way. His definition of quality as fitness for use as perceived by the customer led to the recognition of quality beyond the technical aspect of quality control to the Management arena. His Quality Trilogy combines quality planning, quality control and quality improvement as the three basic processes through which we manage quality provided a big push to the Quality Movement. His approach is based on four elements viz. (1) establish specific goals to be reached (2) establish plans for reaching the goals (3) assign clear responsibility for meeting the goals and (4) base rewards on results achieved.

Fiegenbaum (1983) introduced the Total Quality Control as a management tool with four steps viz. (1) setting quality standards (2) appraising quality with reference to the standards (3) acting when standards are not met and (4) planning for improvement in the initially developed standards. He also emphasized the concept of 'cost of quality' as a way to measure the benefits of adopting the TQC approach. In a precursor to the TQM movement, Fiegenbaum defined Total Quality Control as an effective system for integrating the quality development, quality maintenance and quality improvement efforts of the various groups within an organization so as to enable engineering, production, marketing and servicing at the most economical levels that allow for full customer satisfaction.

Philip Crosby (1984) spoke of the four absolutes of quality viz. (1) Quality is conformance to requirements and nod goodness or elegance, (2) The system of quality is prevention, not appraisal or inspection, (3) The performance standard is zero defects, not 'That's close enough', and (4) The measurement of quality is the price of non-conformance or the cost of poor quality, not quality indices. He also added that 'there is no such thing as a quality problem'. Crosby is also known for his 14-step improvement process adopted by many companies.

Kauro Ishikawa (1985) regarded as the father of quality circles in Japan is famous for the Ishikawa Diagram which has been an effective tool used by a quality circle to analyse and solve a quality problem. And is the most widely used tool to analyse the causes behind a known (adverse) effect. Ishikawa also stressed on enhancing the Quality of Working Life in order to enhance productivity.

Genichi Taguchi (1988) has become a household name in quality analysis and improvement efforts. He defined the 'quality loss function' as the cost incurred or profits foregone relative to a baseline performance, measured as a function of the deviation from a specified target or from an ideal performance level. He provided a system of system, parameter and tolerance designs for carrying out process experiments with signal-to-noise ratio as a response variable to determine optimal process level. He also popularized the use of orthogonal arrays and linear graphs to ensure economic designs of experiments.

Shigeo Shingo advocated to propose his version of zero defects as Poka Yoke or 'defect = 0' by pushing the idea that errors be identified as and when they occur and corrected right away. He emphasized the need for good engineering and process investigation rather than an exhortation or a slogan to achieve zero-defect quality.

Masaaki Imai brought in Kaizens to inject small improvements in many places by anybody within the organization.

Shigeru Mizuno clearly delineated the steps to the practical implementation of Quality Management.

H. James Harrington emphasized the key role of the first-line supervisors besides managers in successful implementation of any quality improvement programme. Richard J. Sconberger stressed on simplicity in designs and manufacturing operations to cut down on manufacturing lead time.

As mentioned earlier, the Malcolm Baldrige National Quality award introduced by the US Department of Commerce in 1987 and several other international as well as national quality awards also provide a big boost to the Quality Movement by encouraging industries to join the competition and qualify for the award. The Deming prize established in 1951 by the Japanese Union of Scientists and Engineers is offered to individuals and groups who have contributed significantly to quality control.

3.3 Recalling Deming, Juran, Crosby and Imai

Deming first visited Japan in 1946 as a representative of the Economic and Scientific Section of the US Department of War. He spoke directly to top industrialists and leaders in Japanese industries and eventually became instrumental to the turnaround of Japanese industrial fortunes. He is famous for his 14 points for management to be adopted and 5 deadly sins to be avoided. He was critical of the prevailing management practices and strongly batted for use of statistical methods and tools to enhance quality and productivity. He was a great philosopher who did not want simply current problems to be solved, but advocated a major transformation of the quality culture in the Japanese industry.

A mere listing of Deming's 14 points for management may not do justice to the management philosophy he suggested, incorporating some radical changes at the corporate level. However, one can definitely read some meaning into each point, stated briefly without any elaboration, as has been attempted in the following.

1. Create constancy of purpose for continual improvement of product and service.
2. Adopt the new philosophy for economic stability.
3. Cease dependency on inspection to achieve quality.

4. End the practice of awarding business on price tag alone.
5. Improve constantly and forever the system of production and service.
6. Institute training on the job.
7. Adopt and institute modern methods of supervision and leadership.
8. Drive out fear.
9. Break down barriers between departments and individuals.
10. Eliminate the use of slogans, posters and exhortations.
11. Eliminate work standards and numerical quotas.
12. Remove barriers that rob the hourly worker of the right to pride in workmanship.
13. Institute a vigorous programme of education and retraining.
14. Define top management's permanent commitment to ever-improving quality and productivity.

Deming referred to lack of constancy of purpose and commitment to plan product and service that will have a market, short-term profit motivation revealed in cutting down corners, cutting down on training or on investment meant to benefit employees, etc., performance appraisals focused on current performance indices and not recognizing ideas and actions likely to bring in long-term benefits, job hopping among managers causing instability in decision-making and use of visible figures ignoring invisible aspects of customer satisfaction, opportunity costs in the absence of employee retraining, inadequate maintenance of plant and equipment, etc., as the five deadly diseases adversely affecting business prospects.

Appreciating the fact the Quality Management has to be a part of corporate management, Deming contributed a lot to modernize management philosophy as such. Some of his remarks bearing on Quality Management are noteworthy in this connection and are reproduced below.

The workers are handicapped by the system, and the system belongs to the management. Putting out fire is not management. Slogans, exhortations and posters with targets to be met (without providing means to meet them) are directed to the wrong people. They take no account of the fact that most of the trouble comes from the system. Abolish the distinction between major and minor defects. A defect will be a defect. Quality and innovation are inseparable: necessary ingredients for achievement of quality are innovation and profound knowledge of variation. There is no process, no capability and no meaningful specifications except in statistical control.

Juran was invited by the Union of Japanese Scientists and Engineers to visit Japan in 1950 to conduct seminars for top- and middle-level executives. Contrasting Juran with Deming, Logothetis remarked 'If Deming is the Old Testament prophet of quality, Juran is the high priest of quality'. But on fundamental principles, they agree a lot. While Deming emphasises principles and methods, Juran is focused on results. Juran believed that senior and middle managers along with workers and supervisors are together responsible for quality. Deming's key issues are variation, continuous improvement, optimization of the overall system. Jura concentrates on the quality trilogy and on a goal-setting approach towards increased conformance and decreased costs of poor quality.

Juran recommends a formula for results which includes four steps viz.

(a) Establish specific goals to be reached—identify what needs to be done. The specific projects to be taken up and completed successfully.
(b) Establish plans for reaching the goals—provide a structural process for going from here to there.
(c) Assign clear responsibilities to individuals and groups for meeting the goals.
(d) Base rewards on results achieved—feed back the information and utilise the lessons learnt and the experience gained.

Juran speaks of a quality breakthrough needed to reduce substantially chronic wastes and to achieve an improved level at which quality should be controlled. In the language of Deming, such a breakthrough can be achieved by innovation and propound knowledge of the system. And Juran gives out a sequence of stages to be passed through for a breakthrough in quality viz. proof of the need, project identification, organization for improvement, the diagnostic journey, remedial action, breakthrough in cultural resistance to change and holding the gains—control at the new level. One can easily find these as early indicators of the five steps involved in the Six-Sigma approach to quality improvement formalized somewhat later.

Crosby, a management consultant, taught thousands of company executives and quality professionals and had has been famous for his concept of zero defects. He starts with the diagnosis of a troubled company using five symptoms viz.

1. The Company has an extensive field service for reworking and corrective actions.
2. The outgoing products normally deviate from the customers' requirements.
3. Management does not provide a clear performance standard, so the employees develop their own.
4. Management denies that is (management) is the cause of the problem.
5. Management does not know the price of non-conformance.

Such an organization, according to Crosby, needs a quality vaccine consisting of three components viz. determination, education (for managers) and implementation with a motivated workforce with adequate guidance. Crosby refers to several absolutes for Quality Management. These are as follows:

Absolute 1   The definition of quality is conformance to requirements, not goodness.
Absolute 2   The system of quality is prevention (and not detection and correction).
Absolute 3   The performance standard is 'zero defects'.
Absolute 4   The measurement of quality is price of non-conformance.

Crosby suggested a 14-step process for quality improvement, where the numbering of steps is not that important and steps can be taken up in parallel.

Imai, a Tokyo-based international management consultant, is best known for his kaizen approach. Kaizen focuses on continual but small improvements, not comparable with breakthrough improvement brought about by innovation. Kaizen quickly became a part and parcel of the Japanese business philosophy, which had already been moulded by the teachings of Deming. Kaizen has been incorporated within the framework of Total Quality Control (a different name for TQM). Kanban, Small Group Activities like quality circles, Poka Yoke or Mistake—Proofing and related tools were all introduced and implemented in the spirit of Kaizen. And Kaizen has since been developed to take care not-so-small improvement projects.

3.4 New Paradigms in Quality Management

Why do we talk about new paradigms in Quality Management?
    The answer lies in the facts that

(a) Human behaviour is changing throughout the world—as producer/server as also as consumer/user.
(b) Pattern of human consumption is changing with the earlier focus on consumption of 'goods' shifting to consumption of 'services'.
(c) Quality has come to be recognized as a management concept and a Management Responsibility.
(d) Quality has become an international business language.
(e) The following propositions regarding quality are now played down as 'myths' quality cannot be measured and quality and costs cannot be harmonized.
(f) Need for measurements and for information-based decisions has come to be appreciated.

Responding to the developments stated above, Quality Management today is

1. an essential component of corporate and strategic management (beyond being absorbed in functional management);
2. oriented to identify and weed out non-value-adding processes and activities;
3. deriving strength from human and technological resources besides techniques for planning, control and improvement;
4. supplemented strongly by an effective cost-management system;
5. aimed at achieving the best return (through value addition) out of quality and not just achieving the best level of quality in products and services with the available or augmented resources.

While discussing Quality Management practices earlier, experts used to make a distinction between the practices prevailing in the West and in Japan by noting the following approaches adopted in these two situations.

West	Japan
Product-oriented	People-oriented
Process-oriented	Society-oriented
System-oriented	Cost-oriented
Specification-oriented	Consumer-oriented

Today, 'best of the world' is the guiding philosophy, knowledge and its applications flow freely across the globe and Quality Management is a transnational activity that blends into itself the niceties of all types of management.

3.5 Quality Integrated with Business

In the industry and business world today, quality has been integrated with business so that quality of products and services are no longer discussed in isolation of business. This transition from a context where quality was taken care of independently of business considerations directly, important business decisions like expanding business markets and business operations would not be directly linked to quality of products and services, which of course would get due attention incidentally on any such decisions being taken has been achieved through the following instruments.

1. A wide umbrella of quality to cover various business processes (much beyond products). Quality has to refer to quality of performance of any activity in any department or division. Thus, the focus should be on process quality, where the process could be responding to a customer request or selecting some vendor(s) for procuring some materials or equipments or services which are not required continuously or the manner in which security staff behaves with an important guest or a consultant invited to render expert advice on a subject. A negative stance at any of these instances may adversely affect business in a broad sense.
2. Emphasis on customer's assessment/perception of quality of products and services and focus on customer's satisfaction. If more customers are satisfied during more transactions, they will narrate their experiences among fellow customers and that may lead to more potential customers or even to more customers. This may be achieved by the organisation's own assessment of quality of its products and services as satisfactory or even excellent. Here customers do include internal customers who should express their frank assessment of quality of all inputs to the processes they handle which come from other individuals or groups within the organization.
3. Involvement of all people in business in quality-related activities in terms of implications of a Quality System. And involvement is ensured only through a conviction that every one in the organization counts in the matter of quality improvement. And this conviction has to be created only through proper awareness programmes.

4. (Top) management involvement in quality review and improvement activities. Quality cannot be left to middle management or to the frontline workers only. In fact, adequate guidance for quality improvement has to come from top management personnel, despite the fact that they might not have knowledge and experience to suggest improvement actions by themselves. They can invoke knowledge management principles to make the best use of 'internal' knowledge —may be not tapped properly. This apart, they should take initiatives to identify and involve external experts whenever called for.

5. Focus on proactive process adjustments and not on reactive product/service rectification/correction/rejection. A diligent comparison of costs of making such adjustments by way of corrections, corrective and preventive actions taken in time to prevent occurrences of defects will reveal that such costs are definitely less than costs of poor quality associated with defects in the final products, detected internally or by the customers during purchase or use. TQM has the greatest focus on controlling costs of poor quality.

3.6   Quality Philosophy, Policy and Practice

Quality management in any organization is explicitly or (more often) implicitly guided by a *QUALITY PHILOSOPHY* which gets reflected in a *QUALITY POLICY*, which, in turn, is implemented through a *QUALITY PRACTICE*.

The **Quality Philosophy**, which the top executive representing all the interest groups or stakeholders may adopt, is revealed by the nature of response to the following questions, pertaining to the felt NEED FOR QUALITY, the CONCERN FOR QUALITY, the EXPECTED GAIN FROM QUALITY and the EXTENT OF ALTRUISM to be pursued. In other words, the Quality Philosophy spells out management's attitude towards quality as also its explicit concern for quality.

On NEED PERCEPTION, attitude of top management towards quality, priority of one statement over the other may be revealing.

1. Quality has to be paid for, insisted upon, conformity to Quality Standards made legally binding versus quality should always be maintained at the optimum level (taking into account factors specific to the organization).
2. Quality takes care of itself and needs no special effort or attention versus optimum quality levels can be achieved and maintained only through a planned and systematic effort.
3. Quality and quantity can never go together versus efforts to build optimum quality also lead to enhanced productivity.
4. Quality and specially reliability (long life) will adversely affect demand, hence production and hence profit versus quality and reliability add to the value of a product/service and value-based pricing more than offsets investments in quality, if such consequences arise.

Quality Philosophy spells out overall intentions of the management in regard to quality—reflecting management's attitude towards quality and also its explicit concern for quality.

Quality Policy gives out broad directions (by way of Quality Plans and Quality Strategies) about quality.

Quality Practice is revealed through detailed processes, procedures and work instructions to implement the Quality Strategies and Quality Plans.

On CONCERN FOR QUALITY, the following statements may be put forth for preference indication.

1. Top management appreciates quality efforts and provides some/all support versus top management is *involved* in quality efforts and provide necessary *leadership*.
2. Quality gets attention only when problems arise or occasions demand versus quality receives continual and regular care.

The **Quality Policy** should spell out overall intentions of the management in regard to quality and broad directions (by way of Quality Plans, and Quality Strategies) to realize these intentions. The following is the hypothetical Quality Policy statement of an Engineering Project Construction Company derived out of a Corporate Vision and a Mission.

Corporate Vision

Company X will continue to achieve excellence in business performance as a leading EPC organization. The company will expand its activities within the area of its core competence as well as in allied sectors. It will be a right-sized, flexible teamworking in harmony with its partners, to achieve business targets.

Mission

We, in Company X, are committed to satisfy, and whenever possible, even delight our customers as also other stakeholders. We will strive continuously to enhance our competence, build and involve all our people, emphasise on value addition in our key business processes, and minimize and eventually eliminate costs of poor quality. We will proactively explore new areas of business and identify new customers. We will implement Total Quality Management embracing all our operations.

Quality Policy

Company X organizes all its quality-related activities within the framework of a Quality System, geared to provide complete satisfaction to all customers, both internal and external, through error-free execution of supplies and services on time at all times.

The **Quality Practice** is revealed through the detailed processes, procedures and work instructions to implement the Quality Strategies and Quality Plans—that are to be followed by all concerned.

The **Quality System** is a vehicle for the transformation

$$Philosophy \rightarrow Policy \rightarrow Practice.$$

A Quality System

It must be noted that quality as is perceived (and judged) by the customer or as is projected to the current/potential customer or as is demanded of the supplier/vendor/sub-contractor is as important as quality as is created, maintained and demonstrated.

3.7 Quality Activities

Traditionally, different activities which are related directly to quality or are supporting quality are identified tasks assigned to different groups and individuals within an organization. These activities need to be clearly identified and subsequently integrated.

Quality of a product manufactured or a process developed or a service rendered/delivered is the outcome of a complex chain of activities like planning, procuring, testing, correcting, evaluating that originates from a demand—internal or external—and ends in delivery. Quality in these activities determines the ultimate quality of the entity that reaches the customer or the user. And these activities have to be so carried out that the quality of the product or the process or the service in which we are interested can achieve a desired level at the cost of a minimum consumption of materials and energy. Further, this quality achievement must lead to a growth in business and in excellence of performance. Quality Management in its broadest connotation has to be an activity in which top management must be involved directly to guide, advise and direct all quality-related and quality-supporting activities along with activities conventionally known as quality-managing activities. These three activity types are to be executed by identifiable groups of people, which—especially in cases of medium and small enterprises—may overlap in terms of persons, but not in terms of responsibilities and authorities.

3.7.1 Quality-Affecting Activities

The following may be taken as an indicative list of activities carried out by different groups of people within an organization that are linked to quality:

1. Perceive quality as is required (stated) and expected (not always stated) by the customers. In fact, requirements relating to materials and components to be used, tests and checks to be carried out, etc., are sometimes explicitly stated in the purchase order or communicated verbally during purchase. However,

beyond those requirements which are communicated orally or in writing, customers will have some expectations about the product purchased or the service received which are not expressly stated. The Marketing Group or better the Market Research Group within that should try to listen to the voice of the customer and to comprehend the perception about quality among the customers.

2. Plan quality through design and development effort, in terms of specifications for materials, processes, equipments, facilities and products **to** meet these requirements and expectations. The Design and Development Group must derive inputs from market research and also interact with procurement and production groups to find out the feasibility and desirability of different alternative designs.

3. Demand quality in supplies by vendors and in jobs outsourced to sub-contractors. Further, in each step in production, quality must be demanded from the previous step in respect of the output of the previous step.

4. Create quality through strict conformance to design specifications during manufacture and inspection. The Production and Inspection Groups must interact with Design and Development Group adequately to ensure that if the design requirements are fully met during production and inspection, the output will meet the customers' quality requirements. It should be always remembered that quality is created, rather infused or built into the product during production, and not ensured subsequently by inspection.

5. Control quality through inspection and detection, review and remedy of non-conformities as and when these arise followed by appropriate corrective and preventive actions to prevent recurrences of non-conformities detected or likely to arise in future. This must be a joint effort of the production and (conventional) Inspection Groups.

6. Demonstrate quality through adequate inspection and test as also through proper documentation of their results. Sometimes, some tests (non-destructive ones) may have to be repeated before own management and before external customers to establish conformity to any quality requirement mutually agreed between the producer and the customer.

7. Improve quality through preventive actions to stem the occurrence of yet-to-be identified non-conformities and through Research and Development effort to raise quality standards of various inputs. Quit often, this will involve properly designed experiments to be conducted and results of experiments properly analysed and interpreted.

8. Project quality through informed marketing effort that focuses on new, novel and non-trivial features of the products and avoids uncalled for increase in customer expectations. It is important to note that the Publicity and Public Relations Group should effectively secure information about product features before advertising for any **product.**

3.7.2 Quality-Supporting Activities

To execute the above quality-affecting activities, some quality-supporting activities should be formally recognized and assigned to some individuals or groups. The important point is not necessarily to have different persons or groups taking care of these activities, but to ensure that none of these activities becomes a casualty.

Quality needs the support of many groups and functions in order to

1. Develop vendor capabilities to meet quality requirements in their supplies, by providing technical assistance and supervision. The same should apply to sub-contractors also. While the Procurement Group normally takes care of activities like vendor rating and vendor selection, vendor development may involve people from Design and Inspection Groups also.
2. Maintain production and inspection equipments on a planned basis—a major responsibility of the maintenance function. The Human Resource Management function should also train and motivate frontline workers to take adequate care of the machines and accessories they use by cleaning, lubricating and making minor repairs to minimize breakdown during operation.
3. Calibrate measuring and inspection equipments periodically against accredited standards. Calibration records should be acted upon by the users in terms of corrections to be made on the measures/values indicated by the equipments.
4. Provide adequate field services for maintaining products at the customers' end. This may not be required in some cases. Application engineers or similar personnel may be involved in providing such services to beget customer satisfaction.
5. Train and motivate people to own the system and operate it. If they find any deficiency in the system of any difficulty in operating the same, they should inform concerned persons to initiate necessary modifications in the system and to acquire necessary skills for operating the same, in case they lack such skills.
6. Use appropriate statistical methods and tools for controlling quality during production, for controlling quality of incoming materials as also of outgoing products, besides using these methods including design and analysis of experiments to improve quality from its current level.

3.7.3 Quality-Managing Activities

A major task of Management is to persuade personnel in different functional areas to have the right attitude of mind towards quality, to behave in a way which is not always the easiest, and to remember that prevention is usually less costly than cure. Management for its part can assist this process by ensuring adequate delineation and communication of its objectives, leading to the establishment of the programme for quality assurance which can provide the basis upon which periodic reviews and evaluations may be made after taking account of essential data feedback.

Irrespective of whether quality specialists or quality departments is established, executive management is ultimately responsible for making balanced judgements, assessing the significance of events and taking appropriate decisions. Management should ensure that each person in the organization

Understands that quality assurance is important to the organisations as well as his future

Knows how he can assist in the achievement of desired quality

Is equipped (in terms of knowledge, skills and other inputs), stimulated and encouraged to do so.

A management representative, preferably independent of other functions, should be appointed with the authority to resolve quality problems. This does not mean that total responsibility for quality is conferred solely on the individual or department. All important tasks for Quality Management would be captured by 4 I's viz.

Identify root causes of errors or non-conformities;

Instruct to show how to do things right first time;

Inspect during the process, emphasis on self-inspection;

Inspire people through motivation, appreciation (recognition) and reward.

Role and Responsibility of (Executive) Management has been clearly indicated in the ISO 9001 standard as follows:

To formulate a Quality Policy (giving out the organisation's overall intentions and directions in regard to quality) and to ensure its understanding by all people across the organization;

To delineate responsibilities in regard to quality for different groups and individuals and to emphasise on interfaces between different groups and functions;

To provide for adequate resources including equipment and skilled manpower for design, manufacture, test/inspection, servicing and related activities;

To appoint a management representative for coordinating, facilitating and monitoring Quality System implementation and

To carry out comprehensive and structured reviews of the Quality System as also its effectiveness at regular intervals.

Apart from these points covered in the generic ISO Standard, some more important roles have to be played by top management in the context of a highly competitive and challenging business world today. Top management has to be driven by an urge to achieve excellence and not to remain complacent with current conformity to standards and contracts. An important focus should be placed on continuous business process improvement and, for that end in view, on new product development and innovation.

Top management has to enthuse executive management to try out some Business Excellence Model or to implement a balanced Business Scorecard or even to enter into contests for excellence awards and accolades.

3.8 Integration of Quality Activities (into a System)

While the above activities affecting quality as well as supporting quality should be separately identified so that none of these becomes a casualty, we must also

1. recognise the fact that deficiency in one quality-related activity cannot be compensated/cured a subsequent activity—can be at the best detected.
2. avoid duplication and wastage of resources and efforts.
3. ensure presence of all elements on identification of missing and non-interacting elements.
4. avoid overlapping and conflicting responsibilities.

In fact, these activities have to be all oriented towards business goals and objectives so that they are integrated into a system and do not remain compart-mentalized, leading to sub-optimal results at the best.

Hence, the need to integrate pre- and post-production activities with production and inspection operations—in regard to quality and thus to develop and maintain a Quality System, to

Meet specific contractual requirements
Win customers' confidence and vendors' cooperation
Provide a basis for Quality Audit—both internal and external
Ensure willing participation of all in the quality effort
Facilitate changes in existing structures and procedures and
Enhance transparency and accountability of all managers and workers.

Such a Quality System has to be documented.

3.8.1 Mechanism for Integration

Activities linked to quality can be integrated through the following steps

1. Appreciation of a Dual Role—as a customer as well as a producer—by each individual or group. A maintenance manager who receives a failing or failed equipment for repair from production area is surely a producer/provider. At the same time, he has to insist on some information about the mode of failure, the time the equipment was last repaired, etc., to be provided by the production people and, that way, is a customer. As a customer, he is entitled to receive

quality information inputs, and as a producer, he is responsible for quality of repair work done at his end.

2. Development of clear and adequate linkage among different functional groups. Each such group besides discharging certain tasks and responsibilities assigned specifically to it has also to interact with its customer group(s) and provide adequate support (by way of materials to be further processed or to be tested/ inspected or in terms of information or guidance) to the latter, as and when needed.

3. Development of methodologies for satisfying internal customers like Quality Function Deployment to link quality perception by customers as conveyed by the marketing group with quality plans to be acted upon by the production group.

4. Resolution of conflicts among objectives pursued by different groups through linking such group objectives with the overall organizational goals.
Securing customer satisfaction through delivery on time is one among the overall goals and the procurement group has to align its procedures and/or the pace at which these are implemented so that the group objective is aligned to the organizational goals.

5. Traceability of any work done at a particular stage by a particular group can be traced through previous stages involving other groups to facilitate remedial action. Non-conformities detected at a current stage may be due to some deficiency at some previous stage(s) and any corrective or remedial action has to be initiated at that (those) stage(s).

3.9 Control, Assurance and Improvement

Quality Management embraces quality control, quality assurance and quality improvement. Quality control is essentially an in-house responsibility and activity, while quality assurance to own management and, more importantly, to customers and other stakeholders like custodians of public interest, involves additional activities like documentation of procedures followed and record of test and inspection results obtained besides third-party audit and even repetitions of some tests and checks to demonstrate conformity to mutually accepted standards in the presence of customers. Quality improvement goes beyond to achieve levels of quality and performance higher than those mutually agreed upon. This entails research and development efforts. Quality improvement activities are essential to ensure continued growth and excellence of the organization in keeping with greater demands for better quality by discerning customers.

Control is a management activity involving

1. Development (or acceptance) of relevant standards;
2. Comparison of actuals with standards;
3. Action on the underlying system based on this comparison
4. Research and Development efforts to

   (a) develop better stricter/more economical standards and
   (b) ensure greater conformity of actuals with standards.

While some people argue that Statistical Quality Control has been a practice of the past, we should bear in mind the fact that both on-line and off-line quality control continue to be not merely relevant but quite a necessity in the modern Quality Management practice. Of course, statistical methods and techniques used earlier have given place to more sophisticated and more efficient ones to suit the present-day needs. And the basic idea behind SQC as was propounded by Ishikawa still demands a close look. Ishikawa took SQC as a triplet of concepts rather than of tools and techniques. He considered the three basic elements of SQC in his unconventional way as

1. SONO-MAMA: 'Is'—ness or 'such' ness of a product, its functional prerequisites. This is to be understood in terms of some generic features that define the product. A mirror has to be a mirror, and not just a piece of glass. A needle has to a needle and not just a pin without a head. Thus, the basic element of quality is that the product in terms of its design and manufacture is the product intended by way of its physical and functional characteristics, at least to some extent.
2. SYNERGY: Composite effect of individual quality characteristics, combined effect of individual efforts to build quality. This is a recognition of the need for involvement of many, if not all, in the organization to build up the desired quality in a product or service to be manufactured or delivered.
3. STATISTICS: Truth revealed by facts and figures relating to quality, factors affecting quality and variations in quality from item process to process as also from item to item within the same process, dependence of item quality on input quality and process quality. The role of statistics for analyzing data and providing a scientific basis for decisions regarding input materials, in-process items and finished goods has now been widely established for both on-line control of products and off-line control to work out improvements in processes and products.

As mentioned earlier, quality assurance goes beyond quality control and involves the entire organization in providing adequate confidence to customers that quality (including price and delivery) requirements as mutually agreed upon have been met. In fact, recent standards developed by national and international bodies for Quality Management System relate essentially to quality assurance.

Today's quality assurance operations have been extended to cover the entire life cycle of products ranging from their planning stage through use or service life to their eventual discontinuation. WE now require preventive measures concerning pollutants and other hazards in the disposal of discarded products. Thus, post-production quality operations—looked upon lightly in the pat—have assumed major importance now. Particularly, important is the step to feed the acquired information for product planning and quality design.

A sound quality assurance system should be developed on the following principles

(a) Standardisation—in definitions and measures, in testing and inspection procedures, etc.;
(b) Innovation—in materials and in design, in manufacturing technologies and control mechanisms;
(c) Adequate use of quantitative techniques—including recently developed techniques like QC-PERT which aims simultaneously at move up of a schedule, assurance of a quality level and cost savings;
(d) Due cognizance of advances in electronics including computers, numerically controlled machines, control softwares for the purpose of in-process control and expert systems for organizing and integrating information systems;
(e) Adoption of a systems approach with due emphasis non-product areas also;
(f) Management support to a continuing process of change for the better.

Quality improvement should be the ultimate aim of all quality activities in a forward-looking organization which looks beyond its current product profile, existing customers and their present-day needs (at least in respect of quality) to new products to be offered to potential customers including the quality-fastidious ones. Fortunately, many efficient methods and techniques and softwares for their convenient applications have been developed in recent times to facilitate and strengthen quality improvement activities.

Quality System Standards in vogue give due stress on corrective and preventive actions to bring about improvements in products and services. The chapter on Quality Systems and System Standards provides some details on such improvement activities.

As can be easily appreciated, improvements in products and services can be effected through improvements in design and production processes. Designs to capture user needs as also to add greater value to products, designs to ensure convenient manufacturability and testability as also ease in maintenance along with practices which reduce deviations from design requirements drastically all lead to process and product improvement. A whole chapter in this book is devoted to improving process quality.

3.10   Levels of Quality Management

Quality Management practices as also their impacts on business performance measured in a holistic manner have been analysed by many research workers and organizations have been placed in several different levels of Quality Management.

This is somewhat comparable to assessments of software development processes in terms of what have been branded as 'capability' and 'maturity' of an organization. In fact, the Capability Maturity Model (CMM) developed jointly by the Carnegie Mellon University and the Software Engineering Institute recognizes five such levels viz. **initial, reproducible, documented, managed and optimized.**

Based on the nature and content of their Quality Management Systems, organizations have been classified as belonging to one of the following categories

Level 1 Uncommitted

These firms are not involved in improvement activities, do not specific Quality Plans, ignore the Total Quality Management Philosophy, have short-term quality objectives and invest little in human resource development activities. Such firms may have an ISO-9000-based Quality Management System just to ensure compliance with the standard in letters.

Level 2 Drifters

More committed to ISO 9000 standards, these firms have started a quality improvement programme, but are yet to chalk out a road map to work out such improvement and to implement TQM in spirit. At this level, differences between ISO 9000 Standard and TQM may be blurred.

Level 3 Tool-Pushers

These firms are more experienced in quality improvement matters and have usually taken recourse to a number of tools and techniques which are widely named as also used, sometimes unimaginatively and out of context. They could even be guided by some model like the European Foundation for Quality Management model or the Rajiv Gandhi National Quality Award model, but the model is not accepted throughout and not in all processes, especially in support processes.

Level 4 Improvers

These firms have been in quality improvement exercise for a long time and have achieved significant progress. The need for a long-term cultural change and the importance of continuous improvement are appreciated. They implement an advanced planning scheme, set the objectives and actions at all levels and focus on employee involvement through work teams, training, incentives and other instruments to motivate and involve people.

Level 5 The Award Winners

These organizations look upon Quality Management as a way to manage their business to their internal and external customers' satisfaction with the participation of all their employees However, they are yet to reach an 'excellence' status.

Level 6 World-Class

The defining features of these firms are the total integration of quality improvement within the firms' business strategies, to the delight of their customers. Their quality strategy consists in reinforcing their competitive advantage by increasing their customers' appreciation of the organization and the attractiveness of their products and services.

The current state of Quality Management prevailing in different organizations can be captured in terms of three variants viz. compliance-oriented (SPC and ISO 9000 Standard), improvement-oriented (TQM) and business-oriented (Six Sigma). According to Conti, Kondo and Watson, 'quality is becoming (at levels 4 and above in the classification mentioned earlier) an integrated system where the best of all approaches are merged into unique Quality Systems that engage the entire business, rather than a single function.'

3.11   Steps in Implementing a Quality System

The phrase 'Quality System' with due emphasis on the word 'system' implies some sequence of steps through which we can implement such a system. While the steps may not be all the same in every organization, the following steps are generally to be gone through, with the proviso that some of the steps indicated here may be skipped or shortened in some cases while a few other steps may also be added in some situations.

1. management *understanding* and commitment management has to

   (a) serve on the Quality Organisation (distinct from the quality/inspection or a similar department);
   (b) establish Quality Goals and Objectives;
   (c) provide the needed resources—physical and organizational;
   (d) provide adequate Quality-oriented training to all;
   (e) stimulate quality improvement through suitable incentives;
   (f) review progress in different quality-related activities;
   (g) recognize excellent work done by individuals and groups;
   (h) revise the reward system from time to time to emphasise the need for quality improvement.

2. Continuing education at all levels

   (a) senior management first;
   (b) different content in training programmes for different types of responsibilities associated with different aspects of quality;

3. Setting up of a Quality Organisation including Quality Councils, Quality Audit Teams, Quality Task forces, etc.

4. Adequate use of relevant procedures and techniques: Statistical process control, Process/machine capability Analysis, Customer Response study, Failure Mode and Effect Analysis, etc.
5. Stress on Innovation in

   Materials and processes;
   Products and services and;
   Use and maintenance.

6. Emphasis on the use of Information and not just opinions, facts and not just imaginations or hunches for making and changing plans
7. Enhanced cooperation to

   Agree on common standards;
   Develop and work with the same vendor(s);
   Educate customers for proper installation, use, storage and upkeep of the products;
   Improve industry image.

3.12 Concluding Remarks

Quality Management today tends to incorporate almost the entire range of management concepts and tools available nowadays as are applicable to the broad field of quality in materials, processes, products and services across all organizations. In fact, attempts to address and resolve problems associated with quality including the vexing issue of quantity versus quality controversy has given rise to methods and tools which have found applications outside the field of manufacturing and the traditional area of quality to be appreciated as problem-solving methodologies like Six Sigma or Eight D.

It is worthwhile to mention that Quality Management today involves a whole lot of ideas, methods and tools which evolved over the years in the realm of Human and Organizational Behaviour and, that way, of Human Resource Management. In fact, people management is a major concern in modern Quality Management. Quality affects business and business processes are currently dealt with in a manner that characterises Quality Management. Although Quality Cost Analysis is not formally carried out on a large scale, resource management and management of finances are quite important in the context of Quality Management. Proponents of Total Quality Management argue that TQM is a strategy for business growth, Quality Management having become an essential constituent of strategic management.

We all appreciate that since the business world has become more demanding and challenging, the role of the Quality Manager (may be glorified as Director of Quality) has gained a lot of significance. Ability to understand and analyse complex business problems and to address both hard and soft issue relating to quality apart from the ability to convince top management about the need for quality improvement in all business processes and to lead a team are now expected of a successful Quality Manager today. In a somewhat unusually expanded role of a Quality Manager today has been indicated by Addey (2004) who says that this person has to be simultaneously (or as when needed) a Salesman, a Teacher, a Psycho-analyst, a Doctor, a Consultant, a Detective, a Policeman, a Social Worker, a Researcher, a Designer, a Strategist, a Lawyer, a Customer and a Statistician. Like it or not, it is thought-provoking at least. This statement speaks volumes on the roles and responsibilities of a Quality Manager today.

References

Crosby, P. (1984). *Quality without Tears*. New York: McGraw Hill.

Deming, W. E. (1986). *Out of the Crisis*. MA, USA: MIT Center, Cambridge.

Fiegenbaum, A. V. (1983). *Total Quality Control*. New York: McGraw Hill.

Fiegenbaum, A. V., & Fiegenbaum, D. S. (1999). New Quality for the 21st century. *Quality Progress, 32*, 27–31.

Imai, M. (1986). *Kaizen*. Singapore: The Kaizen Institute.

Ishikawa, K. (1976). *Guide to quality control*. Tokyo: Asian Productivity Organisation.

Ishikawa, K. (1985). *What is total quality control? The Japanese way*. NJ, USA: Prentice Hall.

Juran, J. M. (1981a). *Management of quality*. CT, USA: Juran Institute.

Juran, J. M. (1981b). *Upper management and quality*. CT, USA: Juran Institute.

Juran, J. M. (1988). *Quality Control Handbook*, New York: McGraw Hill.

Juran, J. M. (1995). *A history of managing for quality: The evolution, trends and future directions of managing for quality*. Milwaukee, WI, USA: ASQ Quality Press.

Logothetis, N. (1992). *Managing for total quality*. UK: Prentice Hall.

Maguad, B. A. (2006). The modern quality movement: Origins, development and trends. *Total Quality Management and Business Excellence, 17*(2), 179–203.

Shingo, S. (1981). *Study of Toyota production system*. Tokyo: Japan Management Association.

Shingo, S. (1986). *Zero quality control: Source inspection and the Poka-Yoke system*. Stamford, CT, USA: Productivity Press.

Wiele, T. V. D., Iwaarden, J. V., Bertsch, B., & Dale, B. (2006). Quality management: The new challenges. *Total Quality Management and Business Excellence, 17*(10), 1273–1280.

Chapter 4
Quality Systems and System Standards

4.1 Introduction

In the context of a market that insists on specified quality, timely delivery and reasonable and competitive price, it has become a bad necessity for any manu-facturing or service organization—right from a big manufacturing house with many distribution and delivery units to a small or medium enterprise that delivers goods and services at the doorstep of each customer—to integrate all its activities that can directly or remotely affect the customer's perception about quality of its products and services into a composite entity that determines the fate of business. And this entity is the Quality (Management) System that is directly looked after by top management and not as a support process.

A comprehensive Quality System that is in sync with the overall Company Business Policy and Strategy, is convincing enough to customers and potential customers, is developed by people within the organization with or without guidance from external experts, and can be implemented by the people within the organi-zation cannot be developed conveniently by all organizations. And it may be desirable, if not essential, to fall in line with some existing standards or specifi-cations for Quality (Management) Systems prepared by some National Standards Body or the International Standards Bodies like the ISO, the IEC (International Electro-technical Commission) or ITA (International Telecommunication Authority). It is true that a Quality System for a particular organization has to suit the management style of functioning and the prevailing work culture (which can be changed only very slowly), while the available standards are usually generic in character—applicable for all organizations or for all organizations within a par-ticular industry yype like automobile.

Fortunately, all standards provide ample flexibility to accommodate specific features of particular organizations and all that is needed by any organization is to clearly comprehend the carefully worded requirement in a standard about a function or an activity that has a bearing on product or service quality. Most of the standards provide illustrative examples and are supplemented, whenever necessary, by other standards relating to processes like Audit or Training of People or Calibration of Measuring Equipments as also to procedures like sampling and Use of Statistical Techniques for Analysis of Data and Information.

Once Quality Systems has been developed and are being implemented, organizations may like to have confidence about the compliance of their systems with corresponding standards and get certified for such conformity to beget customer confidence and gain external visibility as also to qualify for some benefits, which vary from country to country. Starting with only two Certification Bodies recognized by the National Accreditation Council for Certification Bodies in Great Britain, we now have a vast array of certification agencies, with varying degrees of competence and credibility.

Beyond the certifiable standards are some quality awards and some recent approaches to Quality Management which also provide guidance to forward-looking organizations to improve their Quality Systems. In the present chapter, we present some of these standards briefly with some emphasis on ISO 9000 standards.

4.2 A Quality System

A system generically implies a set of inter-related elements (howsoever defined, e.g. individuals, divisions, groups) revealing four fundamental characteristics and all these four have to be identified to make up a system in this sense. The characteristics have to be delineated in different ways for different real-life systems differentiated in terms of their size and type, core and support activities, rights and responsibilities. These four characteristics are expressed in words which do not have their dictionary meanings and are interpreted in terms of a system.

1. Content: ability of the system to identify and adopt its goals and objectives as also the strategies and plans to achieve these. The goals set before the system and objectives to be achieved by its elements have to be appropriately developed in the case of a system which is subsumed within a bigger system and, that way, is really a sub-system.
2. Structure: differentiation among the elements in terms of duties and responsibilities as well as authorities (levels in a hierarchy) of different sub-sets and of different individuals with each sub-set. This corresponds to the delineation of different departments or sections or divisions.
3. Communication: specifying the modes of communication among individuals within a sub-set, between different related sub-sets and between the system or a particular sub-set and the external world:
4. Control: ability to modify any or all the three earlier characteristics whenever found necessary. This characteristic is an authority while the former three are basically responsibilities of a system.

A system, in any context be it a quality assurance or Management System or a Financial Management System, as possessing the above-mentioned characteristics, has to be comprehended in terms of the following four constituents. These constituents have to be appropriately translated when we consider a Quality System. And this is indicated here.

1. An Organisation: a group of individuals integrated through common or shared values, interests and tasks that is guided by a Quality Policy that spells out responsibilities and authorities of different groups/individuals and that also specifies the nature and extent of interactions among such groups/individuals. (These groups are in fact sub-groups of the group.)
2. Processes to be planned and executed by different elements of the organisation, e.g. inspection, maintenance and calibration of test, measuring and inspection equipments; Quality Audit; Training of people at different levels; Control of documents as also of records; analysis of data, corrective and preventive actions to be taken when non-conformities are detected.
3. Procedures to be followed in carrying out the different processes like sampling and preparation of material for test or inspection; calculation of life cycle costs or of vendor-rating index, and the like.
4. Resources to be deployed in connection with the different processes. These could be physical resources like even space to distinguish incoming materials awaiting inspection from materials cleared after inspection for use in production; Material resources like equipments in the manufacturing stages as also for tests and inspection are also required. Human resources by way of people with adequate domain knowledge, experience and right attitude constitute an essential element. And technological resources hold the key for good quality products and services.

In fact, a Quality (Management) System contains ingredients that enable an organization to identify, design, produce, deliver and support products and/or services that satisfy customers. Standards or Models for Quality Systems have been developed over the years at national as well as at international levels.

4.3 Quality System Standards

As globalization takes over, international standards gain importance and acceptability and also popularity. However, there is no model that can provide an ideal, one-size-fits-all solution for all organizational requirements. Some of the more widely used Quality (Assurance/Management) System standards will be indicated in this chapter.

It is worthwhile to consider the four dimensions of standardisation as an activity resulting in a standard. These are

Subject—entity being standardized, could be an unprocessed material, or a process, or a product, or a service or a system;
Aspect—could be a dimension or some property of a product or service, or introduction and sustenance of a system;
Level—could be industry, national, regional or international;
Time—refers to scientific, technological or management knowledge base used in developing a standard.

Thus, a Quality (Management) System Standard is a national or regional or international standard for implementation, assessment and sustenance of a system. It does not relate to specific materials or processes or products or services which normally requires specific standards. A Quality (Management) System standard is expectedly a generic standard.

It is quite important to remember the distinction between requirements for a Quality Management System and requirements for products. ISO 9001 itself does not establish quality requirements or specifications for products. Requirements for products can either be specified by customers or by the providing organization in anticipation of customer/use requirements or even mutually agreed between the customers and the supplier organization. The requirements for products are contained in, for example, technical specifications, product standards and contractual agreements. And technical specification and product standards can be national or regional or even international.

4.4   Quality System Specifications—An Overview

A brief chronological account of the more important standards for Quality Systems and allied documents is presented in a sketchy form below.

MIL-U-9858 for Quality System and
MIL-I-45208 for Inspection System both for USA;
AQAP (Allied Quality Assurance Publication) 1, 4, 9 (4 and 9 for Inspection System);
AQA-1 covered manufacture, inspection and test;
Not accepting AQAP, UK introduced a series of three Defence Standards;
Def. Stan. 05-21, -24 and -29 (for Inspection only);
AQAP aligned with Def. Stan. Later, BS 5750 appeared in 1979.

ISO Committee (with Canada as Chairman) considered many national and international inputs like IS 10201 (Indian Standard Manual for Quality Assurance Systems, first published in 1982) and came out with the ISO series in 1987.

Unlike national standards organizations (and even unlike the IEC), ISO is NOT a certifying agency. It does not have a scheme of certification of the Quality System (or any other system) existing in any organization for conformance to its standards.

ISO 9000 Series of Standards covers quite a number of standards bearing on Quality Management Systems and related processes and procedures. These have been adopted as national standards by many countries including India. These standards have been adopted in India as IS 9000, in the European Community as EURONORMME EN 29000 (1987) and in some 90 other countries. This wide applicability is due to the fact that ISO 9000 standards are neither industry-specific nor country-specific.

The initial set of standards which have undergone several revisions which, in some cases, implied the withdrawal of some of these standards, were

ISO 9000—Quality Management and Quality Assurance Standards—guidelines for selection and use;
ISO 9001—Quality Systems—Model for Quality Assurance in design/development, production, installation and servicing;
ISO 9002—Quality Systems—Model for Quality Assurance in Production and Installation;
ISO 9003—Quality Systems—Model for Quality Assurance in Final Inspection and Test;
ISO 9004—Quality Management and Quality System Elements—GUIDELINES.

Supporting Standards on Vocabulary/Terminology

ISO 8402 (1986) IS 13999 now IS 10201 Part I (1988)

Of the three Standards 9001, 9002 and 9003, the first was meant for organizations engaged in the entire gamut of activities from designing and developing a product or a system through its installation at the customer's site to its servicing during operation. The second was meant for those organizations which were not involved in design and servicing/maintenance activities, and the third was essentially applicable organizations simply procuring products, inspecting them during procuring, storage and delivery and ensuring that quality is not affected during storage at the organisation's end.

In 2000 when the Standards were revised for the second time, the title Model for a Quality Assurance System was replaced by Requirements for a Quality Management System All the three standards 9001, 9002 and 9003 were merged into one ISO 9001. The 2008 revision of ISO 9001 was a big departure in content and presentation.

4.5 ISO 9000 Standards

These standards have been adopted as Indian National Standards and are being implemented on a wide scale in Indian industries in both manufacturing and service sector enterprises. The standard ISO 9001 lays down requirements for a Quality Management System.

The standards (revised four times since the first versions were released in 1987)

1. Relate to a Quality Management System, going beyond quality assurance activities;
2. Are based on a set of eight basic management principles;
3. Are quite close to a TQM model;
4. Are compatible with other management systems;
5. Are comprehensive and, whenever necessary, illustrative;
6. Incorporate processes and procedures for quality improvement;
7. Do not entail additional documentation over and above the necessary part required in the 1987 version and can be completely online;
8. Encourage fact-based decisions using targets and achievements;
9. Involve top management (and not just executive management) in review of and decisions on the Quality System;
10. Highlight the role of quality objectives, quality targets and quality planning;
11. Focus on process management (including process validation, initially as also after any fundamental process adjustments);
12. Emphasise on understanding of customers' requirements and their fulfilments;
13. Stress on measurement of process performance and customer satisfaction and on analysis of available measurements on these two aspects;
14. Draw attention to the role of an effective communication system within the organization as also between the organization and its stakeholders; and
15. Give due attention to resource management (including information as an important resource) and to workplace management.

It may be worthwhile to look closely at ISO 9001: 1987 and then to note how modifications and improvements took place in the revisions made in 1994, 2000, 2008 and 2015. In fact, the 1987 version of the standard had 20 clauses (in fact, these were sub-clauses numbered 4.1–4.20 of clause 4, the other 3 generic clauses covered foreword, applications and normative references), Clauses 4.3 and 4.18 were dropped in ISO 9002 and eight clauses relating to design, production and servicing were dropped in ISO 9003.

Let us examine the contents of the 20 clauses in ISO 9001: 1987 and their implications.

1. **Management Responsibility**
   To formulate a Quality Policy (giving out the organisation's overall intentions and directions in regard to quality) and ensure its understanding by all people, to delineate responsibilities in regard to quality for different groups and individuals and emphasize interfaces between groups/functions, to provide for adequate resources including equipment and skilled manpower for design, manufacture, test/inspection servicing and related activities, to appoint a Management Representative for co-ordinating, facilitating and monitoring Quality System implementation, and to carry out management comprehensive and structured reviews of the Quality System and its effectiveness at regular intervals.

2. **Quality System**

In terms of documents giving out plans, policies and procedures with respect to different aspects of quality and recorded evidence about their implementation as well as about the effectiveness of the implemented Quality System in achieving quality goals and objectives.

3. **Contract Review**

To ensure that customer requirements regarding quality, delivery and service are clearly and adequately defined and that these can be met by the existing capabilities (in terms of raw materials, production and inspection facilities, service mechanisms, etc.) to provide for resolution of differences from tenders and of special problems—if any—as well as to work out modalities for concessions and waivers with authorized representative(s) of the customer(s).

4. **Design Control**

In terms of design planning to decide on the scope and objectives of designing, design input to specify quality requirements to be met, design output to yield relevant specifications and calculated (expected) product quality features, design validation to match design output with design input and with proven designs, design verification in terms of trial runs/prototype tests, etc., and periodic design reviews.

5. **Document Control**

To ensure availability of duly approved (by competent authority) documents at all points of use, to review documents (by involving concerned groups/ functions) and to incorporate necessary modifications/revisions, to ensure timely withdrawal of obsolete documents, to maintain a master index and a set of master copies of all documents, to control distribution of confidential documents, etc.

6. **Purchasing**

To select vendors on the basis of their ability (to meet quality, price, delivery and service requirements), to issue duly approved purchase orders complete with specifications and acceptance criteria, to provide for settlement of disputes with vendors, to carry out vendor performance appraisals regularly, to provide for (quality) verification of all purchases through in-house test/inspection or through test certificates from vendors.

7. **Purchaser Supplied Product**

To be verified (at the time of receipt) for quality, stored and maintained according to established procedures. Any loss, damage or unsuitability for use should be recorded and reported (to the purchaser, if necessary).

8. **Product Identification and Traceability**

To drawings/specification at all stages of production, delivery and installation, so that corrective action—whenever necessary—can be facilitated. Individual product items or batches should have unique identification in respect of customers (destinations), modes of packing, packaging and transportation as well as installation.

9. **Process Control**

   In terms of the use documented work instructions, defined workmanship standards, approved materials, processes and equipment, specifications and acceptance criteria, for utilities and environment, monitoring and controlling variations in parameters of input, process, equipment and output for each operation (affecting quality of the final product), identification and continuous monitoring of special processes, etc.

10. **Inspection and Testing**

    By using appropriate plans and documented procedures for receiving, in-process (first-off and patrol) and final inspection, identifying 'hold' points and non-conforming items, controlling release of uninspected materials/products for urgent use, etc.

11. **Inspection, Measuring and Test Equipment**

    In terms of maintenance of such equipment (including gigs and fixtures as well as test software) through proper handling and preservation as also prevention of unauthorized adjustments, besides calibration of these equipments at regular intervals to ensure traceability to national standards (in terms of precision and accuracy).

12. **Inspection and Test Status**

    To indicate conformity/acceptability or otherwise of various items throughout production by using different marks/labels and allocating different storage areas for items not yet inspected and found conforming, to be further inspected, and inspected and found non-conforming.

13. **Control of Non-conforming Product**

    To avoid inadvertent use of non-conforming material, by using established procedures for identification, documentation, segregation, review and disposition of such products and for issuing notifications to all concerned functions. Possible dispositions of such products could be accepted as it is, re-work and re-inspect, re-grade and scrap (reject) and may involve (in some cases) customer concessions.

14. **Corrective and Preventive Action**

    By investigating the nature, extent and cause(s) of any non-conformity, implementing corrective as well as preventive measures addressed to causes and not symptoms, ensuring that such measures are effective and analysing historical quality records to detect and eliminate potential problems.

15. **Handling, Storage, Packing and Delivery**

    To prevent any damage or deterioration in a product till the supplier's responsibility causes through insistence on authorized receipt and issue, assessment of storage conditions and secured storage of product after final test/inspection.

16. **Quality Records**

    To demonstrate achievement of required quality and effective operation of the Quality System. Pertinent supplier records have to be included, procedures for collection, filing and retrieval have to be specified and retention times have to be defined.

17. **Internal Quality Audit**
    In terms of an independent check of activities for compliance with documented policies and procedures. Planned, comprehensive audits by qualified personnel have to be regularly carried out, documented and acted upon.
18. **Training**
    In terms of procedures to identify training needs and to evaluate results of training undergone (by way of improvement in performance) as well as plans to organize training programmes—in-house as well as external. All training activities should be recorded.
19. **Servicing**
    (Whenever a part of the contract or a part of the marketing activity) In terms of procedures to perform service, to respond to customer complaints and requests and to verify that specified service requirements are met.
20. **Use of Statistical Techniques**
    To verify process capability, to assess material and product characteristics, to develop in-house specifications, to plan experiments for Research and Development, to analyse product failures, to appraise customer satisfaction and company/product image, etc.

4.6   Basis of ISO 9001 Standards

In the 2008 version, certain basic principles and a certain process approach were elucidated to explaining the development of various clauses of the standard ISO 9001. It may be noted that the initial version and the first revision did not explicitly state any such basis which the standards took into account. Further, the earlier versions did not directly match the standard with overall (corporate) management activities and systems.

The 2008 standards are based on eight basic management principles, viz.

1. Customer-focus—Organisations depend on their customers and therefore should understand both current and future customer needs, should meet customer needs and strive to exceed customer expectations. Requirements of customers, both internal and external, should be duly considered in planning each and every business process—both core and support processes. Considering the importance of quality improvement, needs of potential customers currently outside the ambit of the organisation's market have also to be taken into account. And this implies an additional effort on the part of management to identify such potential customers and to track their quality requirements. However, the organization and its processes are not just driven by customers, but are in line with the goals and objectives of the organization.
2. Leadership—Leadership establishes unity of purpose and direction of the organization. Leaders are not just managers or senior executives; they should create and maintain the internal environment in which people can become fully involved in achieving the organisation's objectives. The business processes

must be guided by process leaders to ensure their conformity with plans and targets. Leaders should be possessed of abilities to develop a shared vision in consultation with their team members or develop a vision and diffuse the same effectively among the members. They should have the quality to encourage others to work for the vision. Beyond this, a leader should enable members of the team by organizing knowledge and skill augmentation programmes. Finally, leaders have to empower team members to use discretions and carry out experiments for improving the processes in which they are involved. Management should facilitate emergence of leaders in different process areas and at different levels.

3. Involvement of people—people at all levels are the essence of an organization, and their full involvement enables their abilities to be used for the organisation's benefit. Towards such involvement, management has to implement programmes like quality circles and similar small group activities.

4. Process approach—a desired result is achieved more efficiently when related resources and activities are managed as a process. This approach tries to establish interactions among different elements of a comprehensive process like production. A fragmented look at elements taken in isolation of one another may not yield the desired result.

5. System approach to management—identifying, understanding and managing a system of inter-related processes for a given objective contributes to the effectiveness and efficiency of the organization. This is, in effect, an extension of the process approach to cover the entire organization and the totality of different functions—both core and support—carried out by it.

6. Continual improvement—continual improvement is a permanent feature of any organization. It must be recognized that this feature alone can sustain the organisation's business in the face of increasing customer expectations. This is the cornerstone for business growth, and its absence may lead to undesired consequences in which the organization eventually fails to satisfy its existing customers whose quality requirements become more stringent with the passage of time.

7. Factual approach to decision-making—effective decisions are based on the logical and intuitive analysis of data and information. Decisions based on opinions or hunches or gut feelings can—at least in some cases—be quite disastrous. This principle calls for collection, analysis and interpretation of data on different aspects of business, beyond those pertaining to routine quality control.

8. Mutually beneficial supplier relationships—mutually beneficial relationships between the organization and its suppliers enhance the ability of both organizations to create values. Business partners like vendors of input materials and services, banks and financial institutions providing financial support to the organization, customers for whom the organization exists, regulatory bodies which provide guidelines on several issues and industry bodies which take up the cause of member units and provide support to them should all be treated in a manner that could result in benefits to all and not merely to the organization, at the cost of losses or hardships to others.

A crucial task to be performed by management is to evaluate the Quality Management System in place. When evaluating Quality Management Systems, there are three questions that have to be asked in relation to every process being evaluated, as follows:

(a) Are the processes established and their procedures appropriately documented?
(b) Are the processes implemented and maintained?

4.7 ISO 9001—Some Key Features

The role of executive (as also top) management in the context of any Quality System is paramount. Management responsibilities constitute a major clause in the ISO 9001 standard, in all the versions. Three broad responsibilities are to provide (a) a well-thought-out Quality Policy suitable for the organisation, (b) an organisation to implement the policy and (c) regular management reviews to monitor results achieved and provide necessary guidance and support for an effective Quality System.

The Quality Policy to be developed, documented and disseminated (among all in the organization) should provide the (overall) intentions and directions in regard to quality. The policy should involve, in its formulation, heads of all functions which quality as is produced as also as is perceived by the customers. The policy should not reflect just what is practised currently and, at the same time, should not be too ambitious to be implemented without radical changes.

Intentions must give out both Quality Goals for the system as well as Quality Objectives for the different elements of the systems. Goals and objectives have to be specific for the organization in question and not generic recommendations or advice. These have to be achievable by the organization as it stands now or as it can aspire to be in coming days. Hence, resources—human, technological and physical or material—which can be committed by the organization must be taken into account before goals and objectives are spelt out. Among alternative goals for the organization as a whole, one can consider the following possibilities.

In respect of goals, one can say that the organisation wants to be a Global Leader in quality, or
it will ensure complete conformity to mutually agreed requirements with respect to quality of supplies, or
it will target high customer satisfaction, or
it will stress on increased market share, or
it will focus on continuous improvement in its performance, or
it will (somewhat narrowly) minimize quality losses or
it will take care of societal losses resulting from its operations, etc.

Possible directions to achieve the goals and objectives could be in terms of relative emphasis being planned on

Human resource development to enhance organizational competence;
Vendor development (beyond vendor selection and vendor rating);
Technological upgradation to introduce modern production processes and equipments;
Plant and machinery maintenance to ensure a high uptime;
Customer education to help customers choose the right product and use selected products in the right manner and for the right purpose;
Innovations in materials, processes, control procedures, etc.

Summing up, the Quality Policy should be

Consistent with overall Company Policy or Business Policy and culture;
Forward-looking to take into account the future business scenarios and looking across industries; and
Neither too ambitious nor too modest or complacent.

4.7.1 Documentation

The requirement for documentation in a Quality Management System is driven by the need for communication of intent, constancy of purpose and results achieved being amenable to analysis. Documentation can be easily computerized.

Quality manual—this communicates consistent information both to internal people as also to external customers about processes and procedures followed by the organization, as a matter of course, to fulfil quality requirements (needs and expectations) of the use and maintenance.
Quality manual—this communicates consistent information both to internal people as also to external customers about processes and procedures followed by the organization, as a matter of course, to fulfil quality requirements (needs and expectations) of the customer. The approach includes the Quality Policy, the structure of the organization to assign responsibilities and authorities in relation to quality and the organisation's quality-related processes.
Procedures—more important to customers as well as to internal people are the specified ways in which different activities or processes are to be carried out to ensure products that consistently meet mutually agreed quality requirements.
Records—as objective evidence of activities performed and results achieved concerning, e.g.

Conformity of products to specified quality requirements;
Conformity of activities performed or processes carried out to the procedures laid down in the Quality Management System manual and procedures; and

Conformity of the Quality Management System to the corresponding requirements (as specified in ISO 9001 standard or as spelt out by the organization itself).

The amount and extent of documentation required can depend on several factors such as: size of the organization, complexity of products and processes, competence of personnel and the extent to which it is necessary to demonstrate compliance with the Quality Management System requirements. The production of documents should not be a self-serving exercise but should be a value-adding activity.

4.7.2 Audit

The ISO 9000 and ISO 14000 series of international standards emphasise the importance of audits as a management tool for monitoring and verifying the effective implementation of an organisation's quality and/or environmental policy. Audits are also an essential part of conformity assessment activities such as external certification/registration and of supply chain evaluation and surveillance.

Different types of audits like internal (by self on self), external (by others on self) and extrinsic (by self on others) have been discussed, and details about audit procedures and findings are contained in the international standard IS 19011. Audit has been defined as a 'systematic, independent and documented process for obtaining evidence and evaluating it objectively to determine the extent to which the audit criteria are fulfilled'. It is important to note that audits are not absolute assessments; they are assessments of the existing situation relative to the documented Quality System.

Clause 9.2 on Internal Audit in the 2015 version spells out in Sub-sub-clause 9.2.2 that the organization shall

(a) plan, establish, implement and maintain an audit programme including the frequency, methods, responsibilities, planning requirements and reporting, which shall take into consideration the importance of the processes concerned, changes affecting the organization and the results of previous audits;
(b) define the audit criteria and scope for each audit;
(c) select auditors and conduct audits to ensure objectivity and impartiality of the audit process;
(d) ensure that the results of the audits are reported to relevant management;
(e) take appropriate correction and corrective actions without undue delay; and
(f) retain documented information as evidence of the audit programme and the audit results or findings.

It may be pointed out here that the latest version of ISO 9001 standard is more focused on development of an effective Quality Management System and its sincere implementation rather than on an overdose of audits. Of course, the importance of audits as providing information about the implementation effectiveness of the Quality Management System has not been negotiated.

4.7.3 Measurement, Analysis and Improvement

One of the most important clauses in ISO 9001 (2008 version) is the last clause dealing with measurement, analysis and improvement. The idea, possibly, is to eventually work out improvements in processes, products and services based on appropriate analysis of measurements generated in various stages of the design, procurement, production, inspection and delivery. Management should decide what to measure, what will come out of analysis, and how these findings can motivate and subsequently facilitate and even evaluate improvement efforts.

Analysis of measurements should include, among others, the following

1. Analysis of spread or variability in the feature or property measured;
2. Analysis of trend in the feature over time to indicate probable deviations (excess or shortage) from standards or specifications;
3. Analysis of dependence of the feature on related features separately measured, may be at an earlier stage;
4. Analysis of response–factor relationship, through planned experiments.

Besides measures of process outputs and costs, measures of overall performance in a set of related processes are also important, e.g. measure of customer satisfaction (considering external customers, in particular).

Any measure of any property or any aspect of performance when used repeatedly over time, over units, over processes, etc., reveals variations, among the measured values and hence around the targeted value. Such variations, at least to some extent, are uncontrollable and hence unpredictable. Such random variations along with variations due to controlled factors have to be analysed into components associated with the controllable or assignable factors.

Gap analysis (between targets and actuals), priority analysis (among causes of variation or among consequences of gaps or deviations), process capability analysis (to find out the acceptability of specifications for a process as it is now), etc., are among many other analyses done with the help of relevant statistical tools and software for the analysis types mentioned earlier apart from classical statistical tools.

Quality improvement refers to the actions taken to enhance the desired features and characteristics of products that provide customer satisfaction and even lead to customer delight, and to increase the effectiveness and efficiency of processes used to produce and deliver the products and services.

A process for improvement can include

Definition, measurement and analysis of the existing situation;
Establishing the objectives for improvement, e.g. increase customer satisfaction or reduce wastes or minimize energy consumption;
Search for possible solutions to the existing problems or hindrances;
Evaluation of these alternatives in respect of cost and time of implementation;
Implementation of selected solutions;

Verification, analysis and measurement of implementation;
Normalization of process changes created by the selected solutions.

Improvement processes are continual and not considered as final solutions. Processes are reviewed, as and when necessary, to determine opportunities for further improvement. Audits and reviews of the Quality (Management) System can identify these opportunities.

4.8 2015 Revision

The standard ISO 9001 underwent a radical revision in 2015 with some new clauses and sub-clauses, a new sequencing of clauses and allowing participating organizations ample flexibility in aligning their Quality Management Systems with the standard and enabling them to achieve their desired objectives in regard to performance. The number of clauses has been increased from 8 I 2008 to 10 in 2015, with 7 effective clauses identified as

Context of the Organisation with four Sub-clauses.
Leadership with three Sub-clauses.
Planning including Sub-clauses on Actions to address risks and opportunities, Quality Objectives and Planning to achieve them and, quite interestingly, Planning of Change.
Support including Sub-clauses on Resources covering Environment for the Operation of Processes and Organisational Knowledge; Competence; Awareness; Communication; and Documented Information.
Operation with Sub-clauses devoted to Operational Planning and Control; Requirements for Products and Services; Design and Development of Products and Services; Control of Externally Provided Processes, Products and Services; Production and Service Provision; Release of Products and Services and Control of Non-Conforming Outputs.
Performance Evaluation covering Sub-clauses on Monitoring, Measurement, Analysis and Evaluation; Internal Audit and Management Review.
Improvement incorporating Sub-clauses on Non-Conformity and Corrective Action and Continual Improvement as two Sub-clauses.

Some new clauses and Sub-clauses make a lot of sense and focus on recent ideas, thoughts and practices in modern management. Take the case of Organisational Knowledge part of the standard that can be easily related to Knowledge Management paradigms. This part addresses the need on the part of an organization to determine and manage the stock of knowledge maintained by the organization to ensure the continued operation of its processes and the achievement of conformity of its products and services to agreed performance requirements. These requirements will expectedly safeguard the organization from loss of knowledge, particularly specialized knowledge in some specific areas which may arise from staff turnover or failure to capture and share relevant knowledge.

These will also encourage the organization to acquire new knowledge by way of learning from experience, mentoring by its own veterans and even external advisers as also benchmarking of its processes and results. Thus, this particular Sub-clause infuses into the Quality Management System the idea behind a Learning Organisation enshrined in the balanced Business Scorecard.

Of great significance is the thrust on Planning by devoting a whole clause to the subject. And the very first clause dealing with Risk Analysis and associated actions draws attention to the fact that outcomes of certain processes to be carried out at the supplier's end are not exactly predictable or controllable, giving rise to risks attached to any assurance statements. Risks in efforts to enhance desirable effects and prevent or reduce undesired effects or to achieve improvement as also the likely opportunities that are linked with such efforts have to be analysed and results have to be incorporated in relevant parts of the Quality Management System. The standard does not indicate the need for a formal risk management exercise but provides cautionary guidance in interpreting deviations in outcomes from planned or desired ones at both the supplier and the customer sides.

Sub-clause 6.3 on Planning of Change deserves due notice. Keeping in view the established fact that change is the most difficult entity to manage, this Sub-clause states that whenever a change in some structure or process or procedure is planned for introduction, the organization shall consider

The purpose of the changes and their potential consequences;
The integrity of the Quality Management System;
The availability of resources of various sorts required to implement the changes; and
The allocation or re-allocation of responsibilities and authorities.

Contents of Sub-section 7.1.6 on Organisational Knowledge along with Section 7.2 on Competence and Section 7.3 on Awareness clearly spell out the requirements of knowledge management in all its facets. Sub-section 7.1.6 states 'The organization shall determine the knowledge necessary for the operation of its processes and to achieve conformity of products and services. This knowledge should be documented, maintained and be made available to the extent necessary'. Organisational Knowledge can be based on both (1) internal resources like intellectual property, knowledge gained from experience, lessons learned from failures and successes, capturing and sharing undocumented knowledge and experience, results of improvements in processes, products and services and the like, as also (2) external sources like standards, conferences and workshops attended by its people, knowledge gained from customers or external providers like consultants. This sub-section does not directly speak of knowledge acquisition when needed. However, Section 7.2 mentions the need for actions to acquire competence, when needed, and to evaluate the effectiveness of the actions taken in this regard. Maybe, Section 7.3 on Awareness could have mentioned dissemination of knowledge available within the organization to persons who are in need of such knowledge.

Of course, quite a few other changes over the 2008 version are also equally important for an effective and sincere implementation of the revised standard which marks a distinct improvement in concepts and practices in Quality Management Systems. Just incidentally, no mention is made about a Management Representative in the revised version, without diluting in any way the roles and responsibilities of the organization and its management. In fact, Sub-clause 5.3 on Organisational Roles, Responsibilities and Authorities gives out five distinct areas where top management is required to assign responsibilities and authorities. The only change is that the different roles and responsibilities in connection with the Quality Management System are no longer vested in one person.

Sub-sub-clause 9.1.3 on Analysis and Evaluation spells out the need for appropriate analysis of data and information arising from monitoring and measurement. Results of such analysis (types not specified) shall be used to evaluate

(a) conformity of products and services to corresponding requirements;
(b) degree of customer satisfaction;
(c) performance and effectiveness of the Quality Management System;
(d) effectiveness of implementing plans;
(e) effectiveness of actions taken to address risks and responsibilities;
(f) performance of external providers; and
(g) need for improvement in the Quality Management System.

It is interesting to note that the last (out of 20) clause in the 1987 version of the standard was on the use of statistical techniques, while in the 2015 revision only a NOTE says 'methods to analyse data can include statistical techniques'. However, the standard provides enough opportunity to use statistical techniques, without being explicitly stated within the standard, for risk-based thinking, planning, design development and analysis purposes.

4.9   Other Standards

Apart from the international standards which took due account of prevailing national standards, which are applicable to all types of industries offering 'products', there exist some national standards and some other industry-specific standards which have a relatively large audience within industrial organizations. An important step was taken by the Automobile Industry where the generic standards in ISO 9000 series were found lacking in specifying some critical activities to ensure quality of automobiles and the large number of accessories and ancillaries involved therein. In fact, several of these standards were really based on ISO 9001 standard. In some sense when these other standards came up, they included many additional requirements some of which have since found their entry into the 9001 standard.

Thus, additional requirements such as continuous improvement, manufacturing capability, and production part approval process were introduced in QS 9000. This standard came up as the outcome of an effort taken up jointly by three automobile giants, viz. Chrysler, Ford and General Motors. The idea was to specify standard reference manuals, reporting formats and technical terminology across the automobile industries and their supplier organizations. These standards would help the suppliers of parts, components and accessories to enhance quality of their supplies, at the same time reducing costs. Unlike ISO 9001 that emphasises on 'document what you say and do as you have documented', QS 9000 incorporates demonstration of results achieved at different steps in the production process and, through that, effectiveness of the documented Quality System. Many of the concepts in the Malcolm Baldrige National Quality Award are reflected in QS 9000.

There are two major components in QS 9000: Requirements laid down in the generic ISO 9001 standard and Specific Requirements of customers of both final automobiles as well as of automobile manufacturers. Customer-specific requirements include methods for statistical process control, production part approval process, failure modes and effects analysis, measurement systems analysis, advanced product quality planning and control planning and Quality System assessment. In fact, measurement system analysis, often referred to as r & R studies, is a contribution of this standard that has led to the development of several standards relevant to the measurement process in different experimental work situations.

AS 9000/9001

The aerospace industry was emerging on the scene as an industry that had to lay a greater stress on quality, safety and precision. The concerns for reliability had not been explicitly spelt out in ISO 9001 or any other similar standard. Linked to reliability were issue of safety and maintainability. While design and development management had been there in ISO 9001, explicit requirement about configuration management came to be felt acutely in the aerospace industry. And process performance in such processes as design development, verification and validations also in assembly and inspection processes had to be highlighted in the aerospace industry. All this resulted in the formulation of AS 9001, replacing the initial AS 9000 standard towards the beginning of this century. It may be incidentally mentioned that in view of safety being accorded a very high priority, each part of an aero-engine would be inspected several times not to allow any defect going unnoticed during a single inspection operation.

ISO/TS 16949

With a rapid expansion of the automotive industry in many developed countries and different manufacturers interested to satisfy customers with their own products in their characteristic ways of dealing with customers and meeting the latter's requirements, a need was felt to come up with an International standard that will somewhat more generic compared to QS 9000 and would allow, within a generic framework, a particular manufacturer to control his production and product-related processes including those like product recall or product replacement following his own specific system.

4.10   Quality Awards and Other Incentives

Besides these standards, Six Sigma approach and the Quality Awards provide guidance to industries in their effort to improve business process performance. Unlike standards for Quality Systems to which compliance can be certified by a third party, these awards motivate competing organizations to pull up their quality efforts and to achieve excellence that is reflected in their business performance. Thus, compliance to a system standard has been sometimes regarded as a stepping stone towards success in competition for an improvement-oriented Quality Award. And, then, there are the supporting standards—mostly ISO ones—like those on Auditing, Training, Complaints handling, Documentation. There are some national standards which are quite comprehensive and are widely implemented.

4.10.1   Quality Awards

Among the Quality Awards which stand out prominently as hallmarks of Excellence in Quality, the Malcolm Baldridge National Quality Award established in 1987 by the US Congress with three eligibility categories, viz. manufacturing companies, service companies and small businesses, is the most noteworthy. It takes into account seven major sets of criteria, viz. leadership, strategic planning, customer and market focus, information and analysis, human resource focus, process management and business results. The European Quality Award started by the European Foundation for Quality Management consists of two parts, viz. the European Quality Prize and the European Quality Award, now known as the Business Excellence Model. All government agencies, not-for-profit organizations, trade associations and professional societies are not eligible to apply for this award. In 1996, the United Kingdom Quality Award, the Swedish Quality Award and the New Zealand Quality Award were introduced. The Canadian Awards for Business Excellence introduced in 1997 and the Australian Quality Awards are primarily based on Malcolm Baldridge criteria with some distinctiveness built in. The Deming Prize established in 1951 by the Union of Japanese Scientists and Engineers is awarded to individuals and groups who have contributed significantly to the field of Quality Control. There are other National Quality Awards including the Rajiv Gandhi National Quality Award started by the Bureau of Indian Standards in the year 1994 with four award categories, viz. large-scale manufacturing, small-scale manufacturing, service and best-of-all.

All the above Quality awards really spell out desirable features of a Quality Management System. They are not strictly comparable to the ISO or the National Standards for Quality Management Systems, at least for one reason that they are not certifiable standards. The national awards vary on the dimensions used, the weights accorded to the different dimensions and on the number of different awards spelt out to cover different segments of manufacturing and service industries.

The Malcolm Baldridge National Quality Award which has provided major inputs to many other Quality Awards makes use of the following criteria for award:

Leadership Examines how senior executives guide the organization and how the organization addresses its responsibilities to the public and practises good citizenship.

Strategic Planning Examines organization sets strategic directions and how it determines key action plans.

Customer and market Focus Examines how the organization determines requirements of customers and markets.

Information and Analysis Examines the management, effective use and analysis of data.

And information to support key organization processes and the organisation's Performance Management System.

Human Resource Focus Examines how the organization enables its workforce to develop its full potential and how the workforce is aligned to the organisation's objectives.

Process Management Examines aspects of how key production/delivery and support processes are designed, managed and improved.

Business Results Examines the organisation's performance and improvement in its key business areas: customer satisfaction, financial and marketplace performance, human resources, supplier and partner performance, and operational performance. This criterion also examines organization performs relative to competitors.

Taking a cue from the adage that the proof of the effectiveness of any Quality System should lie in the results produced by the organization, MBNQA has shifted its emphasis on results orientation with a weight of 26 on Financial Results, 21 on Customer Management and satisfaction and a weight of 4.5 on Suppliers/Partners management.

Puay et al. () reported a comparative study of nine National Quality Awards by developing a framework consisting of nine criteria items and 28 sub-items. The major items considered were

1. Leadership—concerns management's behaviour in driving the organization towards Total Quality.
2. Strategy and Policy—concerns how the organization formulates, communicates, deploys, reviews and improves strategy and policy.
3. Resource Management—concerns how the organization manages key resources, such as information, materials, technology and finance.
4. Human Resource Management—concerns management and deployment of the workforce in the organization.
5. Process Quality—concerns how the organization manages, evaluates and improves its key process to ensure quality output.
6. Suppliers/Partners Management and Performance—concern how the organization manages its suppliers/partners to enhance overall performance.

Table 4.1 Summary of weights assigned to framework criteria items

	Framework item	MB	EQA	UK	Brazil	SWQA	New Zealand
1.	Leadership	8	10	10	7	7.5	7
2.	Impact on Society	3	6	6	2	8.5	2
3.	Resource Management	8	8	8	7.5	8	7.5
4.	Strategy and Policy	6.5	8	8	5.5	3	5.5
5.	People Management	15	18	18	14	15	17.5
6.	Process Quality	8	14	14	11	10.5	11
7.	Business Results	26	15	15	25	10.5	18.5
8.	Customer Management and Satisfaction	21	20	20	25	30	25
9.	Supplier Management and Performance	4.5	1	1	3	4	6
	Framework item	Rajiv Gandhi		SQA	CAE	Ave.	Std. Dev.
1.	Leadership	7		12.5	6	8.3	2.1
2.	Impact on Society	13		2.5	2.5	5.1	3.8
3.	Resource Management	10		8	1	7.3	2.5
4.	Strategy and Policy	8		7	12	7.4	3.0
5.	People Management	10		16	20.5	16.0	3.0
6.	Process Quality	12.5		11.5	13.5	11.8	2.0
7.	Business Results	17.5		12	21	17.8	5.4
8.	Customer Management and Satisfaction	21		25	18.5	22.8	3.7
9.	Supplier Management and Performance	1		5.5	5	3.4	2.0

Source Xie et al. (1998). A Comparative Study of Nine National Quality Awards
MB Malcolm Baldridge, *EQA* European Quality Award, *SWQA* Swedish Quality Award, *SQA* Singapore Quality Award and *CAE* Canadian Award for Excellence

7. Customer Management and Satisfaction   concerns ho the organization deter mines the needs of its customers and how the customer relationship is managed.
8. Impact on Society—concerns the organisation's contribution to society and the environment.
9. Results—concerns the results achieved by the organization.

Weights accorded to these nine criteria vary quite visibly across countries represented by the awards. The effort to enhance trans-national image about quality of its products, the current socio-economic conditions and the general management styles along with work culture contribute to such differences. In any case, countries who do not yet have National Quality Awards can definitely gain some insight by looking at the following table that gives out the weights for the criteria in nine national awards, Xie et al. (1998) (Table 4.1).

In some of the awards, the maximum marks that can be given on each criterion to an applicant organization indirectly reflect the weight of that criterion. Weights tend to vary over time to take account of economic, social and technological development in the respective countries.

4.10.2 Six Sigma Model

Six Sigma methodology has brought about a new paradigm in Quality Management which operates on a project mode and may well fit into an existing Quality Management System where Quality Improvement Projects are taken up in different manufacturing or service areas within an organization that has been causing some business problems. Six Sigma is a highly disciplined, quantitatively oriented, top-down approach that has really worked wonders in solving some nagging business problems linked with quality of products and/or services. Integrating Lean Production methods with the Six Sigma approach involving the D (Define)–M (Measure)–A (Analyse)–I (Improve)–C (Control) or D–M–A–D (Develop) V (Verify) sequence, using relevant data has been one more advanced step.

Somewhat different from Six Sigma approach and incorporatingSix Sigma (methodology) implementing some additional steps, Eight Disciplines problem-solving (8D) is a method to approach and resolve a problem, typically employed by quality engineers to identify, correct and eliminate recurring problems. The method has been found useful for both process and product improvement. It establishes a permanent corrective action based on a quantitative analysis of the problem and focuses on the origin and the root causes behind the same. Initially involving eight stages, it was later extended to include an initial planning stage. 8D has become a standard in the auto, assembly and other industries that attempt a comprehensive structured problem-solving methodology using a team approach. The stages in the approach are as under:

The first two steps here are somewhat revealing of a plan to resolve some recurring problems which have already been identified. And a plan cannot operate without envisaging some problem(s) and putting in place some resources including people and their time to take up and eventually resolve the problem(s). The formal steps including these two preliminary ones appear in the following sequence.

Plan: Plan for solving the problem and determine the prerequisites.

Use a Team: Establish a team of people with relevant process/product knowledge. The team may be a cross-functional one.

Define and Describe the Problem: Specify the problem by identifying in quantifiable terms the questions related to who, what, where, when, why, how and how many (usually referred to as 5W2H).

Develop Interim Containment Plan—Implement and Verify Interim Actions: Define and implement short-term actions to mitigate the problem and to isolate it from any customer. The interim containment action is oriented to remove customer problems.

Determine, Identify and Verify Root Causes and Escape Points: Identify all applicable causes that could explain the occurrence of the problem. Also identify why the problem was unnoticed at the time of its occurrence. All causes shall be verified or proved, not determined by fuzzy brainstorming. One can use Ishikawa diagram to map causes against the effect or problem identified at the beginning. Recognising the escape points where problems went unnoticed at the first instances is an additional feature of this methodology, not explicitly spelt out in 6σ approach.

Choose and Verify Permanent Corrections for Problem/Non-conformity: Through pre-production programs, we should confirm quantitatively that the corrections chosen will resolve the problem for the customer on a sustained basis.

Implement and Validate Corrective Actions: Define and implement the best possible corrective actions, using appropriate criteria to identify the 'best'.

Take Preventive Measures: Modify the management systems, operation systems, practices and procedures to prevent recurrence of the problem under study and of similar problems.

Congratulate the Team: Recognise the collective efforts of the team management should at least formally thank the team members and offer some tokens of appreciation.

Again, it should be noted that neither Six Sigma nor Lean Six Sigma nor even Eight D methodology is a substitute Six Sigma (methodology) implementing for any Quality Management System—International or national, Generic or Industry-Specific. These are approaches which can be adopted within the broad framework of a Quality Management System.

4.11 QMS in Public Service Organisations

While Quality Management Systems are essentially generic in character and independent of the nature of the output as also of the type of ownership, services rendered by public service providers have to be more concerned about the customers who are, usually, entitled to receive services of satisfactory quality from the service providers who are branded as Public Service Providers—both in the public and in the private sector. An Indian Standard IS 15700: 2005 provides an idea about the distinguishing features of Quality Management Systems—Requirements for Service Quality by Public Service Organisations. Such an organization provides service(s)—on request or even proactively—to public at large and/or whose activities influence public interest. Government Ministries and Departments, Regulatory Bodies, Public Utility Service Providers and the like provide examples.

As stated in Clause 1 Scope, this standard specifies requirements for a Quality Management System where a public service organization

(a) needs to demonstrate its ability to consistently provide effective and efficient service that meets customer and applicable legal, statutory and regulatory requirements (b) aims to enhance customer satisfaction and (c) aims to continually improve its service and service delivery process.

As expected, this standard has all the features of a national standard and incorporates usual requirements for any Quality Management System. Clause 7 relating to Citizen's Charter, Service Provision and Complaints handling deals with the specific features of the standard, not included in ISO 9001 or other comparable standards.

The Citizen's Charter shall contain

(a) vision and mission statements of the organization;
(b) list of key service(s) being offered by the organization, which should be updated on a continuing basis;
(c) measurable service standards for the service(s) provided in terms of time (to access, to wait, to get delivery, to pay bills, and similar other elements), cost, and other relevant parameters as well as remedies available to the customer for non-compliance to the standard.

The Charter should be non-discriminatory, making no differences among different segments of the public and even between the general public and representatives of the people or of regulatory bodies. It should describe the complaints handling process, giving details about the public grievance officer. It should be periodically reviewed for updation and continual improvement. It should also highlight expectations of the service providing organization from its customers for possible augmentation, continuation and improvement of the service(s) provided.

Clause 7.2 relates to Service Provision and includes, among several other specifications, the requirement to ensure availability and use of suitable equipment, monitoring and measuring devices as also their calibration at specified intervals or prior to use, whenever necessary. Another important requirement is to identify, verify protect and safeguard any customer property or product or equipment for use during some service like repair and maintenance.

The next clause refers to Complaints Handling. It starts with identification of complaint-prone areas in a systematic manner and setting up norms for their redress. Such areas could be linked to some service production or delivery personnel or to some particular aspect of service production or to the absence of some measuring or and monitoring device needed to demonstrate compliance to service standard. The service provider shall provide information concerning complaint handling process, including where and how complaints can be made, minimum information to be provided by the complainant and time limits within which the complaint will be closed to the satisfaction of the complainant. Each complaint should be scrutinized and categorized as critical, major or minor depending upon its seriousness and severity. The standard also includes communication of the decision to the complainant regarding his/her complaint immediately after the decision is taken and a feedback from the complainant obtained. The Standard refers to the nomination of an Ombudsman who could be approached if normal service delivery mechanism does not respond. In some cases, volunteers could be roped in from the community where services are not delivered in accordance with the norms specified in the Service Standard, in the absence of adequate manpower resources.

Under clause 8.2 on Monitoring and Measurement, the standard focuses on monitoring the commitments made in the Citizen's Charter and the complaints handling procedure on a regular basis. It also mentions random checks on the complaints handling machinery.

Clause 8.2.2 related to Customer Satisfaction points out the importance of a suitable methodology like sample surveys for measuring customer satisfaction. Findings of such surveys, if properly conducted and taken earnestly by the customers, can provide inputs for modifications of current policies governing such public services.

Just like the Quality Awards discussed earlier, the **Charter Mark** introduced by the British Government in the early nineties provides an impetus and a guideline for quality improvement in public sector organisations. The Charter mark is based on ten criteria which the service provider has to address. These are

- Set Standards—clear, meaningful and high-performance standards should be set.
- Be open and provide full information—all information regarding the services, facilities, options, etc., should be provided to the user. The user has a right to published standards of services, and they should know how to get the most of these services.
- Consult and involve—users should be widely consulted to ascertain what services they need and they should be involved to find out how to make good use of their ideas to improve services.
- Encourage access and the promotion of choice—services should be available to everyone and users should be offered choices wherever possible, so that the user can choose the option that suits him most.
- Treat all fairly—services should be provided irrespective of any bias to age, gender, religion, etc. The user should get polite and helpful attitude and user-friendly approach from staff.
- Put things right when they go wrong—the service provider must make it easy for people to say when they are not happy with a service. Complaints should be taken in a positive spirit and the service provider should act swiftly to redress the complaint.
- Use resources effectively—resources should be effectively used by budgeting carefully and achieving the 'Best Value' again any resource deployment.
- Innovate and improve—consult users as well own people to innovate and improve. Efforts to improve must be continuous; otherwise even the status quo of the existing service level cannot be maintained. New ideas are necessary to roll on into the future.
- Work with other providers—people often require more than one service at a time and due to lack of harmony among different service providers, people have to face bureaucracy. People want a seamless service resulting from a number of service providers who work in close co-operation.
- Provide user satisfaction—the service provider can demonstrate his efficiency if the user agrees that they have got a really good service.

In 2004, Charter Mark was re-launched as a Standard, more closely aligned with the public service reform agenda. The Charter Mark benefits both the service provider and the user. People feel more important when they are consulted and involved. They have more choices and easier access. They get new and innovative services from a 'single window'. They get courteous behaviour, more information, published standards of service and 'value for money'. Indirectly, loss to society decreases as there is emphasis on efficient utilization of resources.

For the service provider, it provides a win-win situation, improvement and boost to staff morale is evident since work or contribution of every staff member is recognized. It helps the organization with an opportunity to raise its profile and, that way, to generate support from stakeholders and funding agencies.

Unique features of the Charter mark include, among others,

It aims at excellence in services and not organizational excellence.

It is not management-led. Staff motivation and recognition underpin it.

Its focus is on service delivery to the user and not just on the service (production) process.

Charter mark is awarded by an independent and unbiased scrutiny, undertaken directly by a Central Government panel, and not by any certification agency.

4.12   Steps in Implementation of a Quality System

An approach to implementing a Quality Management System that can bring in improvements in performance involves the following sequence of tasks.

Carry out a desk exercise to establish compatibility or otherwise of the documented Quality System and the Quality Policy and Quality Objectives and targets of the organisation. Inadequacy of the Quality System as documented in meeting the policy and objectives cannot be ruled out necessarily and effectiveness audit of the Quality System—as distinct from a compliance audit against the documented system—is always a welcome step in implementing a Quality System.

Once the effectiveness of the system has been established, we have to identify core as well as support processes which are critical to the attainment of the quality goals and objectives. It is possible that performance in several processes affects the attainment of a single objective. Conversely, performance in a single process may influence results in several areas reflected in different quality objectives.

For each critical process, a process plan has to be developed indicating measures of its effective performance, considering the quality objectives influenced by it. The plan must provide requirements to be met during process execution which, if complied with, will ensure attainment of the corresponding quality objective(s).

These effectiveness measures should be applied to the output of each process to assess its current effectiveness to meet the corresponding quality objective(s). The gap between the expected output capable of meeting quality objective(s) and the actual output should be examined.

A comparison between the targeted performance (in terms of the output quality and the cost involved) reveals opportunities for improvement in process performance. A close analysis of the gap through brainstorming and using a fishbone diagram, if necessary, remedial actions must be found out and implemented to bridge the gap.

Remedial actions could relate to materials, machines, methods, operators or operating environments and should include both corrective actions regarding gaps already identified and preventive actions in regard to potential gaps that may arise.

Once remedial actions are in place and gaps have been closed, the system must be observed for its eventual effectiveness to meet the quality goals and objectives. And changes injected in different processes have to be managed for their acceptance by the workforce and sustained implementation of the changes.

If, unfortunately, the system fails to attain the quality goals and objectives even after such changes managed effectively, the initially set goals and objectives may have to be revisited or the entire Quality System has to be re-worked or both.

Management Commitment and Leadership to

(a) establish Quality Goals and objectives;
(b) provide the needed resources—physical and organizational;
(c) provide adequate Quality-oriented training to all;
(d) stimulate quality improvement through suitable incentives;
(e) review progress in different quality-related activities;
(f) recognize excellent work done by individuals and groups;
(g) revise the reward system from time to time to emphasise the need for quality improvement;
(h) continuing education at all levels;
(i) senior management first;
(j) different content in training programmes for different types of responsibilities associated with different aspects of quality;

Setting up of a Quality Organisation including Quality Councils, Quality Audit Teams, Quality Task forces, etc.

Adequate use of relevant procedures and techniques: Statistical process control: Process/machine capability analysis: Customer Response study: Failure Mode and Effect Analysis, etc.

Stress on Innovation in

Materials and processes;
Products and services; and
Use and maintenance.

Emphasis on the use of Information and not just Opinions, Facts and not just imaginations or hunches for making and changing plans.

Enhanced co-operation to

Agree on common standards;
Develop and work with the same vendor(s);
Educate customers for proper installation, use, storage and upkeep of the products;
Improve Industry Image;

4.13   Concluding Remarks

A Quality (Management) System has to be a system first, in the modern connotation of a system in System Science. It has to refer to Quality in products, processes and services and has to be linked up with business to attract due attention and involvement of management. It should not be a static entity to be followed at all times to come. After being developed to suit the special needs of any particular organization, it should be reviewed periodically with respect to its performance and, whenever found necessary, should be modified to keep track of changes in production technology and in market behaviour.

A Quality System should work in tandem with the Environment Management System or Financial Management System and even with the Corporate Social Responsibility profile, wherever applicable. All apparent contradictions among such systems should be discussed and removed right at the beginning.

While organizations will be generally tempted—and for valid reasons—to get their Quality Management Systems, once developed, certified for conformity to ISO 9000 standards to enhance their quality image as also to qualify for some fringe benefits from the government, it must be borne in mind that certification is focused on compliance rather than on excellence. Not too unoften, certification has been a ritual and senior management pays less than adequate attention to quality.

There have been interesting studies on effectiveness of ISO 9000 standards in various countries and in different industry groups and even in individual industries and some of these have been reported in books, journal articles and reports of concerned agencies. In volatile business contexts or where creativity is paramount or where the keys to a company's successful performance lie across complex supply chains and networks of interdependent companies and thus outside the boundaries of the company, the validity of the so-called excellence models and quality awards may not be viewed as goal posts for the Quality System in an organization.

References

British Quality Foundation. (1996). The UK Quality Award 1996 Application Brochure, Information Brochure.
Bureau of Indian Standards. (1994). Rajiv Gandhi National Quality Award 1994 Procedure and Application form.

Conti, T., et al. (2003). *Quality management and future trends in quality in the twenty-first century: Perspectives on quality and competitiveness for sustained performance* (pp. 237–244). Milwaukee, USA: ASQ Quality Press.

European Foundation for Quality Management. (1996). The European Quality Award 1997.

Evans, J. R., & Lindsay, W. M. (1999). *The management and control of quality.* South-western, Ohio.

IS 15700. (2005). Quality Management Systems—Requirements for Service Quality by Public Service Organizations.

ISO 10012. Measurement Management Systems—Requirements for Measurement Processes and Measuring Equipment.

ISO 9000. Quality Management Systems—Fundamentals and Vocabulary.

ISO 10001. Quality Management—Customer Satisfaction—Guidelines for Codes of Conduct for Organisations.

ISO 10002. Quality Management—Customer Satisfaction—Guidelines for Complaints Handling.

ISO 10003. Quality Management—Customer Satisfaction—Guidelines for Dispute Resolution External to Organizations.

ISO 10004. Quality Management—Customer Satisfaction—Guidelines for Monitoring and Measurement.

ISO 10005. Quality Management Systems—Guidelines for Quality Plans.

ISO 9004. Managing for the Sustained Success of an Organisation—A Quality Management Approach.

ISO/TR 10017. Guidance on Statistical Techniques for ISO 9001: 2000.

Juran, J. M. (1995). *A history of managing for quality: The evolution, trends and future directions of managing for quality* (pp. 433–474). Wisconsin, USA: ASQ Quality Press.

Kaplan, F. (1980). *The quality system.* Radnor, PA: Chilton Books co.

Mukherjee, S. P. (1992). Quality System—A Philosophical Outlook EOQ Quality

National Institute of Standards and Technology. (1997). Criteria for Performance Excellence—Malcolm Baldridge National Quality Award.

National Quality Award Foundation, Brazil. (1996). Brazil National Quality Award Criteria.

National Quality Institute. (1997). Entry Guide for the Canada Awards for Excellence.

New Zealand National Quality awards Foundation. (1996). New Zealand National Quality Awards 1996 Introduction and Criteria.

Puay, S. H., Tan, K. C., Xie, M., & Goh, T. N. (1998). A Comparative Study of Nine National Quality Awards.

Singapore Productivity and Standards Board. (1996). The Singapore Quality Award 1996.

Swedish Institute for Quality. (1996). The Swedish Quality Award 1996 Guidelines.

Tan, K. C. (2002). A comparative study of sixteen quality awards. *TQM Magazine, 14*(3), 165–171.

Xie, M., et al. (1998). A comparative study of nine quality awards. *TQM Magazine, 10*(1), 30–39.

Chapter 5
Total Quality Management

5.1 Introduction

Economic growth mandates increasing effectiveness and efficiency of all production processes—in manufacturing as well as in service sectors. In this context, Total Quality Management (TQM) has evolved as an approach to enhance the quality of performance in different business processes, both core and support. In fact, TQM has emerged as a movement spreading across different functional areas in an organization and is not confined to quality of products only. It must be added, however, that a Quality Management System on the lines of ISO 9000 series is a concrete and an important step to initiate Total Quality Management.

TQM was initially known as Total Quality Control (TQC) in Japan. Subsequently, the scope of this approach was expanded to become Company-Wide Quality Control (CWQC). Pillars of TQM include customer orientation, focus on processes, fact-based decision-making, continuous improvement and involvement of all people. Incidentally, these points have been incorporated as basic principles behind the latest version of ISO 9001 Standard. TQM is a company-wide exercise and should be looked upon as a strategy for business growth and not simply for controlling or improving product and service quality. Thus, TQM implies a holistic approach to productivity improvement that use both hard and soft skills.

TQM can be practised in large as well as medium and small industries. Any growth-oriented and customer-focused industry in manufacturing or service sector with a management keen to improve productivity through quality and not just the volume of production can and should practise TQM and can derive benefits proportional to the investment it makes in building up a workforce that appreciates the business importance of TQM and a workplace conducive to TQM implementation.

Volumes have been written on Total Quality Management (TQM)—directly or otherwise, dwelling on the many facets of TQM, describing experiences of a motley myriad of organizations trying out their hands on absorbing and implementing TQM and deriving benefits therefrom, and throwing up a multitude of variants of

© Springer Nature Singapore Pte Ltd. 2019
S. P. Mukherjee, *Quality*, India Studies in Business and Economics,
https://doi.org/10.1007/978-981-13-1271-7_5

TQM with varying degrees of emphasis on soft aspects of human behaviour and hard aspects of quantitative analysis and engineering. A serious reader has to get lost in the deluge of documents to find convincing answers to some questions like the following:

When, where and why did TQM emerge as a new paradigm in Quality Management?
What was the backdrop prevailing in regard to quality then?
Which elements in TQM marked a significant departure?
How could top management be easily convinced to accept TQM?
How did TQM flourish and on which planks?
What role quantitative analysis has been playing in TQM?
How much is TQM concerned with human resource management?
What has been the extent of acceptance of TQM by the industry?
Has TQM been implemented with some success in other organizations?
How has TQM evolved over the years to modify its own character?
Have there been more recent attempts to come up with alternatives?

One does not expect unambiguous answers to these and many other related questions. And the present chapter does not even intend to address all such questions. It simply makes an attempt to introduce the broad issues related to TQM and its implementation with the hope that any one interested to join the TQM movement will be benefitted to some extent by the material contained here along with the references indicated.

Initiated in Japan during the early sixties as 'Total Quality Control'—with an emphasis on the implications of 'Total'—and taken as Company-Wide Quality Control in some other organizations, Total Quality Management has sometimes been regarded—with ostensible reasons—as the Second Industrial Revolution, not in terms of the production process and the technology behind, but reflecting on management practices. TQM has definitely marked a significant milestone in the annals of Quality and Quality Management.

5.2 TQM—A Characterization

Total Quality Management is best characterized as a movement, possessing all the three distinct features of any 'movement'.

Firstly, it has challenged contemporary (erstwhile) concepts, methods and practices bearing on quality. (In a way, it has expanded/augmented/disseminated these.) It extended the borders of quality to pre- and post-production areas to eventually subsume the entire set of organizational activities. In the second place, it has been continuously moving, without being constrained by specifications OR standards, to achieve higher goals of quality (and performance) more efficiently and effectively. Finally, it has integrated Quality Management with overall company management and thereby created a visible impact on business. In fact, Total Quality Management has now emerged as a strategy for business growth.

In the ISO document ISO 9000: 2005, TQM has been taken as a management *approach* of an organization, *centred on quality*, based on the *participation of all its members* (personnel in all departments and at all levels of the organizational structure) and aiming at *long-term success* through *customer satisfaction, and benefits to all members of* the *organization and to society.*

A wonderful exposition, this characterization of TQM is quite concrete in terms of an approach aiming at long-term success—and not just immediate financial gains—that should result in satisfaction to all stakeholders and even in benefits to the society at large.

The document goes on to say that *the strong and persistent leadership of top management and the education and training of all members are essential for the success of this approach.*

The document also points out that the concept of quality relates to the achievement of all managerial objectives, including, but not restricted to, quality of products, processes and services.

5.3 Two Alternative Views

The present author likes to present Total Quality Management in two alternative ways that eventually converge to the same paradigm. The first starts with the phrase Total Quality and talks of managing Total Quality, while the second starts with Quality Management and amplifies it to make it 'Total'.

Total Quality means value-to-price ratio to the customer, who pays a price to acquire a product (before use or deployment) and derives some value (considering some of the three commonly accepted notions of value, viz. use or functional value, esteem or prestige value and exchange or re-usability value. Depending on the nature of the product, these three may have different relative importance measures). Here, Total Quality improves with a higher (perceived or assessed) value for the same price or a lower price for the same value or a higher value against a lower price.

To the customer, the obvious definition is value-to-cost ratio, where, of course, manufacturing cost is the prime consideration and value is as judged by the producer. In any case, this value judgement definitely takes account of customers' perceptions. And, this value is essentially an outcome of the product design and the process capability.

A lower manufacturing cost without compromising on value or an enhanced value against the same manufacturing cost or a combination of both leads to higher Total Quality.

The third definition motivated by a concern for 'societal loss' incurred by a product failing during use or deployment takes care of the consequences of product failure which may affect an entire community and not simply a producer or a customer. To illustrate, a power generating and distributing agency procures a transformer from a transformer manufacturing factory, the transforms fails during operation cause some fire, and the entire neighbourhood suffers from power outage —not just the manufacturer or the power distribution agency (who may not be affected directly in some cases and may have to pay compensations at the best). Lower the societal cost, higher is the Total Quality.

To manage Total Quality, as explained above, we are bound to much beyond the pristine boundary of Statistical Quality Control or to quality-managing activities focused on the production phase or even to classical Quality Management that may take care of some pre- and post-production activities. Considerations of manufacturing costs (including costs of poor quality) of prices to be realized by customers and societal costs which require evaluation of indirect and remote or future costs (not always paid) remain outside the purview of traditional Quality Management. It is true that some of these considerations have been addressed in the Quality Management System Standards. People, policies and actions in the area of procurement do affect manufacturing costs, those in the area of sales and service affect pricing, and designers, people in installation and maintenance and even top management people are involved in the matter of societal costs.

All this justifies Total Quality Management, wherein 'Total' implies

In all activities ranging from receipt and confirmation (on verification of feasibility) of orders, through procurement and deployment of different resources, to pre-shipment inspection and packaging and after sales service, evidently;
By all people in all departments or divisions at all levels (implying a company-wide involvement); and
At all times considering possibilities of stricter quality requirements and greater conformity to such requirements, whenever desired or demanded (emphasizing the need for continuous improvement activities as essential to the programme).

It can be well appreciated that these ways of looking upon Total Quality Management are just two different paths to the same goal—one a little longer but more likely to reach the goal, while the other may be somewhat shorter but less likely to reach the goal.

There have been three different definitions of Total Quality with inherent respective appeals to the three partners in the game, viz. the producer, the customer (user) and the society.

5.4 Total Quality—A Simple Man's Guide to Its Applications

An early characterization of TQM is focused on soft skills, is oriented to the people in the organization and is quite appealing to all in the beginning. His characterization is based on the following requirements to be met by the people in different functional areas at different levels within the organization

1. Being polite on the telephone;
2. Not accepting failures as inevitable;
3. Looking critically at yourself for improving your own performance;
4. Ensuring that judgments and criticism of others are presented in a constructive, not destructive, manner;
5. Recognizing that we all have objectives and aspirations in our jobs;
6. Not blaming other people for your own mistakes;
7. Tackling the source of the problems and not the symptoms;
8. Not making people feel threatened or intimidated; and
9. Avoiding long communication chains, confused accountabilities and endless meetings without clear output actions.

A movement is always goaded and guided by a philosophy that may or may not have an immediate appeal at least to some of the people involved in or likely to be affected by the movement. The philosophy behind Total Quality Management makes it distinct from some other Quality Management Initiatives.

TQM implies more than a Quality Assurance System or a Quality Management System (as are specified in ISO 9000 series of Standards).

TQM rests on three piers, viz.

A belief system—resulting in mutual trust and leading to process ownership by people involved in the processes;
A knowledge system—calling for knowledge about processes and people; tools and techniques for measurement and analysis; prevailing best practices; and
A communication system—ensuring free and fast top-down as well as bottom-up flow of information.

TQM rests on

1. Management commitment and universal participation/involvement;
2. Focus on customers and on processes; and
3. Stress on innovation and improvement

As remarked earlier, TQM is a strategy for business growth. In the same spirit, we should admit that we need to develop a strategy to introduce and sustain TQM in an organization. TQM is not a quick-fix approach to solve productivity-related problems. It is an organization-wide initiative that takes time to build up and has to be sustained over time. Hence, investments in TQM are spread over a reasonable time span; returns are realized on a continuing basis over time.

5.5 An Approach to the Implementation of TQM

There can hardly exists a unique approach to implementing Total Quality Management in an organization, irrespective of its current management style or organizational culture, its customer population and product and/or service profile, its present stock of human resources and a host of other considerations. And TQM cannot be implemented overnight, the stages through which implementation process should go and the time likely to be taken in that would vary from one organization to another. However, several stages have been generally identified and some of the stages that are detailed here may not be required in some cases, while in some others these may have to be further expanded. The stages correspond to some basic issues to be appreciated by top management and the guidance and directions that emerge from such an appreciation.

5.5.1   Stage A

Why TQM? Risks, Costs and Benefits

Management should not plunge into TQM implementation without examining the implications of TQM on management practices, on both core and support business processes and human resource deployment. Further, the costs of implementing TQM—in spirit and not just in letters—by introducing a new work culture and, therefore, making necessary adjustments have to be carefully worked out. Evaluation of benefits—mostly in the long run—is a more difficult task, since it involves a lot of uncertainty in the likely benefits in terms of market expansion, manufacturing cost reduction consequent upon decreased costs of poor quality, improvement in stakeholder satisfaction and the like. Only after a conviction that TQM can play an important role to boost productivity and enhance business results, should the management decide to go for TQM adoption. Otherwise, TQM may be found sometime later as not yielding the anticipated results and may be disbanded. TQM implementation needs investment and investment below a threshold level may not prove to be useful, just as in the case of antibiotics a dose below the threshold fails to bring out the desired benefit.

5.5.2   Stage B

Create a Vision. Develop a Mission Statement and Set Up Quality Goals

This is a crucial step in putting the TQM movement on the right track within the organization. Even if vision and mission statements do already exist, these must be revisited to make them the real driving force behind the TQM movement. Concrete and achievable goals for quality of performance on the important business pro- cesses should be developed with due diligence and shared with and not just communicated to all the concerned people.

5.5.3 *Stage C*

Choose a strategy, involving

1. A systematic identification of areas (covering core as well as supporting business processes) where a visible scope for improvement in performance exists or such improvement is a necessity for business to survive or to grow;
2. A methodical exercise to assess the nature and extent of the problem and to evaluate the consequences of possible deficiencies on manufacturing costs, market shares and profit margins;
3. A comprehensive mechanism to capture required data for the above analysis from different sources within and outside the organization;
4. A detailed data analysis with due segregation for different product lines, different customer groups or market segments, different regulatory regimes, etc.;
5. Development of a comprehensive action plan in terms of activities or tasks to be carried out, the sequence to be followed, the resources human skills and knowledge to be deployed, the supervision and control to be exercised, etc.;
6. A rigorous and time-bound implementation of the action plan;
7. A proactive monitoring of steps in implementation to provide a feedback for any modifications in the plan for implementation, if needed; and
8. Doing it over again to gain confidence.

5.6 TQM—Principles and Actions

An attempt to differentiate TQM from other comparable quality initiatives, in terms of the principles involved and the actions envisaged, was suggested by quite early in the history of TQM. That way the following list of principles and actions as presented below may appear to be too simplistic to comprehend recent developments in the practice of TQM. However, the author feels that this may provide a pragmatic idea about TQM.

Principle	Action
The approach	Management involvement and leadership
The scope	Company-wide involvement after appreciation
The scale	Everyone has some quality responsibility and/or authority
The philosophy	Prevention and not detection
The standard	Right first time
The control	Cost of (poor) quality
The theme	Continuous improvement

The philosophy and the standard speak volumes about their implications. The guiding philosophy is defect prevention in every process element, since the standard set for the system is right first time. Slippages from the standard will add to the costs of poor quality, and this is the entity that is controlled in Total Quality Management. There have some modifications here and there by some exponents of TQM proposing slightly different explanations of some of the principles in terms of the corresponding actions. Even the following principles and actions have been added recently.

Principles	Actions
Ability	Training and education
Communication	Cooperation and teamwork
Reward	Recognition and pride

The framework outlined at the beginning along with the later modifications provides a comprehensive view about what Total Quality Management is. In the next section, an attempt is made to elaborate some of the principles and the corresponding actions. In fact, the scale, the standard and the control principles have been chosen for this purpose, as these appear to be distinguishing features of Total Quality Management philosophy.

It will be useful to consider the operational implications of the 'scale', the 'standard' and the 'control' principles.

5.6.1 Company-Wide Involvement After Appreciation

TQM implies involvement of all within the organization, each individual or team with some specified responsibility and commensurate authority. An important precondition for this involvement is that all the persons are exposed to the idea and the philosophy of TQM, and the likely benefits to the organization from an informed and effective implementation of TQM followed by their participation in sessions to clear any misgivings about TQM so that they come to appreciate TQM as an approach to management which they are ready to adopt. If all within the organization are covered, all processes carried out are incidentally taken into consideration.

Company-wide involvement does not simply mean each individual taking care of the tasks at his/her level. In fact, more important is the involvement of such individuals as members of a team tasked to ensure defect-free execution of a business process. Members of a team may quite often be drawn from different departments and different levels in the hierarchy. The composition of such a cross-functional and cross-level team has to be carefully developed by management on the basis of informed and willing participation.

Proponents of TQM go a little beyond to maintain that such a team should own the process assigned to its care. The concept of process ownership is a key success factor in TQM. It must be remembered that this ownership has to be cultivated through some decisions and actions on the part of top management. The present author recalls his experience in a forward-looking manufacturing industry dealing with synthetics. The management there tried to ensure ownership of machines by those who run those machines by implementing three decisions, viz. (1) getting the worker on a machine to clean the machine, check its jigs and fixtures and correct those, if needed, lubricate the machine in the desired way before starting the machine for regular operation and leave the machine in a run-ready condition at the end of the shift, by providing all the materials required for this purpose right within easy access of the worker; (2) involving the machine operator as a member in a meeting of the purchase committee whenever a new machine is to be acquired, so that experiences of the worker can be taken due note of by the committee and (3) involving a retired machine man to operate the machine he used to operate whenever the machine was idle in the absence of the corresponding operator and required to be run. This was found to yield some benefit. However, mechanisms to inculcate the ownership idea among operators and teams have to be worked out in each case separately, the common strand being a mental orientation.

5.6.2 Right First Time

'Right first time' is the summum bonum of quality improvement and defines an attitude that one who carries out a work item should plan his task in such a manner that none else can identify a non-conformity and a rework is called for to correct the work item. The idea of rework and rejection if rework cannot compensate for the initial defect or deficiency has to be completely done away with.

What is right is often spelt out in the process plan. If not specified there, it should take due cognizance of previous experience to delineate what was accepted without any reservation by the person looking after the next stage in production. Carrying out a job right first time means that the job is not hurried through. And this, in turn, requires adequate time to complete the job in the process plan, and a signal should go the operator that early completion that may beget even a slight defect or deficiency in the output will not be rewarded. In some cases, the completion time does provide for some cushion to take care any possible delay in receiving the inputs and any possible unforeseen delay in execution of the job due to, say, equipment malfunction.

Performance against this standard will be judged by developing metrics like percentage of cases involving rework and reject, percentage of cases involving delayed delivery of the output, average completion time for the job and the like.

5.6.3 Controlling Quality Costs

Measuring quality costs with a view to controlling those at an acceptable minimum is an essential step in TQM. Juran (1988) described the cost of poor quality as 'the sum of all costs that would disappear if there were no quality problems' and presented the analogy that poor quality and the associated costs are 'gold in mine'. Quality cost as including cost of poor quality only as one component and covering costs incidental to initiating quality improvement actions is calculated for any contemplated process improvement plan to judge its feasibility. And, information is used in TQM to indicate opportunities for preventive and corrective actions. In the end, such information is important to assess the strengths and weaknesses of a Quality Management System.

Traditionally, costs of quality have been classified as costs of non-conformance (to desired or targeted quality) and cost of conformance. The first includes primarily costs of failure—detected and acted upon internally besides those found out by the customers. Internal failure costs include costs of rework, rejects and wastes, while external failure costs refer to costs of complaint redressal including costs of replacement and repair and sometimes product recall.

Juran (1988) pointed out that the traditional quality cost analysis looks at the company's costs only and not the manufacturer's. Customers suffer quality-related costs too. Only partly, occasionally and indirectly are costs suffered by customers reflected in the external failure costs to the manufacturer or supplier. Moreover, Total Quality as has been explained earlier in this chapter does take into account societal cost incurred by all due to product or service failure. Thus, quality costing awaits a more comprehensive accounting based on data arising from customers as also from the society at large. And such data are not merely difficult to collect, but may not be amenable easily to quantification and hence may not be poolable with the other costs.

5.7 TQM Models

The literature on TQM contains several models which, as expected, have a lot of communality, but each has some distinctive feature(s).

A three-dimensional model suggested by Kelda (1996) focuses on (1) the human dimension—psychological and political, affecting employee commitment and employee–management relations, (2) the logical dimension emphasizes the rational and systematic aspects of TQM and (3) the technological dimension drawing attention to the engineering aspects and emphasis on processes.

Somewhat remarkable is the Quality-Sweating Model proposed by Kano (1989) where hard work by all people has been stressed. This model includes two alternative approaches, viz. Crisis Consciousness and Leadership make people Sweat for Quality (CLSQ) and Vision and Leadership encourage people to Sweat for Quality (VLSQ).

Another interesting model was suggested by Zairi (1991) called the Building Blocks Model where three blocks, viz. the foundation, the pillars and the roof build up Total Quality. The foundation is laid by three elements, viz. continuous improvement, value addition at each stage of every operation and employee involvement. Various procedures and tools like SPC, workplace design and ergonomics, supplier–customer chain, management control system and the like constitute the pillars. The roof must ensure that the organization is not affected by adverse changes in marketplace and is capable to adapt to new developments. For this, management should take up the responsibility quality planning, leadership and vision for world-class competitiveness.

A more or less similar model called the House of TQM was offered by Kano (1997) in which intrinsic technology becomes the base, motivational approach corresponds to the floor, concepts, techniques and other vehicles are the pillars and customer satisfaction resulting from quality assurance is the Roof of the House.

In a different three-dimensional model for TQM, the dimensions correspond to (1) individuals, (2) teams and (3) organization. Inputs in this model include, among others, a continuing exercise to change the attitude and behaviour at three levels, which are influenced by (a) cognitive factors, (b) motivational factors and (c) socio-dynamic factors. The input exercise has to take care of these factors to be really effective.

The model output (results) is revealed through (1) quality improvement, (2) increase in productivity, (3) enhanced commitment, (4) improved employee satisfaction and (5) enhanced customer satisfaction with product/service quality. All these lead to better business performance, market position and competitiveness and protection against fluctuations in business environment. The underlying idea is that quality improvement is a process of continuous change by individuals and teams.

5.7.1 Role of Management

Management has a much bigger role in the TQM approach to improve performance that is envisaged usually in the context of a Quality System Standard like ISI/IS 9001. In fact, it is often claimed—quite justifiably—that TQM is a management-led movement. Experts in O Organizational Psychology argue that to provide leadership, one has to be first aware of the subject, then become convinced about its feasibility and desirability, then become committed to implement it in one's organization and get involved for a reasonable time to acquire leadership competence Of course, leadership calls for several other attributes.

The direct involvement as leader in TQM implementation is

To challenge the process (as is being currently planned and performed);
To inspire a shared vision to improve the process—envision;
To enable the entire workforce to act through provision of adequate training and of adequate resources—enable;

To model and clear the way to act through necessary authorization to try out new and novel ways and means—empower; and

To encourage the workforce to move forward through proper appreciation of their output and suitable reward—energise.

5.8 TQM Imperatives

To implement TQM, management has to accept that TQM cannot work unless some actions are taken up by all concerned, specially by management.

Management must demonstrate interest in terms of decisions and actions which reflect that management cares for quality, is serious to improve quality and would like the entire organization to get involved in the 'quality' exercise.

Management has to provide resources needed by the people to carry out their tasks without comprising with quality in any way. For this, sometimes 'time allotted' is a resource that should be available to the workers.

One of the most important roles of management is to initiate all steps so that everyone in the organization, irrespective of his level, feels motivated to join the TQM movement. Management has to ensure people involvement through quality circles, quality teams along with policies and programmes to appreciate and reward excellence in performance. This way, the human resource development group has to accept a big challenge to identify any blockades in the path of people's involvement and to initiate actions to remove those. Motivation is often taken as the composite of 'attitude' and 'environment'. Some attitude reorientation programme may have to be implemented, and some care has to be taken to improve the working environment in terms of both physical and psychological parameters.

Management has also to prove that they care for all their stakeholders, with employees first and then customers, regulatory bodies, business partners and surely investors. In some sense, Quality of Working Life has to be enhanced by providing necessary facilities and conveniences, so that workers can feel inclined to put in their best efforts.

Management should create a culture wherein everyone treats complaints as failures (at least to start with). This should apply to both complaints raised by internal customers (workers in preceding or succeeding operations) or by external customers. No complaint should be simply discarded as not worth a consideration and as due to failures on the part of the customer to use the product or in-process material properly. It is true that on proper analysis of some of these complaints will be found to be due to wrong installation or use or upkeep at the customer's end and not genuine failures of the producer.

The best way to ensure this is to provide

Clear and understandable Quality Policy and objectives;
Defined responsibility and authority for groups and individuals;
Laid-down procedures for interdisciplinary (inter-functional) cooperation;

Understandable and unambiguous work instructions; and
Adequate education and training.

Summing it up, imperatives in the context of TQM are

(a) Customer orientation (as distinct from customer-driven or customer-focused) in all decisions and activities;
(b) Human resource excellence in terms of motivation and ability to introduce and 'manage' change;
(c) Management leadership including process and product leadership through technology updating and also providing adequate technology support system; and
(d) Proper use of tools and techniques for problem-solving and troubleshooting.

To highlight the role of management in the context of Total Quality Management, sometimes one finds a question, viz. can there be Total Quality Management without top management? And, the emphatic answer is 'never'. Every successful total quality effort has included the active participation of the top management (in terms of decision-making and executive authority). For quality to become a way of life, top management must carry out specific actions to demonstrate their commitment and culture.

Top management must develop a strategic plan, review and approve the organization's Quality Policy, provide the needed resources, and create and participate in Quality Councils and quality efforts, and advise the merit rating system to include quality measures. Achieving Total Quality demands that top management is leaders, not just cheerleaders.

5.9 TQM Situation

Speaking of an organizational situation reflecting TQM implementation, some authors occasionally mention 4 Cs which get revealed in such a situation. These are commitment, competence, communication and continuous improvement. The first feature, viz. commitment to a cause (satisfying the external or the internal customer, as the case may be) should be exhibited by all in the organization. Commitment, in the context of human behaviour, follows awareness of the TQM approach and conviction about its utility. Competence has to be acquired through necessary education and training as well as through knowledge management to ensure knowledge retention and adequate application. Proper communication facilitates understanding of each other's requirements and ensures avoidance of quite a few mistakes otherwise committed in the absence of relevant information to be passed on by others. And, of course, continuous improvement is the key element of TQM practice and the key result thereof. And this improvement has to be institutionalized in terms of development and use relevant tools and techniques.

For top management as also for visitors, potential customers, representatives of regulatory agencies and of certification bodies, it may be worthwhile to have a feel about what may be branded as a TQM situation prevailing within the organization. In fact, A TQM situation may be briefly characterized by a 'high commitment management strategy' which is revealed through attitudes, plans and actions of managers, supervisors and frontline workers—all striving to perform the best at all times, in every task and on every occasion. In fact, such a situation has characterized by Kelly (1991) in terms of several concrete points.

A TQM situation is visible in the way individuals and teams work within an enabling environment, with clarity about their roles and responsibilities, and with a concerted effort to bring down costs, improve quality and enhance productivity. The defining principles of TQM are evident in the work practices followed at all levels.

Thus, every individual strives to achieve a 'zero-defect' situation where the concept of defect is linked to customer dissatisfaction. In fact, a defect as a deviation from customer requirements (including those of internal customers) or as a deficiency affecting the use of a product (or an in-process unit) is detected even before it reaches a customer to beget his dissatisfaction.

Workers at all levels and managers are well trained in their jobs, and they all develop competence to discharge their assigned tasks by making the best use of their training. Apart from knowledge and skills in their respective areas of operation, each worker or manager is motivated and enabled to own his competence and use the same in a manner he feels the best.

Workers at all levels are also exposed to programmes that motivate them to improve their performance by initiating and/or accepting changes in the way they currently perform. This facilitates management of change worked out by workers themselves or managers or even business leaders. In fact, a repeat visit to any work area in a TQM organization shows up some changes for the better.

In a TQM set-up, individuals are not generally islands looking after only their respective processes or jobs. Most processes require teams to take care of quality and cost issues. Teams could be cross-functional besides cutting across levels or designations. And team spirit—built up through training and team-building exercises—is revealed through mutual help and cooperation among the people. In a TQM environment, no individual will pass the buck on to some colleague for a failure on his part.

Despite teams and individuals working with competence and commitment, their activities are all guided by targets set by management and accepted by all others. Failure to meet targets is taken up seriously, not just to inflict penal measures but to identify and resolve reasons behind a failure. And successes are definitely acknowledged, appreciated and rewarded appropriately.

In this context, people occasionally speak of a vision and a measures matrix—a somewhat dignified presentation of the measures to be analysed and controlled, in each of the situations or vision statements mentioned above. For example, measures against situation 1 would include cost of poor quality, customer perception, customer service levels, e.g. % orders on customer appointed date, responses to staff attitude survey on 'improving quality is part of my normal job.' For the second vision statement, relevant measures are operational zero-defects goals, e.g. 100% calls answered within 15 s, responses to the staff attitude survey. Pertinent to the third situation will be responses to the staff attitude survey on 'How satisfied are you with the recognition you receive for doing your job well?' Similarly, responses to staff attitude survey on 'I have enough information to do my job well.' 'I understand what is expected of me in my job.' are quite important in the context of situation 4. Cost of poor quality as also responses to staff attitude survey on 'I have never been asked my opinion about how to improve quality in our work group' throws light on situation 5. Percentage of units with improvement plans in place, achievement against plans, cost of poor quality along with responses to a survey on 'managers are committed to improve quality' are pertinent measures to monitor situation 6. Measures relating to situation 7 include percentage of managers trained in core skills, percentage of appraisals quoting evidence of demonstrated skills and responses by staff to the question 'How satisfied are you with the training you have received for your present job?'

A TQM situation is not expected to remain static. However, to accept a statement like 'Organization A is a TQM organization', we may be tempted to speak of a threshold situation that can justify such a statement. This threshold is apparent in terms of features like everyone understands, accepts and works for some performance targets: no one complains against a senior or peer for a failure on his part, while requesting for augmentation of his knowledge and skill to avoid failures is welcomed by the organization: customer requirements are always met, resulting in no customer complaints: and consistent growth in business results.

5.10   Integrating TQM into the Business Strategy

Senior management may begin the task of process through seven steps to a self-reinforcing cycle of commitment, communication and culture change. The first three steps are

Secure commitment to change by creating conducive conditions and promoting change in the mindset as a prerequisite to survival and growth in a highly competitive situation marked by rapid and remarkable changes all around.
Develop a shared vision or mission of the business among all concerned—from managers and supervisors to frontline workers. Also, share a vision to implement the desired changes in the way every individual has to function and of the desired changes. Define measurable objectives and set targets. These must be quite realistic and feasible. May be the targets are gradually pushed forward, starting with modes ones.

The remaining four steps comprise

Expanding the mission into more detailed and concrete objectives which are critical to business success. Usually, a list of critical success factors is made out to ensure a shared mission, relating an individual or a group to the success factor they are to be concerned with.

Understanding the critical processes which affect the objectives and the corresponding targets and ensuring that people operating these processes behave as if they own the respective processes. Management should develop appropriate mechanisms to inculcate the feeling of process ownership among process operators. Detailing the critical processes into distinguishable and separately manageable sub-processes along with the activities involved in each and the tasks to be carried out by the concerned individuals and finally.

Monitoring and facilitating process alignments with changes proposed and accepted for implementation. Difficulties in the process of incorporating changes have to be sorted out. If necessary, the proposed changes may have to be examined for feasibility.

Though these seven steps have been generally agreed upon, it is quite plausible that in the context of a particular organization starting with forward-looking and innovative people, fewer steps are called for. What is more important is for management to realize business goals and objectives clearly and the guiding principles adopted to achieve these, through Total Quality Management be design. In such a case, TQM should be incorporated into the overall business strategy.

Ayers (2004) points out the need to take due advantage of Information Technology along with Total Quality Management so that the synergistic effect can lead to profit and growth of an organization. And IT and TQM do reinforce each other and together does improve efficiency and profitability.

5.11   TQM in Services

Applying TQM principles and practices in government organizations including those which provide public services has not met with much success. One problem is that while cost savings are becoming gradually more and more important, there is no clear linkage with profits earned, as exist in the private sector. Moreover, leadership in public enterprises has few incentives to implement TQM; frequent changes on transfers make it difficult to accept a long-term commitment to quality. In some cases, TQM has been implemented only with the weak incentive to meet minimum legal requirements and to initiate programmes to enhance quality on a long-term basis. Reaching quantitative goals set externally by superiors within prescribed time limits supersedes any urge to improve quality.

In the 1990s, the private sector service industry, represented by telecommunication, banking and insurance and other financial services began to introduce TQM to enhance their competitive edge. This created an impact on their public sector counterparts which fell back upon the ideas of TQM to eke out a survival and growth strategy. Members of the public are becoming more and more insistent on quality, price and delivery considerations, and at the same time, governments are feeling the pinch of inefficiency and inability to raise more funds from the public. Not infrequently, leaders in many service organizations—both public and private—take recourse to consultants for introducing TQM in their organizations. Consultants train the people and can enable them to initiate the process, but are unable to motivate them towards a sustained implementation. And TQM is not a quick-fix solution to quality problems.

While the strength of TQM is widely appreciated, the lack of commitment and even the absence of an environment that can sustain such commitment in service organizations—specially those in the government—stand in the way of reaping the benefits of TQM implementation. In fact, cost considerations by a so-called independent wing sometimes reporting to an external body do not allow quality considerations to be duly incorporated in various decision-making situations. Quality gets a back seat in a cost-driven system.

5.12 TQM and Systems Improvement

As a management philosophy, TQM has a lot to do with organizational development and systems improvement. Ayers (2004) mentions several elements that go to consolidate this view.

Dissatisfaction with the Status Quo whatever the present state of customer satisfaction and business growth, quality image and people involvement in quality may be—satisfactory or even high—is the prime mover of the mentality to shake off complacency and push continuous improvement. And the prerequisite for this is to have a clear vision for the future; development whereof needs no tool and is the cheapest exercise.

Customer-driven Philosophy mandates a complete mapping of the customer universe with internal and external customers and their requirements along with a mapping of players in the supply chain and their strengths and weaknesses. This philosophy is backed by effective use of Quality Function Deployment techniques with House of Quality constructs that link customer requirements cascaded down to operator instructions to satisfy the customers. Benchmarking is also helpful to know how such requirements can be met better and with greater cost-effectiveness. Open comparison with better performance by others may help realizing internal deficiencies.

Process Orientation focuses on the fact that a business process often involves wholly or partly a number of functions or departments which have to work in tandem to ensure satisfactory performance of the process. In traditional organizations, departments or functions are first established and processes are designed around them, while in the TQM set-up the organizational structure is changed as soon as processes are changed to suit customer needs and expectations. And this why process owners are created within the organization. Process owners are senior executives who are accountable for the effectiveness, efficiency and maintenance of their processes and not departments to which they might belong traditionally. An efficient owner ensures that the process is constantly improved, tracks the needs of internal and external customers and introduces new technology as and when needed.

Team-driven Change is a key feature of TQM, since teams capture collective knowledge and experience of all the participants and can effectively disseminate the message of ownership. Teams dissect processes and examine the importance of each step in influencing the process output and in drawing up the process plan that should guide process execution. Two types of skills are needed by members of any team. They should acquire and cultivate analytical skills to break down a process into its elements and to establish possible linkages among the elements and subsequently to build on these linkages to work out an optimal or at least satisfactory process plan. Teams can even be trained to use innovation tools to change an existing plan to effect improvement in the process output.

New Metrics to monitor performance in processes. These metrics should be linear wherever possible, should be obtainable from data that already exist or can be conveniently collected and should reveal gaps from expectations and norms and thus motivate improvements. Metrics to deal with customer satisfaction and loyalty, competitive advantage and people involvement, impact on stakeholders including earnings per share or gross value added per worker, etc., are being developed and used by management to take appropriate decisions and actions on the ongoing system.

All these indicate that TQM eventually aims at improving the system as a whole, through integration of different business processes. And that way, TQM does have a visible communality with Balanced Score Card or similar other system improvement philosophies or guides. The Balanced Score Card introduced by Kaplan and Norton (1992) speaks of four components of the organization devoted to customer focus, internal process improvement, learning and innovation and financial results. Mentioned last, financial results are arguably outcomes of the first three components. In fact, one can read into this arrangement as a cause-and-effect chain leading to improved financial performance. Adequate emphasis on the first three aspects can bring about desired financial results.

We can easily notice the emphasis placed on customer focus and on internal process improvement in TQM. And though learning and innovation has not been explicitly mentioned in the context of TQM, there is a great stress on learning and education in TQM implementation. Of course, innovation or new product development do not find explicit places in a TQM strategy, and continuous improvement does always derive strength from Innovation.

5.13 TQM and Knowledge Management

Many articles have been published to highlight some critical factors that contribute to the success of TQM. Quite a few authors agree on a list of ten, viz. top management commitment, adoption of the TQM philosophy, quality measurement, benchmarking, process management, emphasis on product designing, training at all levels, employee empowerment, supplier quality management, customer involvement and satisfaction.

Knowledge management (KM) has defined by some as 'management of organizational knowledge for creating business value and generating a competitive advantage'. We speak of a knowledge chain model by considering four activities within KM, viz. knowledge creation, knowledge conservation, knowledge distribution or dissemination and knowledge application or deployment. In fact, some authors draw a parallel with a value chain.

Both TQM and KM have contributed towards business growth and excellence, and they can well be blended to reap greater benefits than from any one of these two in isolation of the other. Ju et al. (2006) discuss relations connecting the ten TQM critical factors with the four elements in the KM chain, of course, based on experiences in only two firms in Taiwan.

In fact, KM promotes innovation and facilitates new product development and value addition to existing products and services. Several activities which are essential to implementation of TQM in an organization like training at all levels, employee involvement and empowerment can directly lead to knowledge creation and comprehensive documentation of knowledge gained and skills acquired by employees can help the organization to conserve knowledge. This step is quite important in the context of the fact that unless knowledge is documented, the organization will suffer when knowledgeable and skilled persons retire or resign. It should be almost a mandatory practice to disseminate knowledge created within the organization or existing within it should be disseminated among all concerned. And top management should encourage application of knowledge to come up with new ideas or models or designs or practices to enhance business. In fact, KM can be easily integrated into the TQM philosophy.

5.14 TQM Needs a High Commitment Strategy

It is well recognized that the success of TQM is guaranteed only when responsiveness for quality is extended to all levels in the organization, with every individual motivated to accept and discharge some responsibility for quality in some process or product or service. For achieving quality goals as set by the organization and to fulfil customer requirements, employee commitment beyond motivation, a climate of trust pervading the organization and employee participation are essential. Of course, technological and technical aspects of quality have to be taken due care by deriving full advantage from appropriate tools and practices. Hard elements like systematic measurement of processes, performance standards, statistical process control—on-line as well as off-line—are no doubt basic to a TQM strategy.

Towards this, an organization has to reorient its human resource management practice to develop and deploy this type of committed workforce. The organization should remove barriers within the firm to promote a high trust culture in which employees are encouraged to develop a healthy customer relationship, starting with peers as internal customers. This may require some effort on the part of management to organize relevant training and motivation programmes. Contributions of employees to quality should be duly appreciated and rewarded, quality-related synergies available through teamwork should be emphasized, employees should be empowered to be creative or innovative in solving quality-related problems, and a climate of transparency and direct communication should be fostered. Stress must be laid on ethics, on truthful statements about quality and on avoidance of false claims of satisfactory quality and even on penalty for making false claims about quality. Employees at all levels should be clearly told that customers cannot be taken for granted, and once customers detect any quality claim to be false, the loss to the organization is too high to be brushed aside.

For a full implementation of TQM, a lot of initial attention should be paid to developing an organizational climate that is conducive to a quality culture. A sound knowledge management system to acquire new knowledge and to make the best use of available knowledge, an effective communication system that promotes direct and clear communications among all within the organization and a belief system that employees can rise up to the occasion whenever a problem appears to threaten the quality image and that way the business performance of the organization, these three elements are critical success factors for TQM.

Failures of TQM to deliver results have been reported by some organizations, mostly those who carry on with their traditional human resource policies and practices. A simple and glaring difference in such practices relates to differences on criteria used to select employees, viz. criteria focused on knowledge and experience and criteria based on docility and anticipated stability.

5.15 Concluding Remarks

Total Quality Management is an evolving concept or a corresponding exercise or even a system improvement philosophy. No definitive statements about TQM, its implementation, its benefits, its limitations and its communality with other management philosophies or strategies should be made. Enough indications are available on each of these issues. Volumes of case studies reported from varied organizations across locations and cultures speak of impressive success to damp squib. Searches for such consequences have also been studied and reported.

Meanwhile, new developments take place in the arena of Quality Management by way of new concepts, new problem-solving tools and techniques and even in terms of new approaches to management per se. And these are either absorbed suitably within the expanding ambit of TQM or are recognized as distinct quality improvement methodologies, claimed to be superior to TQM. It all depends on how do you look at emerging scenarios, not merely in the context of production and quality, but also in all human endeavours.

As has been pointed out by many, there are two extreme views about TQM which, of course, can be blended to come up with something more meaningful than either extreme. On the one hand, TQM is argued to be depending heavily on human behaviour and that way is prone to subjectivity; on the other side are exponents of TQ who hold it out as a management philosophy which takes due account of involvement of informed and motivated people but also derives strength from analysis of data and data-based metrics. In fact, a holistic view is not to be ruled out and the portrayal in this chapter is on this line of thought.

Some interesting studies reveal some connections between the legal system in relation to product liability and tort prevailing in a country and extent of TQM implementation in that country. In a situation of stringent legal penalty for non-compliance with norms or contracts, TQM can help reducing product liability. Bhat (2006) examines the relation in 49 countries to assert the existence of legal aspects of TQM implementation.

References

Ayers, J. B. (2004). Total quality management and information technology. In V. V Gopal (Ed.), *Partners for profit in total quality management: An introduction*. The ICFAI University Press.

Besterfield, D. H. (1999). *Total quality management*. Englewood Cliffs, NJ: Prentice Hall.

Bhat, V. N. (2006). TQM implementation: Does the legal system matter? *Total Quality management and Business Excellence, 17*(6), 685–690.

Carlos Bou, J., & Beltran, I. (2005). Total quality management, high commitment human resource and firm performance: An empirical study. *Total Quality Management and Business Excellence, 16*(1), 71–86.

Chang, R. Y. (1993). When TQM goes nowhere. *Training and Development, 47*(1), 22–29.

Harari, O. (1997). Ten reasons TQM doesn't work. *Management Review*, 38–44.

Ju, T. L., Lin, B., Lin, C., & Kuo, H. J. (2006). TQM critical factors and KM Value Chain Activities. *Total Quality management & Business Excellence, 17*(3), 373–394.

Juran, J. M. (1988). *Quality Control Handbook*. New York: McGraw Hill.

Kano, N. (1989). Quality sweating theory: Crisis, vision and leadership. *Quality (Journal of the Japanese Society for Quality Control), 9*, 32–42.

Kano, N. (1997). *A perspective on quality activities in American firms in managing strategic innovation and change* (pp. 402–416). Oxford University Press.

Kaplan, R. S., & David, N. (1992). The Balanced Scorecard: Measures that Drive Performance. *Harvard Business Review, 70*(1).

Kelda, J. N. (1996). *Integrating re-engineering with total quality*. USA: ASQC Press.

Kelly, S. (1991). TQM implementation—how will we know when we have got there? *Total Quality Management & Business Excellence*.

Logothetis, N. (1992). *Managing for total quality*. U.K.: Prentice Hall.

Lu, K. K., et al. (1999). Managing for quality through knowledge management. *Total Quality Management, 10*(4&5), 615–621.

Oakland, J. S. (1989). *Total quality management*. Oxford: Heinnemann.

Prajogo, D., & Sohal, A. (2006). The relationship between organisation strategy, total quality management (TQM) and organization performance: An empirical examination. *European Journal of Operational Research, 168*, 35–50.

Ross, J. (1993). *Total quality management: Text, cases and readings*. Fl, USA: St. Lucie Press.

Simmons, D. E., Shadur, M. A., & Preston, A. P. (1995). Integrating TQM and HRM. *Employee Relations, 17*, 75–86.

Snape, E., Wilkinson, A., Marchington, M., & Redman, T. (1995). Managing human resources for TQM: Possibilities and pitfalls. *Employee Relations, 17*, 42–51.

Wilkinson, A., Redman, T., Snape, E., & Marchington, M. (1998). *Managing with total quality management theory and practice*. Hong Kong: McMillan Business.

Yong, A., & Wilkinson, A. (1999). The state of total quality management: A review. *International Journal of Human resource Management, 10*, 137–161.

Zairi, M. (1991). *Total quality management for engineers*. U.K.: Woodland Publishing ltd.

Chapter 6
Quality of Measurements

6.1 Introduction

Measurements are basic tools in any scientific investigation. Many exercises in Science and Technology are aimed at improving the existing state of affairs regarding matter, energy, environment and their interactions—among themselves as also with living organisms. One is reminded of a widely quoted statement made by a twentieth-century German philosopher who runs as follows (Mukherjee)

If I can define it, I can measure it.
If I can measure it, I can analyze it.
If I can analyze it, I can control it.
If I can control it, I can improve it.

Measurements are needed in all scientific investigations to **choose, develop** and **validate models** and used to describe, analyse **predict, control** or **improve** various phenomena. Measurements provide the very basis of all control and improvement actions. Incidentally, one way to differentiate between Science and Technology—if at all one needs to—is to argue that Science is more concerned with *Definition, Measurement* and *Analysis,* while Technology is more engaged in *Control a*nd *Improvement.* However, this differentiation may not be warranted in all cases.

While 'improvement' of any existing 'state' is always our goal, we must remember that 'improvement' comes only at the end of a sequence of actions or process to define the 'state' in an objective manner, to measure the state, to analyse the state in terms of its determinants and correlates and, subsequently, to control the state at a desired level.

In the context of Quality Management, measurements are involved right from quality planning through on-line and off-line quality control and quality assurance to customers and other stakeholders to quality improvement. In a sense, measurements pervade the entire Deming Cycle in terms of Plan-Do-Check-Act operations. And, we have to choose appropriate measures of quality (of incoming materials,

© Springer Nature Singapore Pte Ltd. 2019
S. P. Mukherjee, *Quality*, India Studies in Business and Economics,
https://doi.org/10.1007/978-981-13-1271-7_6

processes, in-process materials, checks and controls, finished products, etc.) and subsequently carry out measurements on physical or chemical or other features or characteristics of the different entities. In case such features and characteristics are not directly measurable, we have to develop suitable proxy measures.

In both the above situations, we need to speak of Quality of Measures (or measurands which are to be measured) as well as Quality of Measurements which are outcomes of the Measurement Process carried out on units of concrete entities. In fact, measures—of productivity, efficiency, dependability, organizational excellence, people orientation, customer satisfaction and similar other concepts— are all based on and derived from several related measurements. It should also be appreciated that any Measure of Quality of a product or process or service entails measurements on a number of pertinent quality characteristics.

While Quality of Measurements has been discussed a lot in recent times, the priority needs to comprehend Quality of Measures and deploy 'good' quality measures for assessing performance has not been fully addressed. A very important consequence of performance is customer satisfaction and, like other latent variables, customer satisfaction does not admit of a unique definition and, obviously, a variety of constructs, models and methods have been in use in various quarters.

We first take up Quality of Measurements and thereafter Quality of Measures, though the reverse order would have been more logical. This has been partly motivated by the fact that national as also international standards have been developed on Quality of Measurements—not, of course, under this nomenclature— and are being used by many laboratories in industries and research organizations.

Quality of Measurements has been discussed in different contexts by different authors and agencies. A good number of national, regional and international standards have been developed over the years to promote the application of con- sensus definitions and measures. Even methods to estimate these measures from repeat measurements have been standardized. Mukherjee (1996) presented the concepts, measures and models relating to quality of measurements in his Platinum Jubilee Lecture in the Section of Statistics in the Indian Science Congress Association and the material that follows is based largely on the content of that lecture.

6.2   Measurements in Quality Management

Measurements play an important role in

- identifying opportunities for improvement (e.g. through measurements of quality cost under different heads like appraisal, prevention and failure (internal as also external) costs) and
- comparing performance of different processes against internal standards (as in process control and improvement) as also comparing performance of processes and results thereof against external standards (as in benchmarking).

The Deming Cycle of continuous improvement—Plan, Do, Check, Act, sometimes modified as Plan, Do, Stabilise and Act—clearly requires measurements to drive it, and yet it is a useful design aid for the measurement process itself. In this cycle, the words, Plan, Do, Check and Act, have been explained in somewhat different ways by different users of this approach. A detailed note appears in Chap. 9. Usually, we accept the following elucidations.

Plan—Establish performance objectives and standards for processes, their inputs and outputs.
Do—Measure actual performance in terms of time taken, quality of output, cost incurred, etc.
Check—Compare actual performance with the objective(s) and standards and determine the gap in between.
Act—Take necessary corrective action(s) to close the gap and make necessary improvements.

It has been often said that it is not possible to manage what cannot be measured. To comprehend a quality or a productivity problem fully in terms of its intensity, frequency of occurrence and consequences, we need to collect measurements on the problem. Even after a provisional solution to such a problem has been developed, we need some measurements on trial runs of the proposed solution before we can establish its effectiveness including economic considerations. Relevant measurements have to be collected and analysed to

* rate vendors for their capability to meet our requirements regarding quality, delivery and price and assess performance of vendors selected on the basis of such a rating

- meet customer requirements as mutually agreed upon;
- set sensible objectives and targets and to compliance with them;
- provide standards for establishing comparisons of performance;
- ensure visibility in terms of a scoreboard for people to monitor their own performance levels against corresponding targets;
- identify quality and productivity problems and prioritise those (in terms of time required and gains expected) for corrective and preventive actions;
- work out costs of poor quality, broken down inti pertinent components;
- determine resource needs objectively, including needs for test, inspection and measuring equipments;
- provide feedback for assessing the improvement exercise. And providing directions for desired modification in the same;
- identify appropriate tools and softwares for carrying out necessary quantitative analysis, keeping in mind the necessity of simplicity in use and of economy.

In order to assess and evaluate process performance as also results accurately, appropriate measurement must be designed, developed and systematically documented by people who own the processes concerned. They may find it necessary to measure effectiveness, efficiency, quality, impact and productivity. In these areas, there are many types of measurement, indirect or direct output or input figures, costs of poor quality, economic data, comments and complaints from customers, information from customer or employee surveys about the extent to which they feel satisfied, etc.

6.3 Panorama of Measurements

Measurement is a process that follows a defined sequence of steps/activities involves physical, material and technological resources and results in a numerical value/a set of numerical values that is assigned to an item in respect of a defined property/parameter/characteristic.

Like any other process, measurement process involves both hard and soft inputs, is carried out under some influencing factors as also some controls and checks, and produces some (soft) output. The hard input is the concrete object on which a measurand has to be numerically evaluated, while the prescribed method of measurement along with the conditions under which the process has to be carried out define the soft input. Ambient conditions of temperature, pressure, humidity, wind velocity, vibration, electromagnetic interference, etc., are some of the influencing parameters. The process is controlled by checks carried out on measuring instruments for their stability, sensitivity, etc., and these are calibrated, as and when necessary. The output is a numerical value or a set of such values which can be ascribed to the input object. Measurement is also the output of a (measurement) process—some numerical value(s).

Measurement is a generic term and the panorama of measurements is enthralling. Measurements are derived from a wide spectrum of sources. In the case of direct measurements, the source is a measuring device in contact with the object being measured in respect of a certain parameter or characteristic. However, photographs, images of various sorts, satellite imageries, etc., are also important sources of measurement.

We have very large measurements like those on interstellar distances (in billion light years) to microscopic measurements of intermolecular separations in solids. On the one hand, we talk of a micro-level measurement like the concentration of a suspended particulate matter in the atmosphere over, say, a paddy field while inter-regional disputes arise over total stocks of such matters in a whole region.

Quite often, macro-level measurements like the latter are obtained by multiplying small micro-level measurements taken on much smaller units. It is not difficult to realize that even a minute error in the micro-level measurement gets largely magnified in the macro-measurement.

Sometimes, the reverse procedure is followed to derive the micro-level measurement as the quotient of a large macro-measurement divided by a usually large number of units. This is the case with, say, per capita national income or per capita annual consumption of active substance extracted from nature.

There are many other distinguishing features of measurements. Thus, we have exact measurements as are yielded by some measuring devices against approximations or estimates. The latter not only correspond to rounding off of measurements to a desired order of accuracy, but also relate to situations where exact measurements are ruled out and estimates have to be made on the basis of limited measurements on related entities and/or some assumptions.

For example, we can speak of the exact quantity of coal raised from a pit and can offer only an estimate of the total exploitable reserve of coal in a coalfield.

It may be of some interest to note the recent revision of the Indian Standard Rules for Rounding off of Numbers, requiring rounding off in one direction only in situations where safety or similar other considerations are expressed in terms of one-sided tolerances/permissible limits.

As Mukherjee (1996) pointed out, measurements carry the charisma of objectivity and there has been a growing tendency among investigators to use measurements as bases for arguments—for and against. It should be remembered that subtle, subjective (individualistic) behaviours, attitudes, aspirations, aptitudes, and similar traits studied in social sciences do not strictly admit of unique measurements.

Though uniqueness and objectivity are not synonymous attributes of measurements, they are quite akin to each other. Hence, the use of measurements in unfolding the vectors of the human mind or in related matters should not be downright denounced, but should be taken with due caution.

6.4 Errors in Measurements

Errors in the observed results of a measurement (process) give rise to uncertainty about the true value of the measurand as is obtained (estimated) from those results. Both systematic and random errors affecting the observed results (measurements) contribute to this uncertainty.

Random errors presumably arise from unpredictable and spatial variations of various influence parameters operating on the measurement process, for example:

- the measurement method employed in case it is not a standard method or has not been validated against a standard method;
- the way connections are made or the system configuration is worked out;
- uncontrolled environmental conditions or their influences;
- inherent instability of the measuring equipment;
- personal equation bias of the observer or the operator;
- judgment or discretion used by the observer or operator in securing the value of the measurand from readings on the instrument, etc.

These cannot be eliminated totally but can be reduced by exercising appropriate controls.

Various other kinds of errors, recognized as systematic, are also observed. Some common types of such errors are

- those reported in the calibration certificate of the reference standards/instruments used;
- those due to different influence conditions at the time of measurement compared with those prevalent at the time of calibration of the standard (quite common in length and direct current. measurements), etc.

It should be pointed out that errors which can be recognized as systematic and can be isolated in one case may simply pass off as random in another case.

6.5   Quality of Measurements

Quality of measurements is comprehended in terms of Accuracy and Precision, based on systematic and random errors respectively that get reflected in repeat measurements. A more recent development takes care of both random and systematic errors and results in a measure of uncertainty about the true value. The Indian Standard IS 5420 Part I describes and illustrates the procedures for calculating accuracy, repeatability and reproducibility of test results. These measures which are linked up with errors in measurements have been quoted, illustrated and explained in several other sources. One can refer to the document by Kelkar which has been posted on the Website of the Maharashtra Pollution Control Board www. mpcb.gov.in.

Accuracy is the critical parameter and is not the same as precision. Accuracy is closeness to the true value (of the measurand), while precision implies consistency among repeat measurements (not always available).

Let X be the measurand and $x_1, x_2, ..., x_n$ be n repeat measurements (on the same object or on exactly similar objects in case measurement involves a destructive test or determination) carried out in the same laboratory, using the same equipment, by the same operator, in the same environment.

Let $\bar{x} = \frac{1}{n} \sum x_i$ be the mean of the repeat measurements and $s^2 = [1/(n-1) \sum (\bar{x} - x)^2]$ be the variance. Also, let T be the true value of the measurand. Then, $|\bar{x} - T|$ is an inverse measure of accuracy, while the standard deviation provides an inverse measure of precision or internal consistency among repeat measurements. Accuracy has a bearing on the equipment while precision has a bearing on control over repeat measurements

Precision is generally measured and reported in terms of
(1) repeatability and (2) reproducibility
Define a quantity r such that

$$\text{Prob.}\big\{|x_i - x_j| > r\big\} < \alpha \quad \text{for any } i = j$$

where α is a pre-assigned small quantity, e.g. 0.05 pr 0.01 Then, r is referred to as the repeatability factor.

In case the repeat measurements were produced in different laboratories (obviously involving associated differences in equipment, operator and environment) and are denoted as $y_1, y_2,...y_n$, we could define a quantity R such that

$$\text{Prob.}[|y_i - y_j| > R] < a.$$

This R is referred to as the reproducibility factor. Factors r and R can be estimated as $r = k_1 s_x$ and $R = k_2 s_y$ where k_1 and k_2 can be determined from the distribution of x or of y. These measures really characterize a measurement process, though these are also used to interpret variations in measurements. These do not involve the unknown true value (μ) of X and hence do not give an idea about the possible range of true values associated with a single measurement (x_i) or a mean value ($\bar{x}$) of several repeat measurements. Indian Standard IS 5420 Part 1 prescribes a common value 2.77 as the value for k_1 and k_2.

The purpose of a specification (one-sided or two-sided) is to fix a limit for the true value of the property concerned. Given that the true value cannot be established in practice, the results will reveal some scattering due to repeatability or reproducibility, depending on the situation. Accordingly, it will be desirable to evolve and accept specification limits taking due account of repeatability and reproducibility of the test method. Thus, the specification range should be equal to at least $3R$ so that the upper (lower) specification limit is more (less) than $1.5R$ from the nominal value specified or intended and some inherent variability if measurements are recognized the case of one-sided specification can be similarly dealt with.

6.6   Measurement System Analysis

The term 'measurement system' refers to the collection of instrument/equipment, operations or processes, procedures, people and software (if involved) which affect the outcome of a measurement process or the assignment of a numerical value to a measurable property (measurand). Measurement system analysis (MSA) is concerned with five parameters viz. Bias (inverse to accuracy), Linearity, Stability, repeatability and reproducibility. The first two are linked up with accuracy and the last two with precision. The third one refers to ability of the measurement system to yield the same measure on the same part/sample/unit tested for the same parameter or measurand at different points in time or after several uses. Unless the process is repeated over different and somewhat separated time periods, we do not check stability and take it for granted within the desired calibration interval for the equipment involved.

Bias is defined as the (absolute) difference between the average of repeat measurements on the same part and the (true) reference value. This really is a measure of the controllable, systematic error in the measurement process.

Linearity corresponds to change in bias over the admissible range of the measurement process in terms of the range of values for the give measurand (possessed by different parts or samples.) In fact, if different parts (with corresponding reference values for the same measurand) and the bias is calculated for each part, we can examine the behaviour of bias against the reference value as remaining constant or increasing (decreasing) linearly (nonlinearly). Even a test for linearity can be carried out.

The remaining two parameters correspond to two distinct components of the total variation observed in an experiment where the process is repeated over several parts or samples, with different reference values for the parameter being measured and involving different operators (possibly using different copies of the measuring instrument).

An MSA study also referred to as a Gauge R&R (reproducibility and repeatability) Study is done by either the tabular method or the Analysis of Variance (ANOVA) method. In the tabular method, variances are estimated by using range, as is done on a control chart for variability where we take R/d_2 as the estimate of standard deviation σ where the value of d_2 depends on the sample size. While the tabular method is simpler, the estimates of variance components based on range do not make use of all the observations directly. This is why the ANOVA procedure is usually preferred.

With the usual Analysis of Variance procedure with two factors, viz. parts and operators, these different components of the total observed variation are estimated and we get a valid idea of the inherent capability of the measurement. A variance component model is appropriate in case of random factors viz. parts and operators, as if the parts considered in the experiment constitute a random sample from the population of all possible parts and similarly the operators involved in the experiment define a random sample from a population of operators. A fixed effects mode or even a mixed effects model focusing only on the selected parts and selected operators or taking the levels of only one of these two factors as a random sample has also been tried out.

Measurement System Capability is expressed in terms of the two metrics viz. Signal-to-Noise Ratio (S/N ratio) and Precision-to-Tolerance Ratio (P/T ratio). These ratios are estimated from the results of the MSA experiment and the Measurement System is accepted as capable provided these ratios satisfy some specified range of values.

A linear model to estimate the different components of variance in an experiment involving l parts P_i and m operators O_j each required to measure each part n times can be presented as

$Y_{ijk} = \mu + \alpha_i + \beta_j + \lambda_{ij} + e_{ijk}$ where y_{ijk} stand for the kth measurement on part I taken by operator j, α_i being the specific effect of part i, β_j the specific effect of operator j, λ_{ij} the interaction (joint) effect of operator j measuring part I and e_{ijk} is the unexplained error component (that corresponds to repeatability). In the usual random effects' model, we denote components of variance due to parts, operators, operator x part interaction and error by symbols σ_p^2, σ_o^2, σ_{po}^2 and σ_e^2, respectively. The total variation in the entire measurement process is often denoted by σ^2. We now have the following relations.

$\sigma^2 = \sigma_p^2 + \sigma_g^2$ where σ_g^2 is the component due to gauge variability (or the measurement process). Further, $\sigma_g^2 = (\sigma_o^2 + \sigma_{po}^2) + \sigma_e^2$. The first part inside parentheses gives reproducibility while repeatability is indicated by the last term.

The two metrics used to assess the capability of a measurement system are defined as

Precision/Tolerance ratio which is taken as $6\,\sigma_g/T$ where $T = $ UTL $-$ LSL is the tolerance range, UTL and LTL being respectively the upper and the lower tolerance limits for the parameter being measured and controlled during production. While this definition has been recommended by Montgomery (1997), the measure of precision recommended by Automotive Industry Action Group (AIAG) is $5.15\,\sigma_g/T$. While Montgomery suggests that

P/T should preferably be at most 0.1, the AIAG recommends that
if $P/T < 0.1$ the gauge is capable
if $P/T > 0.3$ the gauge is not capable while
if $0.1 \leq P/T \leq 0.3$ the gauge may be capable.

Similarly, Signal/Noise or S/N ratio is defined as σ_p/σ_g. Some recommend that S/N ratio should exceed 5 with at least 90% confidence. Confidence intervals for estimated S/N ratio have been derived.

Coming to the desired number of operators and of parts in a gage R&R study, it has been reported that the lengths of confidence intervals for the variance components diminishes significantly with the number of operators, while the number of parts does not affect these lengths that much (Burdick et al. 2003). In fact, the number of operators should be at least 5 or 6. However, some experimenters prefer to increase the number of parts, rather than the number of operators. Incidentally, assumptions of randomization and of replication—basic principles in the design of an experiment—should also be duly taken into account.

MSA studies have been extended to attribute gauges as also to multiple measurands being simultaneously considered. These studies are, as expected, quite complicated.

6.6.1 An Example

Consider an experiment in which five operators are required to measure the diameters (in mms) of three holes punched on a metallic surface, each operator taking four measurements on each hole. The measurements obtained are reproduced below in Table 6.1.

Table 6.1 Measurements on Hole diameters

Operator	Hole 1	Hole 2	Hole 3
A	56 45 43 46	60 50 45 48	66 57 50 50
B	61 58 55 56	60 59 54 54	59 55 51 52
C	63 53 49 48	65 56 50 50	66 58 52 55
D	65 61 60 63	60 58 56 60	53 53 48 55
E	60 61 50 53	62 68 67 60	73 77 77 65

Source Author (during a visit to an industrial unit)

We present the analysis of variance of these measurements in Table 6.2.

As the F-Ratio for the Part × Operator Interaction effect is larger than the significance point, this interaction component is worth noting in the context of our analysis.

It would be appropriate to use the random effects model here, since to examine the measurement system we can obviously go beyond the chosen five operators and the three selected holes. In fact, we can think of a population of holes and similarly a population of operators and consider the set of three holes and the group of five operators as random samples from the respective populations. We then get the estimated variance components as

$$\text{Est } \sigma^2_{operators} = (272.3 - 109.4)/12 = 13.57$$

$$\text{Est } \sigma^2_{Holes} = (54.6 - 109.4)/20 \text{ to be taken as 0 and}$$

$$\text{Est } \sigma^2_{Interaction} = (109.4 - 26.0)/4 = 20.85$$

Thus, 13.57 + 20.85 = 34.42 corresponds to reproducibility, while 26.0 corresponds to repeatability. Their total viz. 60.42 represents gage variability, which along with the variability due to parts make up for the total variability in the measurement process.

Repeatability will be estimated as $2.77 \times \sqrt{26} = 14.12$ approximately while estimated reproducibility in this example works out as $2.77 \times \sqrt{34.42} = 16.25$ approximately.

Table 6.2 ANOVA for data in Table 6.1

Source of variation	D.F.	S.S.	M.S.	F-Ratio
Due to holes (Parts)	2	109.2	54.6	2.1
Due to operators	4	1,089.2	272.3	10.5
Due to interaction	8	875.2	109.4	4.2
Error	45	1,170.5	26.0	
Total	59	3244.0		

Confining ourselves, somewhat unimaginatively, to the observed set of holes (parts) and the selected five operators, the mean squares presented in the table above would directly give us measures of reproducibility and repeatability.

Assuming that reference (true) values of the diameter for the three holes were 56, 57 and 58, respectively, the observed mean values (each based on 20 repeat measurements under reproducibility conditions) came out as 55.3, 57.1 and 58.6, respectively, implying biases of −0.7, 0.1 and 0.6. These figures show nonlinearity of the measurement process.

Estimating variability on the basis of range as is done on a control chart for sample range, the same exercise may be simply carried out as follows.

Calculate the average for each operator to yield values 51.33, 56.17, 55.42, 57.67 and 64.42 for operator 1, 2, 3, 4, and 5, respectively. The range of these five averages is $R_0 = 13.1$ and reproducibility S.D. can be estimated as σ (reproducibility) $= R_0/d_2$, the range being based on five (average) values, $d_2 = 2.236$ yielding the Fig. 5.81. To get the repeatability S.D., we get the range of values for each part (hole) based on 20 values and these come out as 22, 23 and 29 with an average R-bar $= 24.67$ resulting in the estimate of σ (repeatability) $= 24.67/3.735 = 6.605$. This is seemingly larger than estimated repeatability S.D., some consequence of using a range-based estimate of variability without checking for homogeneity of the ranges. The estimate of σ (parts. Holes) can be obtained by getting the average for each part and getting the range of these three averages which come out as 55.30, 67.10 and 58.55 with a range of 3.25, yielding the s.d. estimate as $3.25/1.693 = 1.920$. Components of variance obtained this way will not agree with those given by ANOVA.

6.7 Concept of Uncertainty

It is widely recognized that the true value of a measurand (or a duly specified quantity to be measured) is indeterminate, except when known in terms of theory. What we obtain from the concerned measurement process is at best an estimate of or an approximation to the true value. Even when appropriate corrections for known or suspected components of error have been applied, there still remains an uncertainty, that is, a doubt about how well the result of measurement represents the true value of the quantity being measured.

The true value is indeterminate and unknown, except when given by theory. It is presumed to be the value yielded by the best maintained and used instrument of the desired accuracy class.

Theory of errors is concerned with errors in measurements that can be noted in terms of difference among repeat measurements on the same measurand and that can be explained by a simple model like

$$\text{True value } (X) = \text{Observed Value } (x) + \text{Error } (e).$$

Current interest centers round uncertainty in the true value (as is estimated in terms of a single measurement or a set of repeat measurements). This is understood in terms of the spread of true values where from the observed value(s) could arise. The idea is motivated by the similarity in observed values when different true values of the measurand are considered.

The uncertainty in measurement is a parameter, associated with the result of a measurement, that characterizes the dispersion of the true values which could reasonably be attributed to the measurand. The parameter may be, for example, the standard deviation (or a given multiple of it), or the half width of an interval having a stated level of confidence.

Uncertainty and its evaluation or estimation from repeat measurements and calibration reports using some assumptions have been discussed by many authors and contained in many national standards like NABL 141 in India as also similar standards in other countries required to ensure compliance with the ISO 17025 standard. An important reference could be the document EA-4/02 M rev 01 (2013) published by the European Accreditation Agency. One may also refer to Kelkar which has been referred to by Maharashtra (India) Pradesh Pollution Control Board on their Website. The author dealt with this topic in his Platinum Jubilee Lecture in the Section of Statistics of the Indian Science Congress Association, published in 1996.

6.7.1   Measurement Model

Measurands are particular quantities subject to measurement. One usually deals with only one measurand or output quantity Y that depends upon a number of input quantities X_i $(i = 1, 2, \ldots, N)$ according to the functional relationship.

$$Y = f(X_1, X_2, \ldots, X_N) \tag{6.1}$$

The model function f represents the procedure of the measurement and the method of evaluation. It describes how values of the output quantity Y are obtained from values of the input quantities X_i.

In most cases, it will be an analytical expression, but it may also be analytical expressions which include corrections and correction factors for systematic effects, thereby leading to a more complicated relationship that is now written down as one function explicitly. Further, f may be determined experimentally, or exist only as a computer algorithm that must be evaluated numerically, or it may be a combination of all these.

An estimate of the measurand Y (output estimate) denoted by y is obtained from Eq. (6.1) using input estimates x_i for the values of the input quantities X_i.

$$y = f(x_1, x_2, \ldots, x_n) \tag{6.2}$$

It is understood that the input values are best estimates that have been corrected for all effects significant for the model. If not, necessary corrections have been introduced as separate input quantities.

6.7.2 Estimation of Uncertainty

The standard uncertainty in measurement associated with the output estimate y, denoted by $u(y)$, is the standard deviation of the unknown (true) values of the measurand Y corresponding to the output estimate y. It is to be determined from the model Eq. (6.1) using estimates x_i of the input quantities X_i and their associated standard uncertainties $u(x_i)$.

The set of input X_i may be grouped into two categories according to the way in which the value of the quantity and its associated uncertainty have been determined.

Quantities whose estimate and associated uncertainty are directly determined in the current measurement. These values may be obtained, for example, from a single observation, repeated observations, or judgement based on experience. They may involve the determination of corrections to instrument readings as well as corrections for influence quantities, such as ambient temperature, barometric pressure or humidity.

Quantities whose estimate and associated uncertainty are brought into the measurement from external sources, such as quantities associated with calibrated measurement standards, certified reference materials or reference data obtained from handbooks.

The standard uncertainty in the result of a measurement, when that result is obtained from the values of a number of other quantities, is termed combined standard uncertainty.

An expanded uncertainty is obtained by multiplying the combined standard uncertainty by a coverage factor. This, in essence, yields an interval that is likely to cover the true value of the measurand with a stated high level of confidence.

The standard uncertainty of Y is given by

$$\sigma_y = \left\{ \sum \sum C_i C_j \sigma_{ij} \right\}^{1/2} \tag{6.3}$$

where inputs X_i and X_j have a covariance σ_{ij} and C_i is the sensitivity of Y with respect to variation in X_i. The formula simplifies in case the inputs are uncorrelated. The variances can then be easily estimated if repeat measurements are available on an input; otherwise, these are estimated by assuming some distribution of true values (which could be made to correspond to the same observed value), e.g. normal or rectangular or (right) triangular.

The uncertainty analysis of a measurement—sometimes called an uncertainty budget—should include a list of all sources of uncertainty together with the associated standard uncertainties of measurement and the methods for evaluating them. For repeated measurements, the number n of observations also has to be stated. For the sake of clarity, it is recommended to present the data relevant to this analysis in the form of a table. In this table, all quantities to be referenced by a physical symbol X or a short identifier. For each of them at least the estimate of x, the associated standard uncertainty of measurement $U(x)$, the sensitivity coefficient c and the different uncertainty contributions to $u(y)$ should be specified. The dimension of each of the quantities should also be stated with the numerical values given in the table.

6.7.3   An Example

The tensile strength testing machine in a conveyor belt manufacturing unit is calibrated annually. Tensile strength of finished belts is determined using the equipment involving the tensile value disk and the load cell. Ten repeat measurements on tension in kg/cm^2 were available on a particular belt specimen to estimate uncertainty about the true value. The following information about the equipment was also available for the purpose.

Tensile value disk	
Range used for calibration	0–50 kgf
Accuracy	As per manufacturer's data
Resolution	1 div. = 0.1 kgf
Load cell	
Uncertainty (%) from its calibration certificate	0.37 (A1)

Readings on tension are reproduced below

Reading No.	Tension
1.	153.50
2.	159.78
3.	167.04
4.	161.83
5.	156.10
6.	160.39
7.	187.05
8.	156.12
9.	161.39
10.	160.83

Type A Evaluation of Uncertainty

Mean Reading (kg/cm^2) = 160.40 Standard Deviation = 4.20 kg/cm^2
Standard Uncertainty U_r = standard deviation/$\sqrt{10}$ = 1.33 kg/cm^2
Standard Uncertainty (% U_r) = $U_r \times$ 100/Mean reading = 0.83%

Type B Evaluation
Uncertainty of load cell received from the corresponding calibration certificate. We assume the underlying distribution to be normal so that the coverage factor at 95% confidence level is approximately 2 Thus, U_1 (%) = A_1/2 = 0.37/2 = 0.185% (A_1 considered as the expanded uncertainty U_e = 2 × standard uncertainty). Thus, Uncertainty of load cell U_1 = 0.185 × 160.40 × 0.01 = 0.297 kg/cm^2

Since U_1 = U_1% Mean Reading/100) Thus, the estimated uncertainty of load cell works out as 0.37 × 160.40 × 0.01 = 0.593 kg/cm^2

Combined standard uncertainty U_c = $\sqrt{[U_r \times U_r + U_1 \times U_1]}$ = 1.46 kg/cm^2 and % U_c = 0.91% = $U_c \times$ 100/Mean Reading

Expanded combined uncertainty for approximately 95% level of confidence
U = 2 × 1.46 = 2.92 kg/cm^2
And U% = 1.8%
The uncertainty budget can now be worked out conveniently.

6.8 Improving Quality of Measurements

To improve quality of measurements or to reduce uncertainty (about the true values), we have to reduce chance or random errors (which cannot be completely eliminated) and to remove systematic errors or biases. Broadly speaking, random errors can be reduced by using measuring instruments and maintaining them properly, exercising necessary control on environmental influences on the measurands and training people involved in taking measurements to avoid personal equation biases, etc. Systematic errors are associated with measuring instruments as also with reference measures. This requires instruments to be calibrated regularly against reference standards.

It is generally agreed that calibration of test, inspection and measuring equipments takes care of accuracy, while careful use and maintenance of such equipments lead to improved precision. Both calibration and maintenance are essential to reduce uncertainty about true values. In the following, we provide brief explanations of Calibration and of Good Laboratory Practice that is needed to ensure good quality of measurements.

6.8.1 Calibration—Process and Procedures

The International Standard ISO 10012-1 relating to Quality Assurance Requirements for measuring equipment mentions metrological confirmation system for measuring equipment as 'the set of operations required to ensure that an item of measuring equipment is in the state of compliance with requirements for its intended use'. This includes calibration, adjustment or repair and subsequent recalibration as well as sealing and labelling. However, many practitioners feel that calibration itself covers these different requirements and hence can provide metrological confirmation for the test, measuring and inspection equipment.

Calibration is the set of operations which establish, under specified conditions, the relationships between values indicated by a measuring instrument of measuring system, or values represented by a material measure or reference material, and the corresponding values of a quantity realized by a reference standard. Such a relationship may be used to adjust or correct an instrument or a system, even values of measures or even reference materials, wherever such adjustments or corrections are feasible and desirable. In other cases, these relations provide bases for corrections in or conversions of measurements. The result of a calibration permits either the assignment of values of the measurand to the indications given by a measuring equipment as they are or the determination of corrections with respect to indications. A calibration may also determine other metrological properties such as the effect of influence quantities. The result of a calibration may be recorded in a document, sometimes called a calibration certificate or a calibration report.

Calibration involves checking the operational integrity of a test or measuring equipment or of a measurement standard of unverified accuracy by comparing its performance with that of a standard of known greater accuracy in order to detect, correlate, report or eliminate (by adjustment) any deviation in accuracy, capability or from any other required performance. Calibration gained importance mainly due to stringent requirements in defence supplies and the MIL standards took a lead in formalizing the calibration philosophy and subsequently boosting the calibration practice. As indicated earlier, measurement implies a process as well as the output of that process. The process of measurement needs control and calibration is an important control exercise.

Calibration can be carried out for three possible purposes viz.

(i)   Determining whether or not a particular instrument or standard is within some established tolerance in respect of its deviation from a reference standard.
(ii) Reporting of deviations in measurements from nominal values.
(iii) Repairing/adjusting the instrument or standard to bring it back within the established tolerance.

It is important to note that these three purposes are not mutually exclusive. In fact, all the three may be relevant in a particular situation.

Usually, a hierarchical calibration system is adopted to ensure traceability of measurements given by a measuring instrument to some nationally accepted measurement system through an unbroken chain of comparisons. In the commonest case, we are interested in calibrating a given measuring instrument (which does nor need any material measure to produce a measurement or a value for the measurand of interest) against a certain reference standard.

Calibration procedures vary from one type of measuring equipment to another. For example, in calibrating a micrometer we take sequential measurements of gauge blocks of known size specified by some standard. In the IS, the specified sizes are 2.5, 5.1, 7.7, 10.3, 12.9, 15.0, 17.6, 20.2, 22.8 and 25 mm. Dimensions indicated by the micrometer are noted and deviations from the nominal values recorded. Of course, the accuracy of the gauge blocks themselves has to be ensured or determined as a prerequisite. Alternatively, the length of the gauge blocks can be compared with those of matters of identical nominal lengths.

In the commonest case, we are interested in calibrating a given measuring instrument (which does not need any material measure or reference material to produce a measurement or a value of the measurand) with reference to a certain reference standard. Here also, the same two objectives of calibration—leading to adjustment/correction of the instrument or of the measurements—remain valid, depending on individual situations. We produce n measurements for a measurand (may be n items assessed for the same characteristic) by using both the given and the reference instruments. Let the two series by y, y,...,y and M, M,..., M, respectively. Calibration means establishing a relation—often assumed linear—between the two series of the form $y = \alpha + \beta M$, with numerical values of the parameters α and β determined by the methods of least squares from the two series of measurements.

6.8.2 Calibration System

In a calibration system, the following items shall be defined.

1. Classification of calibration;
2. Standard and levels of standard;
3. Interval of calibration and limit of correction;
4. Procedures of calibration;
5. Action after calibration;
6. Conditions to use measuring instrument;
7. Procedures of measurement.

Table 6.3 gives out relational formulae used in different types of calibration which are used in different contexts.

Table 6.3 Classification of calibration and relational formulae for them

	Type of calibration	Relational formula
a	Calibration with only inspection: Do not correct and take the reading as it is as the measured value	$y = M$
b	Zero point calibration: Conduct the calibration of fixes point by reading of zero point y_0	$y = y_0 + M$
c	Reference point calibration: Conduct the calibration of fixed point by reading y_0 of reference point M_0	$y = y_0 + (M - M_0)$
d	Scale interval calibration: Conduct the calibration of inclination taking the optional point (its reading is y_0) as zero point	$y = y_0 + \beta M$
e	Zero point proportional formula calibration: Suppose the reading of zero point as zero and conduct the calibration of inclination	$y = \beta M$
f	Reference point proportional formula calibration: Conduct calibration of fixed point by the reading y_0 of reference point M_0 and then conduct calibration of inclination	$y = y_0 + \beta(M - M_0)$
g	Linear formula calibration: Conduct simultaneously calibration of fixed point and calibration of inclination with using mean value $\bar{y}$ of reading y and mean value $\bar{M}$ of value of standard M	$y = \bar{y} + \beta(M - \bar{M})$

Source ISO/IEC Standard 17025 on Good laboratory Practice

This table does not deal with calibration by formulae of high degrees of freedom of nonlinear type. In these cases, it is possible to conduct calibration assuming a linear relation within each of several ranges.

In order to take the reading as it is as the measured values, changes of scale or mechanical adjustment may be made. In such cases, correction by change of scale of mechanical adjustment and they are discriminate from no calibration.

6.9   Quality Requirements for Measures

Measures associated with different phenomena or processes and their outcomes should possess some properties in order that we can assess their relevance, appropriateness and dependability for use in making inferences and actions. And these features or properties of a measure really characterise what may be termed as 'quality of the measure'. In this context, we consider these desirable properties for a measure (indicator) of performance. And to be focused on processes which have a bearing on quality, we consider measures of process performance and leave out measures of organizational performance from the scope of the present discussion. In this context, it is worthwhile to mention that 'Process Performance Measurement System' has been discussed by several authors and is regarded as an essential activity to provide inputs for quality improvement.

Kitchenham (1995) and Winchell (1996) have identified the following properties or features as the main requirements for process performance indicators. These were subsequently discussed by Holweg (2000). The list is generic and should not be claimed as exhaustive.

Quantifiability Since a major objective of using measures of performance would be to compare performance across time or units or sections and the like, the measure has to be quantified. If performance indicators are not quantitative by nature, they have to be transformed. For instance, the performance indicator 'customer payment attitude' could be transformed into number of days between 'invoice sent' and 'invoice paid'. This way, qualitative measures may be quantified—though not always uniquely—by using related quantified measures.

Sensitivity Sensitivity expresses how much the performance measure must change before a change in performance can be detected. In fact, a sensitive indicator is able to detect even minor changes in performance. It is well appreciated that improvements in process performance will more often than not be marginal, though continuous. And such marginal improvements will not be detected by a measure that is not sensitive enough. Big changes are obvious and the involvement of a measure is not that critical.

Linearity Linearity indicates the extent to which process performance changes are congruent with the value of a certain indicator. Or, conversely, a small change in the business process performance should lead to a small change in the value of a corresponding performance indicator, whereas an ample performance rise should also lead to strong change in the level of the performance indicator. And this behaviour should be maintained throughout the plausible range of values for the indicator. Otherwise, it will be difficult to interpret the same difference in value of the indicator over different parts of this range.

Reliability A reliable performance indicator is free of measurement errors. To illustrate, if a certain business process has to be rated through a given performance indicator by different experts, the results should not depend on the subjective evaluation of an individual. Inter-rater consistency is an important requisite.

Efficiency Since the measurement itself requires human, financial and physical resources, it must be worth the effort from a cost/benefit point of view. The measure has to reflect changes in process performance faithfully with a minimum of effort.

Improvement Orientation Performance indicators should emphasize improvement rather than conformity with instructions. Therefore, measuring billing errors, number of safety violations, data entry errors and the like do not create an atmosphere where feedback sessions are viewed in a positive, constructive light. Indicators should be so defined and scaled that its values speak of aspects of performance which are directly and not inversely linked up with improvement in process performance.

It should be noted that

1. Performance is not absolute.
2. Performance is multi-dimensional.
3. Performance measures are not independent of one another.

Kueng (2002) points out that even if a process performance measure satisfies the above desiderata, it may not be acceptable by the team that is to make use of the same. That way acceptability by the users of a measure is also quite important.

6.10   Concluding Remarks

Quality of measurements has a crucial role to play in Quality Management. Not too unoften, we are told about disputes between the producer/supplier and the customer not agreeing on the value or level of an important quality parameter of the product under transaction. And in a few of such cases, the very fact that there would always remain some small difference between these values or levels obtained by two parties and a genuine problem should correspond to a d difference exceeding, for example, a multiple of the reproducibility factor R or the length of the expanded uncertainty interval. In this context, Measurement System Capability Analysis becomes a must in situations where a high degree of precision is required or very small measurements are involved.

While the concept of 'uncertainty' about the true value as also methods for estimating uncertainty have been documented by national and international regulatory bodies, its use has not yet been that widespread. More than that, the existing method of estimating uncertainty is not above criticism. It makes use of a measurement model which is not completely objective in the identification and incorporation of all possible inputs, refers to a formula for obtaining the standard deviation of the estimated true value which is applicable to large samples and assumes some probability distribution to convert the range of variation in an input parameter into a corresponding standard deviation. One may genuinely object to the use of an uncertain or asymptotic procedure to estimate uncertainty in a set of measurements or a single measurement. However, the attempt to identify different sources of error in measurements and to quantify their contributions to the overall uncertainty in a measurement should be definitely appreciated.

References

American Association for Laboratory Accreditation. (2014). *Guide G104 for Estimation of Measurement Uncertainty in Testing.*

Automobile Institute Action Group. (2002). *Measurement system analysis manual.* Detroit, MI, USA: AIAG.

Bell, S. (1999). *Measurement good practice guide #11 a beginner's guide to uncertainty of measurement.* Technical report. National Physical laboratory.

Burdick, R. K., Borror, C. M., & Montgomery, D. C. (2003). A review of methods for measurement system capability analysis. *Journal of Quality Technology, 35,* 342–354.

Deutler, T. (1991). Grubbs-type estimators of reproducibility variances in an inter-laboratory test study. *Journal of Quality Technology, 23*(4), 324–335.

Dietrich, C. F. (1991). *Uncertainty, calibration and probability*. Bristol, U.K.: Adam Hilger.

EA-4/02M rev.2. (2013). *Evaluation of the uncertainty of measurements in calibration*. European Accreditation Organisation.

Fridman, A. E. (2012). *The quality of measurements: A metrological reference*.USA: Springer.

Grabe, M. (2005). *Measurement uncertainties in science and technology*. USA: Springer.

Holweg, M. (2000). Measuring cost and performance. Cardiff Business School.

IS 5420 Part 1. (1973). *Guide on precision of test methods—Principle and applications*.

ISO Standard 10012-1 *Quality assurance requirements for measuring equipment*.

ISO/ IEC Standard 17025. Good laboratory Practice.

Kelkar, P. S. (2004). *Quality control for sampling and chemical analysis* (Ref mpcb.gov.in/envtdata/QA-QC-PSK.pdf).

Kitchenham, B., et al. (1995). *Case studies for method and tool evaluation. IEEE Software*, 52–67.

Kueng, P. (2002). Process performance measurement system. *Total Quality Management, 35*(5), 67–85.

Larsen, G. A. (2002–03). Measurement system analysis: The usual metrics can be non-informative. *Quality Engineering, 15*(2), 293–298.

Lilliken, G. A., & Johnson, D. E. (1984). *Analysis of messy data*, Vol. 1 Designed experiments. New York: Van Nostrand Reinhard.

Mukherjee, S. P. (1996). Platinum Jubilee Lectures, Part I. Indian science Congress Association.

Mukherjee, S. P. (2000). Quality of measurements in role of measurements in science and technology. In S. P. Mukherjee & B. Das (Eds.), Indian Association for Productivity, Quality & Reliability and National Academy of Sciences of India.

Pendrill, L. (2008). *Applications of statistics in measurement and testing*.

Tsai, P. (1988). Variable gage repeatability and reproducibility using the analysis of variance method. *Quality Engineering, 1*(1), 107–115.

Vardeman, S. B., & Van Valkenburg, E. S. (1999). Two way random effects analysis and gauge R&R studies. *Technometrics, 41*(3), 202–211.

White, G. H. (2008). Basics of estimating measurement uncertainty. *Clinical BioChemist Review, 29*(Supplement!), 553–560.

Winchell, W. (1996). *Inspection and measurement: Keys to process panning and improvement*. Society of Manufacturing Engineers.

Chapter 7
Analysis of Quality Costs

7.1 Introduction

While no one denies or derates the importance of quality in goods and services produced and delivered, some stakeholders argue that building high levels of quality demands considerable expenses on various heads. At the same time, consumerism provides compensations for poor quality from suppliers or service providers. On the one hand, Crosby would claim that 'Quality is Free'; on the other hand, manufacturers claim that quality has to be paid for. Of course, quality, in this debate, implies something beyond the minimum level of quality which is inseparably present in the product or service. 'Quality without Tears' is the fascinating title of a book on Quality Management. Executives in industries would roll out a whole list of activities linked with quality to build up quality, to demonstrate quality, to control quality during different stages of the procurement–production–delivery process, to improve quality, etc., and would assert that a comprehensive quality activity has to be in place and that necessarily implies considerable costs.

Cost control remains and will continue to remain a major concern for management. And a short-sighted approach to cost control could imply—though not necessarily—spending less on 'building quality'. Management does accept risks and argues that penalty for poor quality would not be a regular phenomenon in the known (to the supplier) absence of adequate checks for quality at the mutually agreed levels by the customers or users.

What is genuinely needed is an honest attempt to identify (1) expenditure to be incurred for maintaining conformity to mutually acceptable quality standards if not for improving quality beyond that, and (2) penalty to be paid for poor quality delivered to customers and detected by the latter, as also costs internally incurred for extra inspection, rework and rejects, etc.

Analysis of quality costs taking into account both these aspects should be an important management activity in the customer-oriented and competitive market today. In fact, an incisive analysis of quality costs—and merely of costs of poor

© Springer Nature Singapore Pte Ltd. 2019   137
S. P. Mukherjee, *Quality*, India Studies in Business and Economics,
https://doi.org/10.1007/978-981-13-1271-7_7

quality as is usually done—is quite likely to come up with suggested actions about various elements in the system where wastes can be minimized, variations can be reduced, delays can be cut down, cheaper but better materials and components can be identified to minimize the total cost of production and after-sales service. Of course, costs of quality should be assessed not merely in terms of expenses directly incurred, but also in terms of indirect costs associated with lowering of image, reneging of customers, penalties imposed by custodians of public interest, etc.

The situation is not one of 'Quality versus Cost', but is one that can be characterized as 'Quality with no cost in addition to cost of production'.

Attention to design for incorporating attractive and value-adding attributes in the product or service can often be done without involving extra costs. During manufacture, due attention of persons engaged should attract no additional expenditure and all that is needed is orienting attitudes towards quality, maybe at a modest cost that would be paid back in more than one ways. A search for new materials or developing a cell for innovation in some sense, if feasible, may bring down costs of materials without sacrificing quality. Inspection amount can be brought down through better inspection at fewer places. Pre-shipment inspection done sincerely will help service costs and costs in attending to customer complaints to come down. A due quantification to track the changes can surely convince management about the slogan 'Quality at no cost in addition to cost of production'.

'Prevention is better than cure' is a proverb too well known, and in the context of quality, prevention is better than appraisal, followed by corrective action whenever needed. However, as reported by Singer et al. (2003), many firms were found to practice the reverse situation, with the primary reason being lack of adequate and convincing data and facts about costs of quality and management being carried away by impressions and even experiences of unscrupulous business not attracting penalties that outweigh cost of conformance to agreed or standard quality requirements.

In the present chapter, the author attempts to briefly consider the alternative schemes for implementing a quality costing practice, the different models followed and their limitations, the paradigm of an 'optimal quality level' and the methods of apportioning total quality cost into components and activities involved.

7.2 Why Analyse Quality Costs?

As is pretty well-known, it was Juran (1951) who first considered cost of quality analysis in the context of Quality Management. The earliest writing on the concept of quality costs was by Juran (1951) wherein he described the cost of poor quality (COPQ) as 'the sum of all costs that would disappear if there were no quality problems' and put it analogous to 'gold in mine'. Fiegenbaum (1956) possibly was the first to classify quality costs into day's familiar categories of prevention, appraisal and failure (both internal and external) costs. Juran later highlighted the traditional trade-off that contrasts prevention plus appraisal costs with failure costs (1962).

'Quality is Free' establishes that quality does not have undesired economic effects, because it is not a productive asset which can be bought or sold. This is the way the economic school explains the title of the book by Crosby. Crosby, however, argues that quality is not a gift and has to build in with some effort that does not entail any extra cost. What actually costs money is something called un-quality and the consequent actions and resources that involve not doing the job right the first time. Thus, the focus in 'Quality is Free' is essentially the need to do things right the first time.

Initially, the idea was that in the absence of any appraisal of processes and products (in terms of in-process or finished goods inspection) and any prevention or improvement activity, both internal and external failure costs would be too high and the total cost of poor quality will be large. If only appraisal is introduced at some cost, failure costs would decline but the total cost including appraisal and failure costs may still be pretty large and even larger than in the first case. Only when prevention of defects during after production is taken up at some extra cost, appraisal cost may not come down but failure costs are expected to fall drastically and the total cost of quality will be smaller than in the earlier cases. In fact, with improvement action taken on hand, even appraisal can be reduced and we expect a further decline in quality costs total.

If we now plot the total cost of quality against some appropriate measure or level of quality (like the fraction non-defective), the total cost of quality would be coming down with increasing quality level and will correspond to a minimum for some high level of quality (may not be 100% non-defective items) beyond which the total cost may even increase, though slightly. In fact, the curve is somewhat convex upwards. This level of quality can be accepted as the optimum level of quality to be maintained till the time further improvements take place to bring down the total cost of quality.

It can be pointed out that with investment in quality less than what is required to initiate appraisal, prevention and improvement activities, return on quality (efforts to improve quality) may not show up. And an analysis of quality costs can help determine the threshold investment which can be judged by top management as being desirable or not.

7.3 Quality Costs Defined

Costs of quality has been defined, categorized and measured in different ways by different exponents of Quality Management. However, we all agree that cost of quality or CoQ is different from cost of poor quality or CoPQ. In fact, the latter should be just a component of the former. However, CoQ is generally understood as the sum of conformance plus non-conformance costs, where cost of conformance is the price paid for prevention of poor quality (e.g. costs of inspection and quality appraisal along with cost incurred in taking corrective and preventive actions) and cost of non-conformance is the cost of poor quality caused by product or service failure (e.g. rework, reject and returns). It is now widely accepted that quality costs are all those costs which are incurred in the design, implementation, operation and maintenance of a Quality Management system; cost of resources committed to continuous improvement; costs of system, product and service failures, and all other necessary costs and non-value added activities required to achieve a quality product or service. The following four are the most often discussed components of cost of quality appearing in the classical P-A-F model. These cost elements and the P-A-F model have been widely discussed, and one can get a comprehensive account of these in Fox (1993).

Prevention Costs

These costs are incurred to reduce failure and appraisal costs to a minimum. The usual elements considered here include costs associated with the following activities

1. Quality planning involving use of planning tools like Quality Function Deployment, benchmarking and related tools;
2. Design and development as also calibration and maintenance of quality measuring and test equipment during production and inspection;
3. Development, validation and verification of designs of products and services even developing specifications for vendor-supplied materials in respect of quality should also be taken care of;
4. Supplier quality assurance including vendor development and motivation;
5. Quality training appropriate for different levels of people in different functions;
6. Quality Audit—both internal and external at periodic intervals;
7. Acquisition, analysis and reporting of data pertaining to quality obtained through inspection, review and audit activities; and
8. Quality improvement programmes.

It may be difficult to apportion costs on some of these heads between those which are incidental to production itself and those which are meant directly for prevention of defects and failures as also to ensure greater homogeneity among units produced. In fact, in most of these activities people from production or maintenance or design functions will be the key persons and to segregate the prevention costs from costs incurred by such people on a regular basis for their usual activities may pose difficulties.

Appraisal Costs

These are the costs associated with activities that aim to ascertain conformance of incoming materials, in-process semi-finished products and finished products to the relevant specifications or other requirements. Such costs are exemplified by costs of stock evaluation, receiving and finished goods inspection as also in-process inspection, evaluation of materials consumed during inspection and testing.

Costs of management review of results of inspection, effectiveness of current Quality Management system customer satisfaction and vendor performance are also in this category.

Internal Failure Costs

These are the costs which occur when a product fails to meet the specified quality requirements, and such failure has been detected through in-process inspection or through finished goods inspection before delivery to the customer. Taking into account internal customers, such costs take place when the output of any process is detected to be unfit for the next-in-line process. Typical examples of internal failure costs are costs associated with scrap, lost productive time and rework. Deficiencies detected during finished goods inspection and honestly withheld from delivery to the customer or to the market at a discounted rate will evidently involve costs of rework, rejection and scrap, maybe taking out the value left in the scrapped units.

External Failure Costs

These are costs involved when a product fails to meet quality requirements, and this failure has been detected after delivery to the customer. Such costs include costs of replacement, warranty claims, costs of attending to service or product recalls, besides loss of image and of sales.

In each of the above cost categories, we have many small but differentiable elements associated with materials, processes and human resources. It is worthwhile to look at Table 7.1 provided by the ASQ Quality Committee about a detailed break-up of each of the above-mentioned cost.

7.3.1 Failure/Appraisal/Prevention Ratio

Typically, failure costs are the largest proportion during the initial phases, then appraisal, with prevention costs the least. The major part of the failure costs may be internal or external, depending on the industry and the market. Gradually, the situation changes to one in which appraisal cost remains more or less the same; more stress is laid on prevention of defects with a significantly increased prevention costs and a much lower failure cots. In fact, as improvement efforts continue, even appraisal costs come down, prevention costs go up, and failure costs come down. This transition is best displayed by the three pie diagrams drawn.

In some cases, appraisal costs may be higher than failure cots. Reasons for such a situation may be some of the failure costs have been overlooked.

Appraisal results in internal failure which cost less (at that stage) than the appraisal activity. However, if undetected and allowed to become external failures,

Table 7.1 Quality cost elements by category

1.0	**Prevention Costs**
1.1	Marketing/customer/user
1.1.1	Marketing research
1.1.2	Customer/user perception surveys/clinics
1.1.3	Contract/document review
1.2	Product/service/design development
1.2.1	Design quality progress reviews
1.2.2	Design support activities
1.2.3	Product design qualification test
1.2.4	Service design qualification
1.2.5	Field trials
1.3	Purchasing prevention costs
1.3.1	Supplier reviews
1.3.2	Supplier rating
1.3.3	Purchase order tech data reviews
1.3.4	Supplier quality planning
1.4	Operations (manufacturing or service) prevention costs
1.4.1	Operations process validation
1.4.2	Operations quality planning
1.4.2.1	Design and development of quality measurement and control equipment
1.4.3	Operations support quality planning
1.4.4	Operator quality education
1.4.5	Operator SPC/process control
1.5	Quality administration
1.5.1	Administrative salaries
1.5.2	Administrative expenses
1.5.3	Quality programme planning
1.5.4	Quality performance reporting
1.5.5	Quality education
1.5.6	Quality improvement
1.5.7	Quality system audits
1.6	Other prevention costs
2.0	**Appraisal costs**
2.1	Purchasing appraisal costs
2.1.1	Receiving or incoming inspections and tests
2.1.2	Measurement equipment
2.1.3	Qualification of supplier product
2.1.4	Source inspection and control programmes
2.2	Operations (manufacturing or service) appraisal costs
2.2.1	Planned operations inspections, tests, audits
2.2.1.1	Checking labour

(continued)

Table 7.1 (continued)

2.2.1.2	Product or service quality audits
2.2.1.3	Inspection and test materials
2.2.2	Set-up inspections and tests
2.2.3	Special tests (manufacturing)
2.2.4	Process control measurements
2.2.5	Laboratory support
2.2.6	Measurement (inspection and test) equipment
2.2.6.1	Depreciation allowances
2.2.6.2	Measurement equipment expenses
2.2.6.3	Maintenance and calibration labour
2.2.7	Outside endorsements and certifications
2.3	External appraisal costs
2.3.1	Field performance evaluation
2.3.2	Special product evaluations
2.3.3	Evaluation of field stock and spare parts
2.4	Review of test and inspection data
2.5	Miscellaneous quality evaluations
3.0	**Internal failure costs**
3.1	Product/service design failure costs (internal)
3.1.1	Design corrective action
3.1.2	Rework due to design changes
3.1.3	Scrap due to design changes
3.1.4	Production liaison costs
3.2	Purchasing failure costs
3.2.1	Purchased material reject disposition costs
3.2.2	Purchased material replacement costs
3.2.3	Supplier corrective action
3.2.4	Rework of supplier rejects
3.2.5	Uncontrolled material losses
3.3	Operations (product or service) failure costs
3.3.1	Material review and corrective action costs
3.3.1.1	Disposition costs
3.3.1.2	Troubleshooting or failure analysis costs (operations)
3.3.1.3	Investigation support costs
3.3.1.4	Operations corrective action
3.3.2	Operations rework and repair costs
3.3.2.1	Rework
3.3.2.2	Repair
3.3.3	Reinspection/retest costs
3.3.4	Extra operations
3.3.5	Scrap costs (operations)

(continued)

Table 7.1 (continued)

3.3.6	Downgraded end product or service
3.3.7	Internal failure labour losses
3.4	Other internal failure costs
4.0	**External failure costs**
4.1	Complaint investigations/customer or user service
4.2	Returned goods
4.3	Retrofit costs
4.3.1	Recall costs
4.4	Warranty claims
4.5	Liability costs
4.6	Penalties
4.7	Customer/user goodwill
4.8	Lost sales
4.9	Other external failure costs

Source Borror (2008), Table 14.1

they would cost much more than appraisal does currently. Therefore, appraisal is deliberately held at a high level to avoid failure in detection of deficiencies internally.

Too much reliance is placed on appraisal rather than prevention—the organization is trying to 'inspect in' quality.

The organization is performing expensive trials and acceptance tests which are not necessary for the organization's confidence in its product, but which are demanded by the customer.

7.4 Generic CoQ Models and Cost Categories

There have been different models to collect, analyse and use quality cost, differentiated in terms of the purpose and the focus. Some models just consider tangible costs associated mostly with poor quality along costs required to prevent the latter. Some others take into account intangible costs like opportunity costs like loss of goodwill and of customers. Costs to improve quality are recognized in some models. The focus in some cases is on costs incurred for different aimed-at and/or realized outcomes, while in certain other models individual activities along with the deployment of material and human resources in those activities are accounted for.

Jaju et al. (2009) present a consolidated table of contributions made by different researchers on the subject, pointing out limitations of each and mention that the models proposed need validation by analysis of real-life data. They also feel that such researches possibly mislead people concerned by suggesting that huge reductions can be achieved at little extra cost. The table, of course, gives a reasonably comprehensive account of different approaches to measuring quality costs.

The following is a commonly accepted profile of CoQ models along with the cost categories involved in each. These models find place in many publications, e.g. Fox (1993) and Schiffaeurova and Thomson (2006). Other models have also been reported in the literature, e.g. the one suggested by Chen and Tang (1992) where costs are put into two broad domains, viz. those incurred at the supplier's end and those incurred in transactions with customers.

Generic model	Cost/activity categories
P-A-F model	Prevention + appraisal + failure
Crosby's model	Conformance + non-conformance
Opportunity or intangible	(1) Prevention + appraisal + failure + opportunity
Cost models	(2) Conformance + non-conformance + opportunity tangibles + intangibles
Process Cost Model	Conformance + non-conformance
ABC models	Value added + non-value-added

Components of detailed metrics for cost needed for use of CoQ analysis include

Cost of assets and materials used (tangible and direct);
Cost of labour engaged in prevention (tangible and direct);
Cost of labour engaged in appraisal (tangible and direct);
Cost of defects per 100 pieces (partly tangible and direct);
Cost of late deliveries (partly tangible and partly direct);
Time between service calls (tangible and indirect);
Number of non-conforming calls (tangible and indirect); and
Number of complaints received (tangible and direct).

Based on different components and corresponding to different requirements, some global metrics should be computed to establish comparisons of quality and productivity in relation to CoQ over time within the same organization and across similar organization for the same time period. Some of the common global cost metrics are the following

$$\text{Return on Quality} = \frac{\text{Increase in profit}}{\text{Cost of Quality Improvement Programme}}$$

$$\text{Quality Rate} = \frac{[\text{Input} - (\text{quality defects} + \text{start - up defects} + \text{rework})]}{\text{Input}}$$

$$\text{Process Quality} = \frac{\text{Available Time} - \text{Time for Rework}}{\text{Available Time}}$$

$$\text{First Time Quality} = \text{Percentage of product with no rework}$$

It should be remembered that these global metrics are not synonymous with figures of merit discussed in a later section and used to compare cost of quality with other financial parameters to help top management to assess CoQ analysis and to modify the existing practice in anticipation of better business results.

7.5 More About CoQ Models

Quite a few models describing relations among the various cost components have been reported in the literature, and a good review has been attempted in Ittner (1994, 1996) besides Plunket and Dale (1988). Schiffaeurova and Thomson (2006) provide a comprehensive review of cost of quality models as also best practices of such models judged in terms of benefits to organizations practicing these.

Plunkett and Dale classified the various models into five groups. Group A models indicate the principle of reducing failure and appraisal costs by increasing expenditure on prevention activities. All data are plotted on a linear time base rather than some measure of quality performance in Group B models. In Group C models, quality costs are plotted against stages in quality development, e.g. no appraisal and only external failures, appraisal and internal failure along with somewhat reduced external failures, appraisal and prevention along with reduced failure especially external failure and finally reduced appraisal together with improvement resulting in much reduced failures. Group D models accurately reflect the relationships between the major cost categories and the cost savings to be expected from prevention and appraisal activities. Group E models indicate that the rate of change of cost with quality is much greater and the cost optimum level of quality is positioned much closer to perfection.

The importance of opportunity costs and other intangible costs has been taken into account more recently. While costs of poor quality as usually calculated are directly observable and relate to the present, there are costs that affect business indirectly and in future and, in some cases, quite significantly. Thus, goodwill lost among customers who came up with complaints of poor quality and were compensated in terms of replacements or free reworks or other freebies really implies an opportunity cost. Intangible costs can only be estimated indirectly rather than calculated directly in terms of, say, profits not earned because of lost customers and reduction in revenue because of non-conformance of products and services. Some authors go beyond lost goodwill and trace opportunity costs to three sources, viz. under-utilization of installed capacity, inadequate material handling and poor delivery of service. Costs of inefficient resource utilization and design costs are also considered as components of quality costs by some analysts. These all illustrate areas where a manufacturer or service provider can definitely improve its current performance level and avoid costs of opportunities lost through inefficiencies or inadequacies in resource utilization.

The Process Cost Model which takes into account costs of conformance to process plan requirements and costs of non-conformance leading to rework, rejection and delayed delivery of products and services in a particular process. The analysis of costs associated with this process will help in finding out ways to improve the process. For improvement in organizational performance, to the extent it is influenced by such a process, criteria to assess improvement can be decided on (a) percentage of on-time delivery and (b) average time taken, post execution of the process, resolve a customer complaint (may be internal or external) and (c) costs of rework and replacement. It must be noted that cost of conformance can be reduced only through improved process planning and execution and this may involve off-line quest for new materials or processes or equipment or knowledge and the like. With the current process plan, conformance costs can be calculated by people in planning, while supervisors and shop-floor managers can provide time estimates to be included in the plan.

Process cost elements are derived from and can be recorded under

People:	Salaries and wages
Equipment:	Usage and depreciation
Materials:	For work as also rework
Environment:	Cost of control

Cost of conformance (COC), in this model, is defined as the cost of operating the process as specified in a 100% effective manner. This does not imply that the process is efficient or even necessary, but rather that the process, when operated within its specified procedures, cannot be achieved at a lower cost. This is the minimum cost for the process as specified, worked out in terms of standard times, standard costs, etc.

Cost of non-conformance (CONC) is the cost of inefficiency within the specified process, i.e. over resourcing or excess costs of people, materials and equipment arising from redundancy in design, unsatisfactory inputs, errors made, rejected outputs and various other modes of waste. These are considered non-essential process costs.

The complete analysis of a company's activities into interlinked 'processes', accurately and without duplication (and consequent double-counting of costs), maybe more onerous than the traditional categorization of quality costs. Furthermore, the inclusion into the COC of the running costs of 'inefficient or unnecessary' processes but running exactly according to the plan appears to hide the inefficiency. A repair station with no waiting time, no wasted repair materials or no ineffective repair would incur no COC, yet its only reason for existence would be a copious flow of rejected work from the line. And standard costs of yields or times would also allow hidden yield losses.

Jaju et al. (2009) present a consolidated table of contributions made by different researchers on the subject, pointing out limitations of each and mention that the models proposed need validation by analysis of real-life data. They also feel that such researches possibly mislead people concerned by suggesting that huge reductions can be achieved. Existing accounting systems are usually considered as poorly fitted to generate reports on quality measurements. They fail to bring out benefits resulting from improved quality. Although most CoQ measurement methods are activity/process oriented, traditional accounting establishes cost accounts by categories of expenses instead of activities. Thus, many CoQ elements need to be estimated or collected by other methods. There is no consensus on how to allocate overheads to CoQ elements and no adequate method to trace quality costs to their sources.

An activity-based costing (ABC) model was developed by Cooper and Kaplan (1988) to address this issue. ABC can be defined as 'the tracing of overhead and direct cost to specific products, services or customers'. Tracing involves two stages. Stage one assigns resource costs based on the amount of each resource consumed in performing specific activities. Stage two assigns activity costs to products, services or customers based on actual consumption of the activities. Prior to performing the two-stage ABC tracing process, an organization must identify its resources and map its activities.

Under ABC, accurate costs for various cost objects are achieved by tracing resource costs to their respective activities and the cost of activities to cost objects. Not a CoQ model by itself, ABC can be used to identify, quantify and allocate quality costs among products. The long-term goal of ABC which can be integrated with the Process Cost Model conveniently is to eliminate non-value-adding activities and to continuously help in improving processes and activities so that the defect rate reduces. ABC has also been used to identify costly steps that could be eliminated through more robust product and logistical designs. ABC facilitates evaluation of the cost to serve a certain customer, profitability of a particular product, or whether a particular function of process should be outsourced.

Gryna (1988) points out several problems that have caused cost of quality approach to fail. Quality cost analysis looks at the company's cost only, not the customers' cost. The manufacturer and suppliers are definitely not the only stakeholders who incur quality-related costs. Towards this, Taguchi's societal cost concept plays an important role. Cost of poor quality should be calculated by taking into account all the stakeholders, including costs incurred by the state to consider and redress quality complaints by customers and users.

7.6 Figures of Merit

When comparing quality costs in different plants, or on different occasions, a figure of merit enables us to make comparisons on a similar basis. The particular ratio to be used depends on the factor to be emphasized, e.g. quality costs in relation to labour utilization, sales value, production costs. Within a particular organization, a figure of merit is required to provide a baseline against which to set goals and to measure improvements. Such figures of merit differ from the global CoQ metrics discussed earlier in this chapter and serve different purposes.

Some of these ratios are

Labour-based	Internal failure costs/(direct labour costs)
Cost-based	Total failure costs/(manufacturing costs)
Sales-based	Total quality costs/(net sales billed)[a]
Unit-based	Total quality costs/(# of units produced)
Added-value based	Total quality costs/(value added)

[a]Some like to use net sales realized instead of billed

It should be noted that the first two ratios account for only failure costs or cost of poor quality, while the other three consider the entire costs of quality. The fourth ratio should be used with caution for cross-industry comparisons since the cost of manufacturing or the sale value of a unit may differ grossly from one industry to another. A labour-intensive service organization which wants to improve labour utilization of personnel and reduce human error will direct attention to trends in the labour-based ratio. However, a manufacturing company going to introduce automation in its processes will not like this ratio at all. Such a company would better focus on cost-based or the unit-based ratios.

It is important to remember that tracking improvement in terms of a figure of merit over time, the numerator and denominator in a ratio must relate to the same reference period. For example, the numerator in the second ratio will include costs of external failures reported during a subsequent period with costs of repair or replacement incurred in a later period, making the numerator and the denominator not strictly comparable. A similar problem may arise with the last ratio where 'value added' will be calculated for units produced during a reference period, while the denominator may easily spill over a subsequent period. The problem with the third ratio may be even deeper.

Table 7.2 lists a few successful organizations who have used different models for Quality Cost Analysis and have derived substantial benefits from this analysis. This came out as the finding of a survey carried out by Schiffaeurova and Thomson (2004) on models and best practices of cost of quality.

7.7   Establishing a System for Reporting Cost of Quality

For the P-A-F model explained in BS 6143 Part 2, collection of cost data takes place in five steps, viz.

Step 1. Calculates those costs that are directly attributable to the quality function.
Step 2. Identifies costs that are not directly the responsibility of the quality function but which can be counted as part of the Total Quality-related cost of the organization.

Table 7.2 Some successful use of CoQ models and methods

Company (reference)	Method model	Gains
P-A-F model		
UTC. Essex Group (Fruin 1986)	CoQ = P + A + F CoQ is calculated as a % of total manufacturing cost	• CoQ reduced from 23% to 17%
AT&T Bell Lab (Thompson and Nakamura 1987)	CoQ = P + A + F (I + E)	
Hydro Coatings (Purgslove and Dale 1995)	CoQ = P + A + F (I + E) CoQ is calculated as a % of annual sales turnover CoQ is also expressed as a % of raw material usage	• CoQ reduced from 4.1% to 2.5% in 4 years • Investment in quality paid back in the first year
Electronic Manufacturer (Denzer 1978)	CoQ = P + A + F (I + E)	
Crosby's model		
Solid State Circuits (Denton and Kowalski, 1988) BDM International (Slaughter et al. 1998)	CoQ = CoC + CoN CoQ is expressed as a % of the revenue CoQ = CoC + CoNC	
Opportunity and alternative costs models		
Xerox (Can 1992)	CoQ = CoC + CoNC + OC	• CoQ reduced by $ 53 million in first year
Pharmaceutical Company (Malchi and McGurk 2001)	CoQ = Operating Cost + CoNC + Alternative Cost	• 11% reduction in CoQ
Process model		
CEC Alsthon Engineering Systems (Goulden and Rawlins 1995)	CoQ = CoC + CoNC	
ABC model		
Hewlett Packard (Jorgenson and Enkerlin 1992)	ABC (activity-based costing) CoQ = Process (Quality + Board Test + Repair + Bench Test + Defeat Analysis	

CoQ cost of quality; *COC* cost of cost of conformance; *P* prevention cost; *CONC* cost of non-conformance; *A* appraisal cost; *OC* opportunity cost; *F(I + E)* failure cost (internal and external failures)
Source Survey conducted by Schiffaurova and Thomson (2006)

Step 3. Identifies and enters in the memorandum the internal cost of 'budgeted failures'.
Step 4. Identifies the internal costs of failure not allowed for in Step 3.
Step 5. Identifies the cost of failures after change of ownership.

A 12-step process to establish a CoQ reporting system is outlined below:

Step 1. Obtain management commitment and support. Top management might initiate establishment of a quality cost system. Alternatively, if the idea comes from accounting or quality assurance personnel, top management must be convinced of its necessity. An estimate of cost of poor quality (CoPQ) will usually ensure management support.

Step 2. Set up a team comprising individuals (including product managers, engineers, line workers, customer service representatives and others) throughout the organization. Active participation of the users of information is also recommended.

Step 3. Select an organizational segment to start the programme. The initial segment could be a specific product, department or plant, believed to have high measurable quality costs.

Step 4. Obtain cooperation and support of users and suppliers of information. Supplier's non-cooperation can force delays in reporting information and make the system useless.

Step 5. Define quality costs and quality cost categories. The operational definition of each cost category should be developed and disseminated among all users and suppliers of quality cost information.

Step 6. Identify quality costs within each cost category. To start with, ask users and suppliers to identify and define specific costs incurred because of poor quality.

Step 7. Determine the sources of quality cost information. Data may not be readily available in existing accounting systems. Quality costs must be visible and not hidden within other accounts. If some data are not available, the team must determine how much extra effort is necessary to collect such data or whether reasonable estimates from available information are good enough.

Step 8. Design quality cost reports and graphs to meet the needs of users. More detailed information may be required for lower levels of the organization. Appropriate stratification of the information by product line, department or plant aids further analysis. Also, the choices of quality cost indices should be determined.

Step 9. Establish procedures to collect quality cost information. Individuals must be assigned with specific tasks. Forms may be designed with the help of computer systems personnel to simplify the task.

Step 10. Collect data, prepare and distribute reports.

Step 11. Eliminate bugs from the system. In early trials, issues such as unreliable or unavailable data, employees who feel uncomfortable in collecting the data or interpreting the results, or computer system problems may require attention and resolution.

Step 12. After the initial project has succeeded, plans should be developed to expand the system to other segments. Rotation of membership on the team broadens the base of persons who understand the system operation. Also, the system should be reviewed periodically and modified as necessary.

7.8   Distribution of Quality Costs

A break-up of the total cost of quality across different functional areas within a manufacturing or service enterprise and according to some classification of costs depends a lot on the product/service profile of the enterprise as also its current focus on quality. Going by the classical P-A-F classification, a typical situation has been illustrated in Logothetis as shown in Table 7.3.

Table 7.3 Classification of visible quality costs

	R&D	Engineering and maintenance	Quality	Logistics	Manufacturing
Preventive costs					
Audits	5	10	25	10	25
Training	10	10	35	5	80
Quality planning	300	200	220	30	45
Process capability	10	10	10	–	10
Calibration	–	–	10	–	5
Vendor assessment	–	–	40	30	–
QIP teams	25	20	20	5	50
Prevention total	350	250	390	80	215
Appraisal costs					
Inspection of product	200	–	700	–	400
Supplier monitoring	–	–	140	60	–
Product audits	–	–	80	25	–
Test materials	10	–	100	–	–
Checking procedures	40	10	30	10	30
Appraisal total	250	10	1050	95	430
Failure costs					
Isolation of causes	60	10	80	20	20
Reinspection	–	–	50		30
Customer returns	–	–	70	–	–
Concessions and downgrading	20	–	30	–	–
Scrap disposal	–	–	70	–	270
System failure	10	–	20	20	40
Materials supply	40	–	–	170	30
Manufacturing losses	–	–	–	–	2100

(continued)

Table 7.3 (continued)

	R&D	Engineering and maintenance	Quality	Logistics	Manufacturing
Manpower failure	–	–	–	–	80
Process equipment	–	610	–	–	230
Failure total	130	620	320	210	2800

	Assembly and packing	Sales	Finance	Administration	£ thousand
Preventive costs					
Audits	15	–	25	10	120
Training	50	75	–	–	300
Quality planning	10	–	–	–	805
Process capability	–	–	–	–	40
Calibration	–	–	–	–	15
Vendor assessment	–	–	–	–	70
QIP teams	10	10	5	5	150
Prevention total	85	85	30	15	1500
Appraisal costs					
Inspection of product	200	–	–	–	1500
Supplier monitoring	–	–	–	–	200
Product audits	5	110	–	–	220
Test materials	40	–	–	–	150
Checking procedures	20	–	60	30	230
Appraisal total	265	110	60	30	2300
Failure costs					
Isolation of causes	10	–	–	–	200
Reinspection	–	–	–	–	80
Customer returns	–	340	–	–	410
Concessions and downgrading	–	10	–	–	60
Scrap disposal	160	–	–	–	500
System failure	20	30	80	–	220
Materials supply	–	–	–	–	240
Manufacturing losses	310	–	–	20	2430
Manpower failure	60	–	–	–	140
Process equipment	80	–	–	–	920
Failure total	640	380	80	20	5200

Source Logothetis (1992)

7.9   Use of Quality Cost Analysis

One among the traditional uses of Quality Cost Analysis has been to examine the increase in total cost of quality with increasing levels of conformance to accepted levels of quality and even to examine how is this level of quality being aimed at by the Quality Management System related to total cost of quality. An analysis in terms of a graphical display can lead us to find out the optimum level of quality to be maintained as the level corresponding to which the total cost of quality indicates a minimum.

A close look at the costs of poor quality or quality deficiencies—restricting ourselves to only the visible or the revealed costs—will tell us how serious efforts to bring down costs of internal and external failures should be assigned the greatest priority.

Coming to practice of Quality Cost Analysis, the situation is really not encouraging. Measuring return on quality which alone can motivate investments in quality improvement is not a common practice. Spending money on quality improvement programmes without ever estimating the expected benefits leads to investment with little or no impact on the bottom line. Smaller firms do not generally have any quality budget and do not attempt to monitor quality costs. Larger companies claim to monitor quality costs, but quite often underestimate the same, by ignoring invisible costs as also opportunity costs that are really visible with a delay. And even when costs are measured, not much analysis to compare the total cost of quality to net sales billed or total manufacturing costs and the like is taken up, nor is the information channelized to identify situations demanding improvement.

Studies by the American National Institute of Standards and Technology have reported the following relation between Sigma Quality Level and related cost of quality (Kaidan 2007)

Sigma level	% Non-defective	PPM/DPMO	Cost of quality As % of sales
3	93.32	66,810	25–40
4	99.379	6,210	15–25
5	99.9767	233	5–15
6	99.99966	3.4	Less than 1

As the sigma level of quality increases or variability in quality diminishes, the need for testing and inspection goes down, work in progress declines, cycle time reduces, and customer satisfaction level goes up.

7.10 Concluding Remarks

To derive real benefits from a conscious and comprehensive Quality Cost Analysis, management should first be convinced excessive costs due to poor quality is a business problem that justifies a big attention. A quality cost study when coupled with a pilot quality improvement project is quite likely to reveal the need for quality improvement and for a continuing quality cost monitoring activity. It must be noted that a quality cost analysis by itself does not improve quality or boost business. It only indicates opportunities for improvement in processes and practices where the cost of poor quality is too high. Management has to act on its findings—not just cost figures, but situations where improvement is a bad necessity—and fix clear responsibilities for diagnosing and removing problems that cause high costs of poor quality. In fact, the aim should be to unearth redundant processes and non-value-adding activities and wastes, to eliminate sorting inspection and, this way, to estimate and reduce opportunity costs and revenue losses that are not directly related to non-conformities. Quality Cost Analysis carried out for each process can be an effective continuous reminder to the process owners about cost consequences of any failings on their part.

References

ASQC. (1987a). *Guide for reducing quality costs* (2nd ed.). Milwaukee: ASQC.

ASQC. (1987b). *Guide for managing supplier quality costs*. Milwaukee: ASQC.

Bhat, K. S. (2002). *Total quality management: Text and cases*. Mumbai: Himalaya Publishing House.

Borror, C. M. (2008). *The certified quality engineer handbook* (3rd ed., pp. 109–110). Milwaukee: ASQ Quality Press.

British Standards Institute. (1990). *BS 4163: Guide to the economics of quality: Part 2 prevention, appraisal, failure model*. London: BSI.

British Standards Institute. (1992). *BS 6143 Guide to the economics of quality: Part 1, process cost model*. London: BSI.

Campanella, J. (1990). *Principles of quality costs*. Milwaukee, USA: ASQ Press.

Cooper, R., & Kaplan, R. S. (1988). Measure cost right: Make the right decisions. *Harvard Business Review, 66*(5), 96.

Crosby, P. (1967). *Cutting the cost of quality*. Boston, USA: Industrial Education Institute.

Fox, M. J. (1993). Ways of looking at quality related costs: The P-A-F model. *Quality assurance management* (pp. 312–327). New York: Springer.

F4ss.org https://f4ss.org

Giakatis, G., & Rooney, R. M. (2000). The use of quality costing to trigger process improvement in an automotive company. *Total Quality Management, 11*(2), 155–170.

Grieco, P. L., & Pilachowski, M. (1995). *Activity based costing: The key to world class performance*. Florida, USA: PT Publications.

Hwang, G. H., & Aspinwall, E. M. (1996). Quality cost models and their applications: A review. *Total Quality Management, 7*, 267–281.

Ittner, C. D. (1994). An examination of indirect productivity gains from quality improvement. *Production and Operations management, 1*, 153–170.

Ittner, C. D. (1996). Exploratory evidence on the behaviour of quality costs. *Operations Research, 44*, 114–130.

Jaju, S. B., Mohanty, R. P., & Lakhe, R. R. (2009). Towards managing quality cost: A case study. *Total Quality Management and Business Excellence, 20*(9–10), 1075–1094.

Juran, J. M. (1951). Directions for ASQC. *Industrial Quality Control.*

Kaidan, V. (2007). Why quality, cost and business excellence are inseparable. *Total Quality management and Business Excellence, 18*(1–2), 147–152.

Larson, P. D., & Kerr, S. G. (2007). Integration of process management tools to support TQM implementation: ISO 9000 and activity based costing. *Total Quality Management and Business Excellence, 18*(1&2), 201–207.

Logothetis, N. (1992). *Managing for total quality.* U.K.: Prentice Hall.

Malchi, G., & McGruk, H. (2001). Increasing value through the measurement of the cost of quality (CoQ)—A practical approach. *Pharmaceutical Engineering, 21*(3), 92.

Morse, W. T., Roth, H. P., & Poston, K. M. (1987). *Measuring, planning, and controlling quality costs.* New Jersey: National Association of Accountants.

Plunkett, J., & Dale, B. G. (1987). A review of the literature on quality-related costs. *International Journal of Quality and Reliability Management, 4*(1), 40.

Porter, L. J., & Rayner, P. (1992). Quality costing for total quality management. *International Journal of Production Economics., 27*, 69.

Punkett, J., & Dale, B. G. (1988). Quality costs: A critique of some 'economic costs of quality'. *International Journal of Production Research, 26*, 1713.

Pursglove, A. B., & Dale, B. G. (1995). Developing a quality costing system: Key features and outcomes. *Omega, 23*, 567.

Rooney, E. M., & Rogerson, J. H. (1992). *Measuring quality related costs.* London: The Chartered Institute of Management Accountants.

Sandoval-Chavez, D. A., & Beruvides, M. G. (1988). Using opportunity costs to determine the cost of quality: A case study in a continuous process industry. *Engineering Economist, 43*, 107.

Schiffaeurova, A., & Thomson, V. (2006). A review of research on cost of quality models and best practices. *International Journal of Quality and Reliability Management, 23*(4), 1–23.

Singer M., Donoso, P., & Traverso, P. (2003). Quality strategies in supply chain alliances of disposable items. *Omega, 31*(6), 499–509.

Tatikonda, L. U., & Tatikonda, R. J. (1996). Measuring and reporting the cost of quality. *Production and Inventory management journal, 37*, 1.

Tsai, W. H. (1998). Quality cost measurement under activity-based costing. *International Journal of quality and Reliability Management, 15*(6), 719.

Chapter 8
Measurement of Customer Satisfaction

8.1 Introduction

Customers occupy the centre stage of business and industry today. Customers present a wide array of interests and preferences, requirements and expectations, purchasing powers and demand profiles. In the current globalized and liberalized market economy, understanding customers' requirements and fulfilling these requirements followed by an assessment of customer satisfaction are the key functions in any business or industry, where quality has come to stay as an international language.

Subsection 9.1.2 under Clause 9 on Performance Evaluation in ISO 9001:2015 states that 'the organization shall monitor customers' perceptions of the degree to which their needs and expectations have been fulfilled. The organization shall determine the method for obtaining, monitoring and reviewing this information.' There is an illustrative note in which examples of monitoring customer perceptions through customer surveys, customer feedback on delivered products and services, meetings with customers, market-share analysis, compliments, warranty claims and dealer reports have been mentioned.

Measurement of customer satisfaction has been attempted in various ways by different groups, primarily based on responses in a properly designed (sample) survey of customers or potential customers (or their representatives) seeking their opinions or ratings of features and functions of a supplier that beget customer satisfaction or otherwise. These responses to various items in a questionnaire (some of which could be questions and some others could be statements with which a respondent may agree or disagree partly or fully) are converted to scores and their weighted average is accepted as a Customer Satisfaction Index (CSI). Lots of

Much of the material presented in this chapter are from the author's work on Measurement of Customer Satisfaction (2002).

157
S. P. Mukherjee, *Quality*, India Studies in Business and Economics,
https://doi.org/10.1007/978-981-13-1271-7_8

variation remained in sampling, in questionnaire design and administration, in scaling of responses, in assignment of weights, etc.

Of late some national standards to derive CSI from a comprehensive model that includes drivers and consequences of customer satisfaction have gained adequate recognition. One such model is the American CSI Model.

The American Customer Satisfaction Index (ACSI) is the only uniform, national, cross-industry measure of satisfaction with the quality of goods and services available in the USA. Established in 1994, the ACSI is both a trend measure and a benchmark for companies, industries and sectors of the household consumer economy. Research that makes use of the database from the first four years of the ACSI shows that this index is predictive of both companies' financial returns and national economic performance.

ACSI modelling of survey data provides satisfaction indices (on a 0–100 scale) and indices of drivers and consequences of satisfaction with the products and services of specific companies and industries within seven economic sectors, chosen to be broadly representative of the national economy.

The USA is the third country to establish a national measure of customer satisfaction with quality. Other countries are now in start-up phases. The first national Customer Satisfaction Index was the Sveriges Kundbarometer (Swedish Customer Satisfaction Barometer or SCSB) established in 1989 and designed by Dr. Claes Fornell and faculty at NQRC, University of Michigan Business School. The ACSI uses the modelling and survey methodology of the SCSB, which was refined during the period 1989–1993 and pretested in the USA in 1993.

Based on the recommendations from a feasibility study and by the work provided by the ECSI Technical Committee, a new framework for measuring customer satisfaction and customer loyalty known as the Extended Performance satisfaction Index (EPSI) was designed around the turn of the century. The EPSI is recognized by the European system for measuring customer and employee satisfaction, as well as organizations' social responsibilities and management effectiveness. A huge number of European industries are surveyed under the EPSI rating scheme.

8.2 Satisfaction and Its Measurement

It is important to start with a widely accepted definition of terms like 'customer' and 'satisfaction' like the ones given in the Standard ISO 9000: 2005. Or, better, ISO 10004: 2012 'Customer' refers to an organization or person that receives a product or service. Examples are: consumers, end-users, retailers, beneficiary and purchasers. A customer can be internal or external. However, satisfaction is generally measured in respect of external customers as a formal and useful exercise.

The Standard defines 'Customer Satisfaction' as 'Customer's perception of the degree to which the customer's requirements have been fulfilled'. It goes on to state that customer complaints are a common indicator of low customer satisfaction, but their absence does not necessarily imply high customer satisfaction. Further, even when customer requirements have been agreed with the customer and fulfilled, this does not necessarily ensure high customer satisfaction (this is the case when customer requirements as originally felt by the customer had to be modified during interaction with the supplier on price or delivery or some other consideration).

'Satisfaction' has been defined as a judgement that a product or service feature or the product or service itself provided a pleasurable level of consumption-related fulfilment, including levels of under- or over-fulfilment. Satisfaction with a product or service is a construct that requires experience and use of the product or service. Thus, customer satisfaction is about user satisfaction rather than about buyer satisfaction. Satisfaction is a feeling and admits of some lower and upper threshold values or levels.

The basic objectives of measuring customer satisfaction can be to

(1) identify areas (features of products and services) of customer dissatisfaction which call for appropriate corrective action;
(2) identify areas wherein customers are currently satisfied and where efforts should be made to retain or even to enhance customer satisfaction;
(3) link customer satisfaction to performance in processes carried out internally and to employee satisfaction. In fact, it has been often argued—and rightly so—that unless employees within the organization are themselves satisfied with the work environment or quality of their working life, they cannot make customers feel satisfied;
(4) Determine customers' perceptions and priorities in regard to the different facets of the product or service, viz. on-time and secure delivery, warranty provisions, justifiable cost, maintenance issues, convenient ordering and billing procedures, product recall procedures, response to special requests or suggestions.

Thus, a measure of customer satisfaction developed with due care and applied with due consistency can definitely help any organization improve its performance to enhance customer satisfaction and thereby to grow its business.

8.3 Models for Satisfaction Measurement

Though macro-models are sometimes used to measure customer satisfaction, micro-models make more sense. Of the several models developed in different contexts, the following are more commonly used.

Expectation Disconfirmation Model—Consumers' pre-consumption expectations are compared with post-consumption experiences of a product or service through measurement of an attitude of satisfaction/dissatisfaction on a scale. Expectations originate from beliefs and past experiences about the performance of a product or service.

Perceived Performance Model—The performance of the product or service as realized by the user dominates over the prior expectations to determine the level of satisfaction. This model is specially useful in situations where the product or service performs so positively that expectations get discounted.

Affective Model—Emotion, liking or mood influence the feeling of satisfaction or its absence following the use experience. Here we go beyond the rational process in situations where services, in particular, are consumed by users with special likes and dislikes or with particular mental make-ups.

8.4 Approaches to Customer Satisfaction Measurement

Any attempt to measure customer satisfaction has to start from determining customer requirements, not just about some products or services delivered—a little while ago or a long time back—but about the organization(s) manufacturing the product and involved it its installation (whenever applicable), delivery and maintenance. In this context, customers are basically users and not just distributors or dealers. Of course, the latter category of people remains in direct contact with users generally, in case of goods and services for domestic consumption and can provide some information about customer satisfaction or otherwise.

Determining customer requirements calls for identification of quality dimensions that appeal to customers and on which customers can offer their opinions, complaints and even suggestions for improvements. All this is often captured through what is referred to commonly as the Voice of the Customer (VoC). Another approach is the Critical Incident Approach in which a customer describes clearly some aspect of the organizational performance with which the customer comes into direct contact. That way, this approach provides information mostly about quality of services (may be related with some products procured or otherwise).

Though customer satisfaction should be continuously measured and monitored as an input to product/service improvement or new product development, we can think of several situations which generally call for customer satisfaction measurement.

Event Driven shortly after a sale or delivery of a service is completed, with questions limited to the specific event, put to people involved or informed about the event. The sample of participants is usually small, and a simple analysis of responses is carried out. Results of the survey or actions taken on that basis may or may not be communicated to the respondents but are made use of in initiating necessary corrective and/or preventive actions to enhance customer satisfaction.

Periodic Performed more or less at regular intervals, with a broad range of questions covering related facets of the organization's performance, probing for in-depth information and suggestions for improvement of products and services. Usually, a larger sample is covered, and comprehensive data analysis is carried out. Feedback is provided to participants in the exercise, and inputs are derived about satisfiers and dissatisfiers among product or service features.

Continuous performed as a routine with a specified periodicity coupled with exercises needed to deal with a new product release or a big complaint from an important customer.

8.5   Quality Attributes and Customer Satisfaction

Quality attributes of a product are features which are evaluated by customers (not always consistently or quantitatively or even objectively) and which influence the level of satisfaction of customers. These features may be related to use value or esteem value or salvage value and could be present in terms of product augmentation or extension. This way, quality attributes constitute a subset of features of an augmented or extended product.

Quality attributes have been categorized by Kano on the basis of their contributions to customer satisfaction. This classification helps the supplier or provider to focus attention on those attributes which contribute significantly to customer satisfaction and that way to business growth. Kano puts quality attributes of products and processes into five broad classes and identifies four areas for the supplier or provider in terms of the actions needed.

Kano's attractive attributes cover two extremes, viz. (1) highly attractive attributes which are product/service features that can easily attract potential customers and that way could be offered by the organization in terms of a strategy to expand market or to claim a bigger share of the market and (2) less attractive attributes which are not essential to satisfy customers and thus can be dropped if cost considerations so require. However, the issue of retaining these if cost considerations so permit versus the question of discarding these attributes to reduce costs which indirectly imply less price and hence a greater demand should be dealt with in terms of the organization's business policy. Similarly, the question of adding the highly attractive features at extra costs has also to be resolved as a business problem.

Kano's one-dimensional attributes are again in terms of two extreme situations according to the value addition a feature makes to the product or service under consideration. Some are high-value-added quality attributes, making a high contribution to customer satisfaction. These features can generate better revenues and hence should be incorporated in the products or services offered. Pitted against such features are low-value-added quality attributes which contribute less to customer satisfaction, but cannot be ignored by the providers since their absence can definitely lead to customer dissatisfaction. The firms also need to avoid providing too less a level of these attributes to dissatisfied customers to stop losing such customers altogether.

Kano considers another classification of attributes depending on the extent these are a must. In fact, these attributes are often referred to as Kano's 'Must Be' Attributes classified as (a) critical quality attributes which are essential to customers and in their absence customers will leave the organization. Hence, firms must provide sufficient fulfilment of these attributes to customers and (b) necessary quality attributes which have to be there at levels that will avoid customer dissatisfaction.

1. Kano goes on to have another finely tuned classification of features into two categories. These are referred by Kano as indifferent quality attributes. The first category comprises potential quality attributes which will over time become the attractive attributes and organizations can consider providing these as strategic weapons to attract customers in the future. The second category includes indifferent quality attributes about which customers are generally indifferent. If necessary on cost considerations, firms need not offer these attributes.

Besides these categorizations of quality attributes in relation to customer satisfaction or its reverse, Kano provides a classification of areas to organizations for appropriate actions in order to achieve a high level of customer satisfaction. And in this context, he mentions four different areas. These are

1. Excellent area

The attributes located in this area are those that customer considered to be important, and for which the actual performance of the firm has been judged as satisfactory to customers. Retention of customers requires that performance in these attributes be continued at least at their present levels.

2. To-be-improved area

Here are attributes which are considered by the customers as important, but for which actual performances have not met expectations—at least fully. Customers are not totally dissatisfied. The company must focus on these attributes and make improvements immediately.

3. Surplus area

This contains attributes which are not rated by customers as very important, but the perceptions of customers about the supplier/provider are quite satisfactory. The firm can put aside these quality attributes. If the firm needs to cut costs, these are the attributes that can be ignored or dropped without incurring a significant negative impact on customer satisfaction. Again, a business policy issue crops up.

4. Care-free area

These quality attributes are such that customers have a lower satisfaction level, but at the same time they also rank these attributes as being less important. The firm does not need to worry about these quality attributes because these items have less impact on the overall quality evaluation processes by the customers. However, providers must maintain these attributes at cost-optimal levels to have an edge over competitors.

8.6 Attributes of Customer Satisfaction

The main attributes or elements are

- image of the provider/producer/supplier in terms of overall beliefs and impressions, business practice, ethics, social responsibility;
- expectations from service life, availability, maintainability, cost per unit of usage, product and service parameters;
- cognizance of customer voice as reflected in easy access and communication, satisfactory and timely redressal of complaints;
- loyalty in terms of intentions or decisions for repurchase, recommendations to others for purchase in contrast to intention or decision to change;
- satisfaction reflecting perceived quality and value compared with expectations and ideal perceptions, economic considerations;
- linked up with customer satisfaction as manifest variables is customer loyalty. Hayes (2010) has distinguished among Advocacy Loyalty, Purchasing Loyalty and Retention Loyalty and has proposed an index for each.

We should develop and use a standardized (may be at the national or industry level) procedure or instrument to measure customer satisfaction, like the ACSI or ECSI model. With the growing importance of a CSI model, we must also determine which Structural Equations Model (SEM) can better work in a CSI model. Cassel et al. suggested two criteria for choosing an SEM technique for a CSI model, including (1) the SEM technique should determine the CSI score to make the comparison possible and (2) the SEM technique should exhibit good statistical properties like robustness.

8.7 ACSI Methodology and Model

Measures of customer-perceived value, customer satisfaction, customer loyalty, customer complaint behaviour and related constructs used in the widely accepted ACSI model as also the methods for estimating the relations connecting different variables are quite general and do apply to all types of products and services including public services across the general milieu of customers even in highly competitive markets. It is true that different market segments or customer groups differentiated in terms of product and service brands preferred and purchasing power justify different measures of the constructs, though the methods for estimating relations among the variables could remain the same.

Measures of the various constructs used come from responses to questions in well-designed surveys that are inputs to the model and thus define the exogenous variables in the model. And the variables are manifest, though related to respondents' perceptions. And that way these are not as stable as directly measurable manifest variables. Responses to several related questions are often to be collated to derive any such manifest variable input to the model. The relationship in the model and the variable measures used to estimate these relationships apply to public services and competitive product markets alike. The model does include some latent variables which are treated as endogenous variables.

The ACSI model is virtually a structural equations model, linking endogenous and exogenous variables—both manifest and latent—in terms of cause–effect relations. Thus, although the formal representation of this model involves a set of linear relations, these relations are not strictly speaking linear regression equations wherein variations in some dependent variable(s) are explained through a number of independent explanatory variables. In fact, in the ACSI model some variables appear as both independent and dependent in the set of equations defining the model.

The primary objective in developing and solving the ACSI model was to explain customer loyalty which is a very important construct in marketing management in terms of two contributory latent variables, viz. Perceived Product Quality and Perceived Service Quality. And the main problem remains with developing good proxies for these two in terms of manifest variables.

Model validity in situations customer satisfaction measurement really implies predictive validity, established when the model can result in reasonable predictions. Of course, nomological validity of this model as a form of construct validity to affirm that a construct or a proposition behaves as it is expected in a system of related constructs has been established by Bogozzi (1980).

As explained earlier, satisfaction is a multi-dimensional latent variable (construct) with each such dimension derived from a number of manifest indictors (revealed through responses to the corresponding questions). Any one concrete measure of satisfaction such as a single survey question rating is at best a proxy for latent satisfaction (Simon 1974). In effect, each indicator serves as a proxy for some dimension of satisfaction. Instead, the ACSI uses a variety of proxies or benchmarks that customers use to evaluate their overall consumption experience. These proxies are combined into an index to operationalize the concept of satisfaction as a latent construct.

The structural equations model behind ACSI is indeed a system of linear equations connecting manifest and latent variables (Mukherjee 2002). Some of the variables involved are exogenous while some others are endogenous. Manifest variables are scaled responses to questions in the survey while latent variables like customer satisfaction, resulting from perceived value (another latent variable-experienced through use of the product or service), and resulting in customer loyalty (a latent variable of primary interest).

- The equations involved in the model are essentially like linear regression equations with variables not classified as independent or explanatory variables (regressors) and dependent variables to be explained (regressands) as in classical regression analysis.

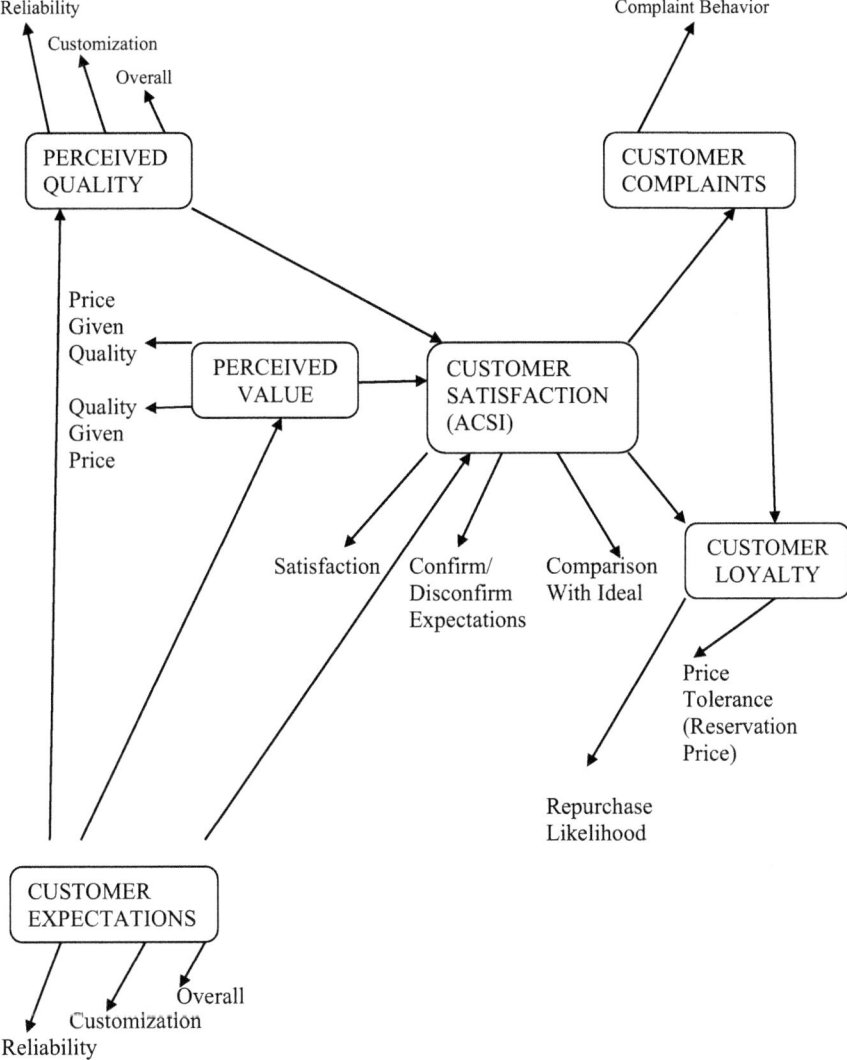

Fig. 8.1 ACSI model *Source* Anderson et al. (1994)

- The primary objective in estimating the model is to explain customer loyalty, a construct of universal importance in the evaluation of current and future business performance. The ACSI model is shown in Fig. 8.1.
- Since the number of variables and the associated coefficients may be larger than the number of data points at least on some occasions, Partial Least Squares method has been generally used.

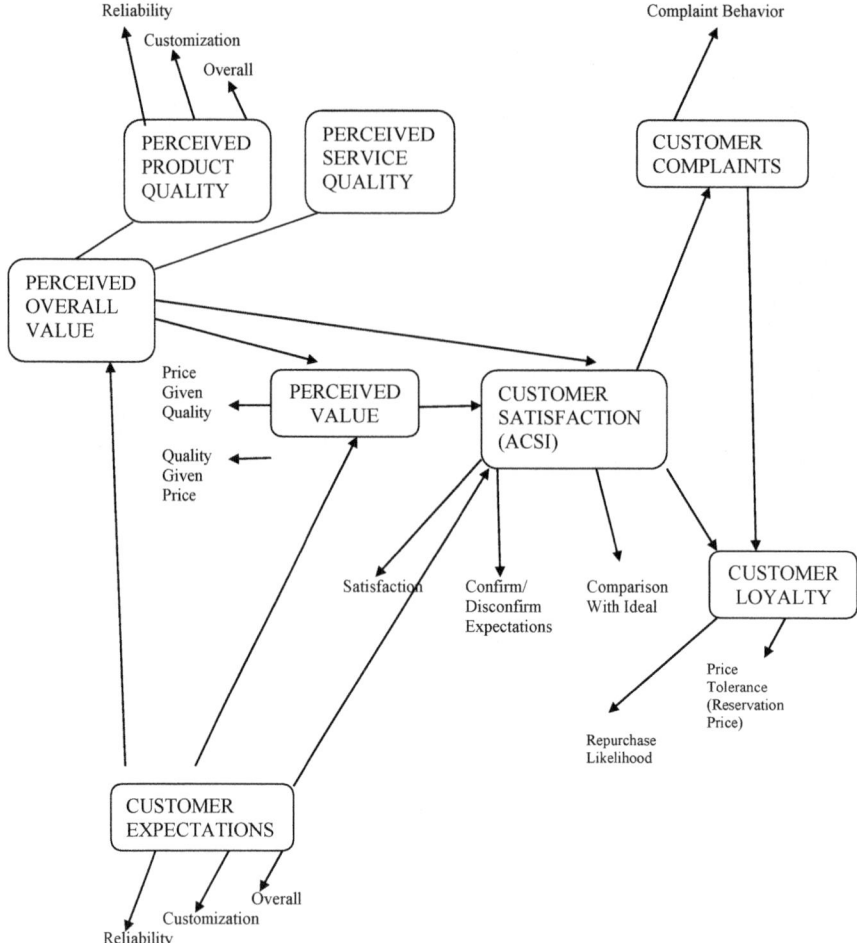

Fig. 8.2 ACSI model for the manufacturing/durables sectors *Source* Fornell et al. (1996)

The model was expanded experimentally in 1996 to produce the latent variable Perceived Overall Quality as the combination of two latent variables (constructs), Perceived Product Quality and Perceived Service Quality. Thus, four variables, viz. image (of the organization as revealed to the customer), expectation (from the product and associated service), perceived quality and perceived value (taking into account the price paid) are taken as exogenous latent variables in the expanded model.

Figure 8.2 shows the expanded ACSI model now used for measuring satisfaction with industries in the manufacturing/durables sector.

The EPSI framework uses an additional exogenous latent variable, viz. transparency about products and services as perceived by the customer. Customer's assessment of the perceived value of a product or service is likely to be influenced by the adequacy or otherwise of the information provided by the supplier. Information asymmetry caused by a perceived inadequacy of information about the product or service by the customer may lead to under-assessment of the customer's perceived value and that way may adversely affect customer satisfaction. Thus, in the EPSI framework, perceived value depends on image, expectation, product quality, service quality and, in addition, transparency. While the first four directly affect customer satisfaction, besides influencing perceived value, transparency influences only perceived quality, at least directly.

The formal expression of the model depicted in Figs. 8.1 and 8.2 can be written as in a structural equations model in terms of a set of linear equations estimated by partial least squares (PLS). The systematic part of the predictor relationships is the conditional expectation of predictands for given values of predictors. The general equation is thus specified as

$$E[\eta|\eta, \xi] = B\eta + \Gamma\xi \tag{8.1}$$

where $[\eta' = (\eta_1, \eta_2,..., \eta_m)$ and $\xi' = (\xi_1, \xi_2, ..., \xi_m)$ are vectors of unobserved endogenous and exogenous variables, respectively; $\mathbf{B}(m \times m)$ is a matrix of coefficient parameters for η; and $\Gamma(m \times n)$ is a matrix of coefficient parameters for ξ. This implies that $E[\eta\xi'] = E[\xi\xi'] = E[\xi] = 0$, where $\xi = \eta - E[\eta|\eta, \xi]$.

The equation that relates the latent variables in the model shown in Fig. 8.1 (which is used for all sectors except Manufacturing/Durables) is[1]:

$$
\begin{bmatrix} \eta_1 \\ \eta_2 \\ \eta_3 \\ \eta_4 \\ \eta_5 \end{bmatrix} =
\begin{bmatrix} 0 & 0 & 0 & 0 & 0 \\ \beta_{21} & 0 & 0 & 0 & 0 \\ \beta_{22} & \beta_{32} & 0 & 0 & 0 \\ 0 & 0 & \beta_{43} & 0 & 0 \\ 0 & 0 & \beta_{53} & \beta_{54} & 0 \end{bmatrix}
\begin{bmatrix} \eta_1 \\ \eta_2 \\ \eta_3 \\ \eta_4 \\ \eta_5 \end{bmatrix} +
\begin{bmatrix} \gamma_{11} \\ \gamma_{21} \\ \gamma_{31} \\ 0 \\ 0 \end{bmatrix} \xi +
\begin{bmatrix} \xi_1 \\ \xi_2 \\ \xi_3 \\ \xi_4 \\ \xi_5 \end{bmatrix}
$$

where

ξ Customer Expectations
η_1 Perceived Quality
η_2 Perceived Value
η_3 Customer Satisfaction (ACSI)
η_4 Customer Complaints
η_5 Customer Loyalty

---

[1]In the expanded model used for the manufacturing/durables sector (Fig. 8.2). Perceived overall quality is composed of the latent variables. Perceived product quality and perceived service quality. The equations presented here refer specifically to the model used for all other sectors (Fig. 8.1).

The general equation for relating the latent variables to empirical variables is:

$$y = \Delta_y \eta + \varepsilon$$
$$x = \Delta_x \xi + \delta$$

(8.2)

where $y' = (y_1, y_2, \ldots, y_p)$ and $x' = (x_1, x_2, \ldots, x_p)$ are the measured endogenous and exogenous variables, respectively. Δ_y $(p \times m)$ and Δ_x $(q \times n)$ are the corresponding regression coefficient matrices. By implication from PLS estimation (Fornell and Bookstein 1982), we have $E[\varepsilon] = E[\delta] = E[\eta\varepsilon'] = E[\xi\varepsilon'] = 0$. The corresponding equations in the model are:

$$\begin{bmatrix} x_1 \\ x_2 \\ x_3 \end{bmatrix} = \begin{bmatrix} \lambda_{11} \\ \lambda_{21} \\ \lambda_{31} \end{bmatrix} \xi + \begin{bmatrix} \delta_1 \\ \delta_2 \\ \delta \end{bmatrix}$$

and

$$\begin{bmatrix} y_1 \\ y_2 \\ y_3 \\ y_4 \\ y_5 \\ y_6 \\ y_7 \\ y_8 \\ y_9 \\ y_{10} \\ y_{11} \end{bmatrix} = \begin{bmatrix} \lambda_{11} & 0 & 0 & 0 & 0 \\ \lambda_{21} & 0 & 0 & 0 & 0 \\ \lambda_{31} & 0 & 0 & 0 & 0 \\ 0 & \lambda_{12} & 0 & 0 & 0 \\ 0 & \lambda_{22} & 0 & 0 & 0 \\ 0 & 0 & \lambda_{13} & 0 & 0 \\ 0 & 0 & \lambda_{23} & 0 & 0 \\ 0 & 0 & \lambda_{33} & 0 & 0 \\ 0 & 0 & 0 & \lambda_{14} & 0 \\ 0 & 0 & 0 & 0 & \lambda_{15} \\ 0 & 0 & 0 & 0 & \lambda_{25} \end{bmatrix} \begin{bmatrix} \eta_1 \\ \eta_2 \\ \eta_3 \\ \eta_4 \\ \eta_5 \end{bmatrix} + \begin{bmatrix} \varepsilon_1 \\ \varepsilon_2 \\ \varepsilon_3 \\ \varepsilon_4 \\ \varepsilon_5 \\ \varepsilon_6 \\ \varepsilon_7 \\ \varepsilon_8 \\ \varepsilon_9 \\ \varepsilon_{10} \\ \varepsilon_{11} \end{bmatrix}$$

(8.3)

where

x_1 Customer Expectations about Overall Quality
x_2 Customer Expectations about Reliability
x_3 Customer Expectations about Customization
y_1 Overall Quality
y_2 Reliability
y_3 Customization
y_4 Price Given Quality
y_5 Quality Given Price
y_6 Overall Satisfaction
y_7 Confirmation of Expectations
y_8 Distance to Ideal Product (Service)
y_9 Formal or Informal Complaint Behaviour
y_{10} Repurchase Intention
y_{11} Price Tolerance (Reservation Price)

Kristensen and Eskildsen (2005) have shown that the structural equation models behind ACSI, ECSI and EPSI are quite robust against the following anomalies:

Exogenous variable distributions not showing traditional patterns;
Multi-collinearity among latent exogenous variables;
Indicator validity being at stake;
Indicator reliability not shown to be high;
Structural model S difficult to specify;
Specification errors having non-traditional distributions;
Sample size being not that large;
Number of manifest variable present in each block;
Missing values cannot be ruled out.

This robustness coupled with the construct validity has meant a wide use of the ACSI model and of the Customer Satisfaction Indices derived therefrom.

8.8 The Indices

The general form of the ACSI as an index of customer satisfaction is as follows:

$$\text{ACSI} = \frac{E|\xi| - \min|\xi|}{\max|\xi| - \min|\xi|} \times 100 \qquad (8.4)$$

where ξ is the latent variable for Customer Satisfaction (ACSI), and $E[.]$, $\min[.]$ and $\max[.]$ denote the expected, the minimum, and the maximum value of the variable, respectively. And this index can vary between a very small positive number and 1.

The minimum and the maximum values are determined by those of the corresponding manifest variables

$$\min[\xi] = \sum_{i=1}^{n} w_i \min[x_i]$$

and

$$\max[\xi] = \sum_{i=1}^{n} w_i \max[x_i]$$

where x_i's are the measurement variables of the latent Customer Satisfaction, w_i's are the weights, and n is the number of measurement variables. In calculating the ACSI, unstandardized weights must be used if unstandardized measurement variables are used.

In the ACSI, there are three indicators for Customer Satisfaction that range from 1 to 10. Then the calculation is simplified to:

$$\text{ACSI} = 100 \times \left[\sum w_i x_i - \sum w_i\right]/9 \sum w_i \qquad (8.5)$$

where w_i's are the unstandardized weights.

Both ECSI and ACSI have strengths and weaknesses which have been considered by some authors to come up with a few modifications. The relative strength of ACSI over ECSI is the inclusion of a construct on customer complaint. However, customer expectation in the model has been found to be somewhat less important and Johnson et al. (2001) and Martensen et al. (2000) replace 'customer expectation' by 'corporate image'. Corporate image refers to the brand name and the kind of association that customers derive from it. With a high image of the company from which a customer buys, a customer usually has a higher satisfaction level and repurchase. Questions reflecting on corporate image include internal consistency, emphasis on public affairs, trustworthiness and caring about the customers' needs.

An important aspect not to gloss over is the fact that in any customer satisfaction and loyalty survey, demographic differences do exist in the opinions and even in the veracity of statements or figures and such differences should be taken due care of in basing conclusions on the indices obtained through survey data.

8.9   Customer Satisfaction Surveys

Measures of customer satisfaction and of loyalty have been often derived from responses to questions in a survey where a properly designed questionnaire is canvassed among a sample of respondents representing the customer group of interest. In Chap. 4 of his famous book on the subject, Hayes (2010) provides a detailed account of item section, rating format choice, sampling of items for inclusion in the final questionnaire as also sampling of respondents. As stated by him, items should appear to be relevant to what is attempted to be measured, should be concise and unambiguous, should contain only one thought and should not contain double negatives. Either a Yes–No checklist format or Likert-type agreement–disagreement format should be used to record responses. The questionnaire should be preceded by a well-written introduction spelling out the purpose of using the questionnaire and providing help to comprehend the questions and to provide responses. He also provides guidance about sampling of items from the list of questions initially developed and for evaluating the items finally retained by carrying out factor analysis and checking item-total correlations (in case several similar items are merged into one) as also testing internal consistency by using Cronbach's alpha coefficient. Indeed, Hayes devotes a whole Chapter to illustrate Customer Satisfaction Questionnaires for various services like dental care, library service, restaurant service.

Findings from customer satisfaction surveys may not always be credible. It may be impossible to cover the entire customer base for the purpose of a customer satisfaction survey, given the limited resources available and the constraint on time by which survey results can be processed to yield the appropriate measure(s). It may not be necessary either to get a credible measure of satisfaction. The problem in random sampling is that quite often a sampling frame or a list of all customers along with their details required for access does not exist. Sometimes, a quota or judgement sampling is done from the population of customers who may not be conveniently canvassed. If the selection of customers is just left to convenience, the estimate yielded by the survey may not be that biased. However, a judgement sample is often affected by the choices of the investigator and that will introduce bias in the estimated result.

Sometimes, an analogue of multi-stage sampling may be used where the first-stage units (like distributors or retail outlets) are not really users and hence are not the eventual respondents. A proper selection of such units can be easily done. However, to reach second-stage units within each selected first-stage unit may be pretty difficult. It is just natural that the population of users is not homogeneous in respect of their expectations and previous experiences and, that way, their requirements. Evidently extent of satisfaction with different facets of the product or service offered to them will vary. Unless possible clusters of relatively homogeneous users or customers are identified and adequately represented in the sample to be canvassed, findings from the survey unduly loaded by some cluster or the other will fail to bring out the true picture about customer satisfaction.

Quite often, the customer population is divided into some natural groups or strata depending on features of customers that can possibly influence their satisfaction or otherwise with the product or service under consideration, e.g. gender, socio-economic condition, lifestyle and the like. From each such stratum, a sub-sample of an appropriate size is selected, usually randomly or systematically, to constitute the overall sample. Stratified sampling is likely to yield more precise estimates of population parameters than simple random sampling.

A second problem with responses to questionnaires arises from sources like improper administration of the questionnaire leaving not enough time to the respondent to think over an item, tendency on the part of the respondent to avoid uncomfortable responses, inability on the part of the investigator to explain the items properly to a respondent, etc. While the first source can be tackled by adequate training of the investigators, the second is linked up with the inherent attributes of the respondent which is likely to be influenced by the introduction to the survey as presented by the investigator.

Non-response and response biases and errors vitiate findings from such a survey. In several situations where some confidential information is sought in a survey or when the respondent is asked to tell which of several brands of a consumer item he (she) uses, the response may be biased due to a feeling that the correct response may not convey the desired impression about the respondent's consumption behaviour in terms of purchasing power or concern for quality in preference to price, etc. To account for such errors and biases, randomized response techniques have been developed and are being used in market research surveys.

Sample size in a survey has to be adequate for sampling error in the estimates to be small. It is possible to calculate the sample size required to give an estimate of a population parameter with a specified (small) relative standard error, and hence, a confidence interval with a specified length provided some guess values from some previous surveys or some other sources are available. However, resources available say the final word regarding sample size. It may be noted that the sample size to start with should be given by (sample size actually needed)/(response rate), the latter being the proportion of selected persons eventually responding.

8.10 Design and Administration of the Questionnaire

Developing the Customer Satisfaction Questionnaire with its generic and specific aspects has been comprehensively discussed by Hayes (2010). Clause C.4 in ISO/ DIS 10004 also treats this subject. In fact, as in any international standard, this one also gives out detailed advice on each aspect of questionnaire and its administration.

Since responses from 'concerned and informed' persons form the basis for any credible measure of customer satisfaction, it is imperative that a lot of attention has to be devoted towards ensuring that the questions included in the questionnaire are ungrudgingly accepted by the respondents, are not difficult or time-consuming to answer, do not encroach on privacy and are just necessary and sufficient for the purpose. Responses provided should be pre-coded wherever possible to facilitate analysis. Provision should exist for responses that do not fit into any of the response categories listed in the questionnaire. Administration must suit the respondent's convenience in regard to time and place of interview.

Most customer satisfaction surveys are expected to be anonymous if the purpose of the survey is to come up with an index of customer satisfaction. However, if the objective of the survey is also to inform the organization about problems with customers who are not fully or partly satisfied with the products and/or services of the organization so that appropriate measures can be initiated to address such problems, details of the respondent should be sought and possible kept confidential, not to be put on the public domain. Starting with identification particulars of a respondent and the nature and extent of his/her transactions with the organization, questions which are easy to answer should be put first, followed by questions which may take some time to be answered, may be involving quite a few alternatives. The questions should follow a natural and logical sequence. To illustrate this point, a question like 'Were you dissatisfied with some products or services you procured from this organization?' followed by a pair of questions 'Which products and/or services you procured from this organization?' and 'Which of these products and/or services failed to satisfy you' is a wrong sequence and we should have only the second pair in place. Usually, sensitive questions should be avoided. If responses to some sensitive or confidential questions are genuinely needed, one can avoid asking such questions directly and take recourse to Randomised Response Technique. One should also avoid difficult or complex questions. In fact, it may be sometimes helpful if several elementary questions are asked to derive—somewhat indirectly—the response to a complex question.

The organization should organize the questions in a logical sequence where possible, and use more than one question to make it easier for the respondent, if the response involves numerous alternatives.

Administering a questionnaire should not be taken as an easy task. In fact, the value of the response to a question depends a lot on how the question was comprehended by the respondent and the latter depends on how the investigator explained the question and/or helped the respondent to answer. Extremes alike 'take it as you understand' or 'provide an answer like this…' should be avoided. No attempt should be made to put words into the mouth of the respondent. Similarly, several respondents should not be interviewed together to assess their individual perceptions.

Items in a questionnaire may have answers which are numbers or quantities; otherwise, these could be categorical or even binary. In assessing customer satisfaction, we may have to use questions or statements relating to some underlying latent variable. When accessing attitudes, a 5-point scale for categories across a continuum is often used. For example, a statement may appear in the questionnaire about some aspect of the product or service either claimed by the provider or artificially asked to elicit the respondent's reaction to the statement. This reaction may be sought as one of five different positions like 'Strongly agree', 'Agree', 'Neutral', 'Disagree' and 'Strongly disagree'. Where greater discrimination is required, a wider scale, e.g. a 9-point scale, might be used. Usually with an odd number of response categories, the central one corresponds to 'neutral' or 'undecided'.

If the need is to compel the respondent to take a position and avoid a neutral response, the questionnaire can make use of an even number of scale points (e.g. 4 or 6). In such a case, the response categories could be 'Very satisfied', 'Satisfied', 'Dissatisfied', 'Very dissatisfied'.

A 'pretest' or a 'pilot test' is a preliminary survey conducted with a small, but representative, set of respondents, in order to assess the strengths and weaknesses of the questionnaire. Findings may help us to identify training needs for the investigators so that they can elicit the desired information from the respondents. It is a strongly recommended practice, even though it might not be feasible if the number of respondents is limited.

Where possible, all key aspects of the questionnaire should be tested, using the same methods (e.g. by mail or telephone) as in the actual survey. This should be repeated with each significant revision of the questionnaire.

Generally, the pretest results are analysed to assess the reaction of interviewees particularly in terms of their being bored by too many questions, some of which appear to be irrelevant to the respondents, evaluate the quality of responses recorded, estimate the time taken on an average to interview a single person, and note the problems of selecting the interviewees as already done. On the basis of such inputs, the questionnaire and its method of administration may be modified, operational definitions of terms and phrases used in the questionnaire may have to be revisited, investigators may be given a further dose of training, In fact, the 'pretest' or the 'pilot test' often leads to a thorough change in the entire exercise.

The relative advantages and limitations of different survey methods, as summarized in the DIS, are reproduced in Table 8.1.

Table 8.1 Comparison of survey methods

Method	Advantages	Limitations
Face-to-face interview	- Possibility of complex and directed questions which may need some instant clarification - Flexibility in conducting interview to suit the convenience of the respondent - Immediate availability of information - Ability to verify apparently inappropriate information	- Takes more time, therefore slower - More costly, especially if interviewees are geographically dispersed - Risk of possible distortion through interviewer influencing the response or even indirectly prompting the same
Telephone interview	- Wider coverage of respondents than face-to-face interview - Flexibility of time and mode to suit respondents' convenience - Ability to verify information - Greater speed of execution - Immediate availability of information	- Non-verbal responses cannot be observed (no visual contact) - Risk of distortion caused by deficiency in extracting information from responses as deciphered by the interviewer - Information limited by relatively short duration of interview (20 min to 25 min) - Customer reluctance to participate in telephone conversation
Discussion group	- Lower cost than individual interviews - Partially structured questions - Spontaneous responses resulting from group interaction	- Requires experienced facilitator and related equipment - Outcome depends on participant's familiarity with technique - Difficult if customers are dispersed over wide region
Mail survey	- Low cost Can reach a widely dispersed geographic group - No distortion by the interviewer - High level of standardization - Relatively easy to manage	- Response rate might be low - Self-selection of respondents might result in skewed sample that does not reflect the population - Possible difficulty with unclear questions - Lack of behaviour control in answers - Longer time for data collection

(continued)

Table 8.1   (continued)

Method	Advantages	Limitations

On-line survey (Internet)	- Low cost
	- Previously prepared questions
	- No distortion by interviewer
	- High level of standardization/
	  comparativeness
	- Fast execution
	- Easy evaluation
		- Low response rate
		- Lack of behaviour control in answers
		- Delay in availability of data
		- High probability of interruption in
		  case of unclear questions
		- Assumes customer has the equipment
		  and is familiar with the technology

Source ISO/Draft International Standard 10004

Some of the points given out above may have a different status now, e.g. response rate in Web-based surveys has increased over time and such surveys are gaining ground particularly when respondents are educated and expected to tell the truth in the absence of an interviewer.

8.11   Measuring Customer Loyalty

The impact of customer satisfaction on business prospects and performance has sometimes been attempted through the Net Promotion Score (NPS). This is derived from the response to a single loyalty question, viz. 'How likely are you to recommend us to your friends/colleagues?' A 0 to 10 Likert scale is used where a score 0 implies 'not at all likely' and 10 implies 'extremely likely'. Customers are segmented into three groups (a) Detractors (ratings from 0 to 6), (b) Passive (ratings 7 and 8) and (c) Promoters (ratings 9 and 10). The Net Promotion Score is then computed as

$$\text{NPS} = \text{Proportion of Promoters} - \text{Proportion of Detractors}$$

While NPS has been reported by some organizations as being the best predictor of growth, recent studies show that NPS is no better than other measures of customer loyalty like those based on overall satisfaction or repurchase decision or recommendation. This has led to loyalty indices derived from responses to four questions, viz.

1. Overall, how satisfied are you with our organization?
2. How likely are you to recommend our organization to friends or colleagues?
3. How likely are you to continue purchasing the same product or service from our organization?
4. If you were selecting an organization (within the industry of interest) for the first time, how likely is it that you would choose our organization?

Based on the average rating for these four questions, one gets what is sometimes referred to as the Advocacy Loyalty Index. Hayes (2010) speaks of three measures of loyalty based on a factor-analytic approach. These are

Advocacy Loyalty Index—reflects the degree to which customers will be advocates of the organization (average across satisfaction, recommendation, repeated choice and repurchases decisions)
Purchasing Loyalty Index—reflects the degree to which customers will increase their purchasing behaviour (average across purchase of different items, purchase increase, increase in purchase frequency)
Retention Loyalty Index—reflects the degree to which customers will remain with a given organization (defection from the organization, response coded inversely).

8.12   Concluding Remarks

Customers have the last word in the context of Quality and business evolves around customer satisfaction. Of course, the term 'customer' has a wide connotation and includes anyone who is entitled to make a statement on quality. While consequences of customer satisfaction in terms of repurchases or recommendations to others, its determinants are not all manifest and can be quite intriguing. There exist situations where responses relating to satisfaction or dissatisfaction have to be sought from some authorized spokesperson in the customer organization who may not always have a correct idea of how satisfied the actual users within the organization are about the product or service under consideration.

Indices of customer satisfaction are sometimes considered as elements in some Quality Awards without necessarily establishing strict comparability among such indices reported by different contestants. Sometimes, the methods of computation differ, some times data analysed are inadequate in quantity and poor in quality, and sometimes values obtained are not properly interpreted. Responding to problems arising, various national and regional standard bodies have developed standards for measurement of customer satisfaction and in some other countries such standards are currently being developed.

Measures of customer loyalty as a manifest reflection of customer satisfaction have been derived on the basis of the traditional mover–stayer model based on the purchase behaviour of a fixed panel of users/customers over time and noting proportions of respondents who have stuck to the same brand in two different periods of time. However, this early Markov Chain-based model is not much used these days, because of many limitations including the inability of this model to account for entry of new brands and withdrawal of old ones in between the two periods of time and for genuine changes in brands taking place.

Emphasis on customer satisfaction has led to several techniques for listening to the Voice of the Customer and for incorporating customer requirements in designing the product or service, in specifying its parts or components, in developing process plans and ultimately framing operator instructions. This constitutes Quality Function Deployment, and the aim is to enhance customer satisfaction as also competitiveness in that direction.

Appendix

See Table 8.2.

Table 8.2 Survey questions used in the ACSI model

Question number	Measurement variable description	Latent variable
1	Overall expectation of quality (pre-purchase)	Customer expectation
2	Expectation regarding customization, or how well the product and service fits the customer's personal requirements (pre-purchase)	
3	Expectation regarding reliability, or how often things would go wrong (pre-purchase)	
	Overall evaluation of quality experience with product (post-purchase)	Perceived product quality
5P	Evaluation of customization experience, or how swell product fits the customer's personal requirement (post-purchase)	
6P	Evaluation of reliability experience, or how often things have gone wrong with product (post-purchase)	
4S[a]	Overall evaluation of quality experience with service (post-purchase)	Perceived service quality
5S[a]	Evaluation of customization experience, or how well the service fits the customer's personal requirements (post-purchase)	
6S[a]	Evaluation of reliability experience, or how often things have gone wrong with service (post-purchase)	
8	Rating of quality given price	Perceived value
9	Rating of price given quality	
10	Overall satisfaction	
11	Expectancy disconfirmation (performance that falls short of or exceeds expectations)	Customer satisfaction (ACSI)
12	Performance versus the customer's ideal product and service in the category	
13	Has the customer ever complained either formally or informally about the product/service	Customer complaints
15	Repurchase likelihood rating	Customer loyalty
16	Price tolerance (increase) given repurchase	
17	Price tolerance (decrease) to induce repurchase	

[a]Used only in modelling satisfaction for manufacturing/durables

References

Anderson, T. W., & Fornell, C. (2000). Foundations of the American satisfaction index. *Total Quality Management, 11,* 869–882.

Anderson, E. W., Fornell, C., & Lehmann, D. R. (1994). Customer satisfaction, market share, and profitability. *Journal of Marketing, 58*(July), 53–56.

Bogozzi, R. P. (1980). *Causal models in marketing.* New York: Wiley.

Cadotte, E. R., Woodruff, R. B., & Jenkins, R. L. (1987). Expectations and norms in models of consumer satisfaction. *Journal of Marketing Research, 24*(August), 305–314.

Chandler, C. H. (1989). Beyond customer satisfaction. *Quality Progress, 22*(2), 30–32.

Eskildsen, J., & Kristensen, K. (2007). Customer satisfaction: The role of transparency. *Total Quality management & Business Excellence, 18*(1& 2), 39–48.

Fornell, C. (1992). A national customer satisfaction barometer: The Swedish experience. *Journal of Marketing, 56*(January), 6–21.

Fornell, C., & Bookstein, F. L. (1982). Two structural equations models: LISREL and PLS applied to consumer exit-voice theory. *Journal of Marketing Research,* (November) 440–452.

Fornell, C., Johnson, M. D., Anderson, E. W., Cha, J., & Bryant, B. E. (1996). The American customer satisfaction index: Nature, purpose and findings. *Journal of Marketing, 60*(October), 7–18.

Goodman, J. (1989). The nature of customer satisfaction. *Quality Progress, 22*(2), 37–40.

Hannan, M., & Karp, P. (1989). *Customer Satisfaction: How to maximize, measure and market Your Company's ultimate product.* New York: American management Association.

Hayes, B. E. (1997). *Measuring customer satisfaction.* Milwaukee: ASQ Quality Press.

Hayes, B. E. (2010). *Measuring customer satisfaction and loyalty.* New Delhi: New Age International Publishers.

Hsu, S.-H., Chen, W.-H., & Hsueh, J.-T. (2006). Application of customer satisfaction study to derive customer knowledge. *Total Quality Management, 17*(4), 439–454.

ISO 10001. (2007). Quality Management—Customer Satisfaction—Codes of conduct for organizations.

ISO 9001. (2008). Quality Management System Requirements.

ISO/TS 10004. (2010). Quality Management—Guidelines for monitoring and measuring Customer Satisfaction.

Johnson, M. D., & Fornell, C. (1991). A framework for comparing customer satisfaction across individuals and product categories. *Journal of Economic Psychology, 12*(2), 267–286.

Johnson, M. D., Gustafsson, A., Andreassen, T. W., Lervik, L., & Cha, J. (2001). The evolution and future of national customer satisfaction index models. *Journal of Economic Psychology, 22* (2), 217–245.

Kristensen, K., Kanji, G. K., & Dahlgaard, J. J. (1992). On measurement of customer satisfaction. *Total Quality Management, 3,* 123–128.

Martensen, A., Gronholdt, I., & Kristensen, K. (2000). The drivers of customer satisfaction and loyalty: Cross-industry findings from Denmark. *Total Quality Management, 11,* 544–553,

Morgan, N. A., & Rego, L. L. (2006). The value of different customer satisfaction and loyalty metrics in predicting business performance. *Marketing Science, 25*(5), 426–439.

Mukherjee, S. P. (2002). Measurement of customer satisfaction in statistics for quality improvement. In A. B. Aich, et al. (Ed.). Calcutta: IAPQR.

Selivanaova, I., et al. (2002). The EPSI rating initiative. *European Quality, 9*(2), 10–25.

Simon, J. L. (1974). Interpersonal welfare comparisons can be made and used for redistribution decisions. *Kyklos, 27,* 63–98.

Zanella, A. (1998). A statistical model for the analysis of customer satisfaction: Some theoretical and simulation results. *Total Quality Management, 9,* 599–609.

Chapter 9
Improving Process Quality

9.1 Introduction

By far the most important task as well as target for Quality Management is to achieve visible (even quantifiable) improvement in process quality. This implies improving the performance of all business processes—both core and also support. In fact, any forward-looking organization striving for excellence must have in place an effort to improve processes on a continuing basis. Continuous process improvement is the motto, and with improved processes leading to new or improved products and services, the organization can offer such new or improved products and services over time. [It must be noted that effort to improve has to be continuous, while the results of improvement by way of new products and services are bound to appear at discrete intervals of time].

Business process improvement is fundamental to business development, quality improvement and change management. It is the core of the various models of organizational excellence in use like the EFQM Excellence Model, the Malcolm Baldridge National Quality Award Model and the Deming Prize Model.

In practical terms, much business process improvement activity is basic, consisting of simple process mapping and analysis, leading to greater process understanding, ownership and some redesign.

While many organizations have commenced the early stages of business process mapping and analysis and have obtained certification to ISO-9001: 2015, few are seriously applying strategic process improvement approaches such as Six Sigma or Lean Organization or adopting some general model for the purpose.

Incidentally, Total Quality Management has been aptly described as a principle, a systems approach, a way of working to introduce continuous improvement in all operations—manufacturing as well as pre- and post-manufacturing—in order to achieve Total Quality. No doubt improvement has to be stabilized and acted upon to come up with improved quality products for some time till the next phase of improvement. As remarked earlier, we have to look at all business processes, going

S. P. Mukherjee, *Quality*, India Studies in Business and Economics,
https://doi.org/10.1007/978-981-13-1271-7_9

beyond the core process of the organization, in order that improvement leads to business growth. It will be proper to start with some generic definition of a process, then have some suitable definition of process quality in relation to some process plan and then proceed to discuss different approaches for quality improvement in processes.

Even in the more recent Quality Management paradigms emphasizing on quantitative analysis in the Six Sigma or the Eight-D methodology, process improvement gets its pre-eminent position. In the DMAIC (Define–Measure–Analyse–Improve–Control) or the DMAIS (Define–Measure–Analyse–Improve–Stabilize) methodology, the ultimate objective is improvement, followed expectedly by control or stabilization of the process at the improved level. The focus is not that direct in the DMADV (Define–Measure–Analyse–Develop–Verify), partly because this approach applies to the initial phase of new product or process development.

As remarked earlier, we have to look at all business processes, going beyond the core process of the organization, in order that improvement leads to business growth. And we should take due advantage of tools and techniques which prevent the occurrence of defects or errors in a process, either during the process planning phase or during an exercise to review the existing process plan.

It will be proper to start with some generic definition of a process, then have some suitable definition of process quality in relation to some.

9.2   Process Defined

A process is a set of interrelated and interacting activities or operations to be carried out in a given sequence along with the corresponding set of resources to be committed, through which some input is converted into some output. The activities are all in relation to a particular task to be performed and/or a particular outcome to be achieved. That way, the activities are interrelated. Further, these activities are interacting in the sense that any of these is affected by some preceding activity(ties) and affect some subsequent activity(ties). A simple process may appear to involve only one operation. However, it is possible to identify the activity elements even in a simple process. An activity is carried out by a man with or without a machine or a similar device using some material or physical resources to convert some input hard or soft—into some output. Sometimes, a machine by itself carries out an activity.

The following is a user-friendly process model that can facilitate process analysis which is essential for process control and subsequent process improvement.

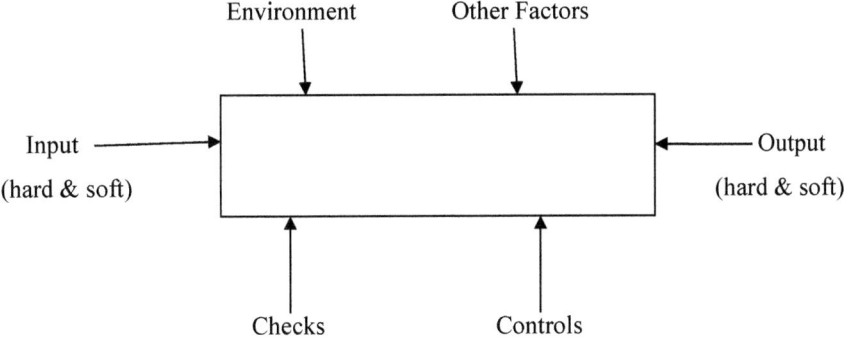

This model is equally applicable to manufacturing and also other business processes. Some of the characteristics mentioned in the model are technology-determined, some are input-driven, some others are equipment-dominated, while a few could be operator-decided.

Items of information about market requirements or stocks of raw materials or work-in-progress or finished goods awaiting final inspection or delivery schedules for different customers and the like may be processed (summarized or tabulated or plotted) with the help of a calculator or a graph paper to result in an actionable report illustrates an activity. However, one can break up this simple activity into elements like examining the input data for consistency or completeness, deciding on the type of summary to be prepared, drawing the table or the graph, etc.

A step in manufacturing like assembling some components into a sub-system:

Designing a product to possess some desired features and functional properties
Calibration of production or test or inspection equipment against duly certified reference standards
Inspection of incoming materials or in-process semi-finished items or finished products
Maintenance of equipments and facilities, including inspection, repair of failed parts and replacement of failing parts
Carrying out a market survey to assess the demand for a new product
Providing services to customers on request or as a part of service contract
Securing payment against a pending bill

Any process can possibly be segmented into sub-processes, each to be separately analysed. However, for the purpose of process improvement, the process chosen should not be too narrowly defined, since in such a case any improvement in the process will hardly have any impact on the overall process concerned or on business. At the same time, if the process chosen for analysis and improvement be too wide in its coverage, it will be pretty difficult to work out a significant improvement. Thus, the process to be considered should strike a balance between these two situations.

The assessment of a process in relation to the need for improvement arises from the fact that a process can be looked upon as a sequence of value-adding operations or tasks and a process plan is expected to spell out the nature and extent of value addition to be effected by a process in terms of functional value or aesthetic/prestige value added to the output of the process or to the final product through this output. The process plan takes due care of customer requirements, considering both internal as well as external customers. And requirements, particularly of external or final customers, may not be always explicitly stated and may remain implicit as expectations over above explicitly stated needs. And a process has to meet both needs and expectations of the customers, through appropriate improvements, whenever needed.

9.3   Process Quality

Process Quality is generally assessed in terms of the quality of process performance as compared to requirements spelt out in the process plan. And performance takes care of both the quality of conformance in terms of the agreement between process execution and corresponding specifications in the process plan as well as in terms of the output measures or values compared with the targets set up in the plan.

Analysis of costs associated with the execution and outcomes of a process yields a fair idea about process (performance) quality. Basically, we consider costs of conformance with the process plan and also costs of non-conformance and of the consequent outcomes. Costs of conformance depend partly on the stringency of requirements in the process plan, and we should check whether we are overdoing in this matter. In fact, stricter requirements and high conformity to those involve costs which may not be off-set by the direct and also the indirect gains from better quality and may not always be justified. That way deciding on the level of quality to be achieved becomes a very important managerial issue. It must be borne in mind that costs of conformance have to be incurred as something basic and our focus will be on reducing, if not eliminating, costs of non-conformance. In this context, quality of a process has been generically measured by the following quantifiable aspects of a process, broken up into work elements or items. There are seven generic ways of measuring process quality in relation to the corresponding process plan. These are

1. Defect—resulting from non-conformity with plan requirements and leading to process output not being accepted for the next-in-line process (or for packaging and dispatch). Defects may require.
2. Rework—to rectify the non-conformity so that process outputs are defect-free. In some cases depending on the number and nature of defects in a work item we may have to.
3. Reject—the work item and do it over again. These three measures relate to the outcome of the process as can be detected during process inspection.

Associated with process execution, we may encounter situations characterized by

4. Work Behind Schedule or Delayed Start—of the process element beyond the point of time stipulated in the process plan Such a situation could arise because of rework or rejection followed by repeat execution of the previous process element. In the absence of sufficient cushion time, such a situation is likely to cause.
5. Late—of the process output either in terms of the execution time being more than the specified time or because of work behind schedule caused most of by delayed arrival of the work item from the previous segment of the process.
6. Lost Items—items of work executed but corresponding outputs not properly preserved and not available for use at the times these are wanted. Sometimes, jobs done on a computer may get lost in the absence of a simple command 'save'.
7. Items Not Required—repeat performance of the same work item like a count or a check more than once, particularly in situations where one count or check is adequate, or some redundant work items not essential in carrying out the process and not adding any value to the process output are not too uncommon.

Apart from these seven generic ways to assess process quality—in an inverse fashion—process quality is also revealed in the process cost, considering both cost of conformance to process plan as well as costs of deviations or non-conformance leading to poor quality.

In some situations, process quality should go beyond yield or cost or wastes to cover energy consumed or emission per unit of product.

9.4 Model for Process Analysis

The selected business process (not necessarily a manufacturing or even a service process) should be first mapped properly to delineate its

(a) initial and terminal boundaries;
(b) hard and soft inputs fed into the process as inputs and hard as well as soft outputs coming out of the process;
(c) suppliers or providers of inputs—internal and also external, depending on the process—and customers—internal as well as external—to receive and act upon the output(s);
(d) the process of transformation of the input into the output;
(e) specifications for the inputs and targets for the output(s).

This exercise is usually referred to as Process Mapping.

Let us consider the case of a machine failing during production and taken for repair (involving replacement of some part or parts) to the repair facility within the production area. For the repair process, the hard input is the failed machine, the soft input includes information about the time the machine failed last, loading on the machine when it failed, nature and extent of routine care of the machine by the operator, and the like. The soft input could also include some advice about the time by which the repair process should be completed. The repair process will involve thorough inspection of the machine to identify the failing components and also the failed ones, if any, repair work using required materials follows. The hard output would be obviously be the repaired machine and the soft output could be in terms some advice to the user about routine maintenance and load to be put, a report on the time and cost involved in repair to be forwarded to management

The process mapping exercise exemplified above is often referred to as the SIPOC model and can be shown in terms of a customer–supplier relationship diagram. SIPOC is the acronym for Supplier, Input, Process, Output, Customer. A detailed process map can reveal value-adding and non-value-adding activities in the process. Such a detailed process map can provide four different perspectives about a process being considered for possible improvement, as indicated below.

- Perceived Process—What you think the process is.
- As Is Process—What the process really is currently.
- Should Be Process—What the process should be to satisfy the customer(s).
- Could Be Process—What the Process could be through appropriate actions.

While the 'Should Be Process' may be difficult to achieve, we have to reach the 'Could Be Process'. The extent to which the gap between these two entities can be bridged depends the nature and extent of resources which can be committed to planning, executing and monitoring and also correcting the process as is now.

Vale stream mapping also known as a material and information flow mapping is a variation of process mapping that looks at how value flows through a process and to a customer and how information flow facilitates the work flow. A process (particularly processes which involve quite a few steps or activities) is essentially a flow of activities. In this context, value is what customers want or need, and are agreeable and even able to pay for. On the other hand, waste is an activity that consumes resources, but creates no value for the customer.

Activities in a process can be classified into three categories, viz. (1) value added (or, value-adding) which have to be performed (2) non-value-added or wastes (also muda) which have to be either completely eliminated to at least reduced without affecting the process and (3) activities which do not directly contribute to any value but have to be retained for the sake of business, e.g. inspection, audit, use of control charts—sometimes referred to as business-value-added activities. In cases of new products or services, it may be difficult to distinguish between value-added activities and non-value-added activities, and we should carry out some focus group surveys or some similar exercises and should not depend on internal customers.

A value stream consists of all activities, both value-added and non-value added, required to bring a product from raw material into the hands of a customer, a customer requirement from order to delivery, and a design from concept to product launch. A value stream consists of product and service flows as well as information flow. Non-value-added and unnecessary steps that can be eliminated without consequences to business and the customer are sometimes referred to as muda of type II. The purpose of using a value stream as it exists now along with a desired value stream where non-value-added activities are absent or rarely come across is to pinpoint steps or phases where muda exists and effort should be made to eliminate or reduce such steps or phases, resulting in a cheaper and faster process that meets customer requirements.

The following is a process model in which the roles of internal resources committed and external influences affecting the process and its output, and also the checks performed on the process and the controls exercised to make the process achieve its targeted output, are clearly indicated.

For the purpose of understanding, controlling, improving or optimizing a process, we need to deal with data on inputs and outputs as well as on the conversion operations. Thus, we need data on input parameters like quality and quantity of input (hard inputs relate to materials, components or products while soft inputs relate to information regarding quality of input materials or about the state of an equipment when it failed etc.)

Process parameters including those of

(1) equipment to be used and the relevant parameters and
(2) environment under which the process is carried out.

E.g. feed-rate of input materials, speed of some machine, alignment between jaws in a cutting tool, time for the operation(s), ambient temperature and humidity etc. It is well known that the sample size and the acceptance number are the process parameters in a single sampling inspection process. Besides, the features of an item checked during inspection and the method of checking are also important process parameter in inspection.

Output parameters like yield, cost, quality, residual life or availability of the equipment used should also be clearly defined and measured. For the repair process, time to the next failure of the repaired machine could be an important output parameter. For an Inspection Process, number or fraction of defective items left in the lot and found during manufacture and causing problems would be a natural output parameter.

For each important process, there has to be a process plan identifying these different parameters, the way pertinent data have to be collected, and also the target and tolerances for each such parameter. Process Planning is an elaborate task involving analysis of data relating to the past or to the experiences of other organizations engaging similar processes as well as requirements of internal and external customers. In fact, unless a process is completely specified in a process plan, it will not be possible to assess its quality and, thus, to initiate steps for its improvement.

9.5   Approaches to Improvement

It should be appreciated that the plan for a process to be improved may itself require improvement, besides improvement in its execution. In fact, process planning in keeping with the objectives and targets and in conformity with constraints on resources is a very important task in process improvement. And as is true for planning in general or for any targeted feature, several concepts, tools and techniques are available for the purpose of planning a business process, To illustrate, it is better to go outside the realm of manufacturing or production to some critical process like realization of the billed amount from a customer. The plan here should specify when does one need to first remind the customer's concerned executive about non-payment, how quickly and effectively does one respond to any queries or some missing documents like test and inspection results or to some complaints raised by the customer, how and when should the producer's representative invoke any provision of penalty for late payment without any ostensible reason for holding back the payment, and the like.

Organizations which frequently use some of the following practices tend to have better performance manifested in improved quality. However, most organizations do use these approaches only occasionally. These practices are: Cycle-Time Analysis, Process Value Analysis, Process Simplification, Strategic Planning and Formal Supplier Certification programme.

One of the earliest and most discussed approaches for improving process quality is contained in the famous Deming Cycle, also referred to as the Deming Wheel. Below we provide a brief outline of this approach.

Deming Cycle (PDCA)

PLAN   Establish performance objectives and standards
DO   Measure actual performance (NOT JUST execute)
CHECK   Determine the gap between actual performance and performance objectives/standards
ACT   Take appropriate action to bridge the gap and make the necessary improvements

This representation relates to a given process to be executed by an organization. A somewhat more comprehensive explanation of the terms has sometimes been preferred. And in that context, the tasks denoted by the four words are as follows.

PLAN   Establish the objectives and the processes necessary to deliver results in accordance with customer requirements and organizational policies
DO   Implement the processes
CHECK   Monitor and measure process features against policies, objectives and requirements for the outputs and report the results obtained
ACT   Initiate actions to continually improve process performance

Let us consider the context of a given process, since the spirit of the more comprehensive approach can always be clubbed with the process-specific approach. The four words—or, better, constructs—in the Deming Wheel (as it is sometimes called) can be interpreted in different ways, possibly depending on the context. Of course, there has to exist a reasonable balance of emphasis on the four constructs.

The step 'Plan' in the context of process control, means to specify how a given process should be performed and how deviations from such a plan can be assessed quantitatively. The process should be controlled at the specified levels or values in respect of inputs, equipments, working environment and also output. However, in the context of process improvement to result in a greater conformity to the desired specifications or standards, 'Plan' means to work out ways and means to improve operations by finding out what is going wrong, causing non-conformity with standards and *proposing* solutions. Similarly, the word 'Do' does not simply mean to execute the process but really implies that one solves the problems causing deviations from the standards on a small or experimental scale first, to see whether the changes suggested in the solution will work or not. It is learning by doing, using trial-and-error to develop a *perception* about what will eventually lead to an improvement in process performance. 'Check' means to see whether the changes are achieving the desired results and modifying or refining them, if necessary. This word was later changed to 'Study'. Both 'check' and 'study' have connotations of putting changes into effect in an organization, a sense of *pulling people* in the direction of making quality improvements. 'Act' means to implement changes on a larger scale, to make them routine, to roll them out to the whole organization, a *push* activity.

Some authors point out an imbalance among the four constructs. 'Do' and 'Act' seem to be more active, while 'Plan' and 'Check/Study' are more investigative.

At the 'Plan' stage, one has to select the problem, schedule the activities, understand the current situation, set the target, analyse the problem and plan countermeasures. At the 'Act' stage, if the objectives are not met, one has to analyse the results further and plan countermeasures; if the objectives have been met, one has to formalize and standardize the measures adopted. At this stage, the whole exercise is to be reviewed and measures for further improvements in process are to be planned.

PDSA cycle has been dovetailed into the seven phases of corrective action which also phases are dependent on previous phases and define a cycle. Phase 1 is to identify the opportunity for improvement followed by the second phase to analyse the selected process and the third phase to develop the optimal solutions correspond to P in Deming Cycle. D in the cycle is the fourth phase to implement the solutions found in the previous phase. Phase 5 involving a study of the results on implementation of the improvement actions corresponds to Do in Deming Cycle. The last element in Deming Cycle covers two phases, viz. standardize the solution for sustained use and plan for the future.

For understanding and solving a quality problem towards achieving the twin objectives, the essential approach is to study the pattern of variation in the pertinent area, and properly quantify this variation. For this purpose, the following seven simple tools of statistical process control can be taken advantage of

- Cause-and-Effect Diagram or Fishbone Diagram or Ishikawa Diagram
- Check sheet to properly and comprehensively record variations.
- Pareto Diagram to prioritize the causes in terms of their effects;
- Histogram to get a visual display of the pattern of variation;
- Stratification to segregate data according to relevant groupings;
- Run Chart to detect recurrences of some type of defect or abnormality;
- Scatter Diagram to comprehend the nature and extent of cause–effect relations.

However, analysis of the patterns of variations and the causes behind those are more efficiently detected by appropriate statistical tools like Analysis of Variance (ANOVA) to identify groupings of the observations or measurements which differ significantly among themselves in respect of the mean values, Analysis of Covariance (ANCOVA) to identify significant covariates or concomitants which provide some explanation of variations revealed through ANOVA, Regression Analysis to find out the role of explanatory covariates and the like. Dealing with joint variation in several quality characteristics, we can gainfully make use of appropriate tools for multivariate data analysis like clustering or principal component analysis or multi-dimensional scaling.

9.6 Continuous Process Improvement (CPI)

Continuous process improvement is now recognized as the 'best' or possibly the 'only' way to come up with products and services that can be successfully marketed to customers with high expectations and easy access to a wide array of alternatives. This is a systematic approach to make both incremental as well as breakthrough improvements in processes. Through this, we take a hard look at the existing process in all relevant details and discover ways to improve the process elements. The end result is a faster, better, more efficient and cost-effective way to produce goods and services.

CPI is not a one-time investment. If the goal is to achieve total customer satisfaction, both internally and externally, CPI has to be an ongoing affair, a way of life.

Every quality philosophy speaks of process improvement and several distinct approaches to continuous process improvement though not differing much among themselves in the essential content. Possibly the PDCA (Plan-DO-Check-Act) Cycle due to Deming has been the most discussed and deserves a special mention. However, the role of customers (internal or external) has not been explicitly mentioned in the Deming Cycle, though a process that fails to provide full satisfaction to its customer(s) should be taken up as the opportunity to improve and to 'plan' for appropriate corrective action. The only point to remember as a difference between a process to be regarded as an opportunity to improve and one that fails to satisfy the customer(s) is the fact even a process that currently satisfies the customer (s) may be less efficient than expected and should be improved upon to enhance its efficiency. And there could be some process that fails to satisfy the customer(s) but cannot be taken up for improvement conveniently if a third party like a supplier is involved and an extra effort may be needed to cause improvement in such a process.

It must first be admitted that improvement takes place through corrective actions and preventive actions. Preventive actions are proactive in nature and should result in an improved process plan which then should be followed during process execution, while corrective actions are to be taken during process execution whenever deviations from the process plan are detected. A robust process design that prevents deviations during process execution exemplifies a preventive action to improve process quality.

Of the several approaches towards CPI, a very workable one is provided by the SAMIE model (Chang 1993). SAMIE is an acronym for Select, Analyse, Measure, Improve and Evaluate. Some may argue that the phases Analyse and Measure should be taken in the reverse order since Analysis will involve measurements. Each phase in this model consists of steps that will guide the potential user through a CPI cycle. Of course, the SAMIE model can be adapted to suit one's own improvement efforts and organizational requirements.

Phase: Select
1. define key requirements for core customers;
2. identify the process to be improved on a priority basis;

Phase: Analyse
1. document the process 'as it is' now
2. establish process performance measures

Phase: Measure
1. gather 'baseline' process performance data
2. identify and quantify process performance gaps

Phase: Improve
1. set process improvement goal(s)
2. develop and implement improvement measures on a' trial run' basis

Phase: Evaluate
1. assess impacts of process improvements
2. standardize process and monitor ongoing process improvements.

One can easily see that SAMIE does not differ much from the Six Sigma methodology using the DMAIC (or DMAIS) approach. The apparent difference that shows up is that SAMIE explicitly involves the step to evaluate or to assess cost-effectiveness of a proposed improvement (really a change or a solution to the problem analysed) before proceeding to standardize process at the changed level. Even in DMAIC, the last stage of controlling or standardizing does involve this assessment, though not explicitly emphasized.

Not all these steps need be followed in all situations, and sometimes several steps may overlap one another. The phase Select may not apply to a case where a particular process is known to be the first candidate for any possible improvement. Sometimes we may retrace our steps and revisit some earlier step. Thus, while documenting a process to be improved, we may find tat we need to look back and redefine the customer's output requirements before embarking on the Analyse phase.

The SAMIE approach need not be confined only to processes that are currently not performing satisfactorily. It can be applied even for processes which are currently productive. Here our target could be to achieve excellence.

The most important step is development of improvement measures. For each step, the actions to be taken and the common tools to be used are also indicated. As expected, most of these run through all improvement exercise plans in some form or the other.

9.7   Toyota Model for Process Improvement

Process improvement is concerned with making a process *better, cheaper or faster*. Better in that it delivers *higher levels of satisfaction to its stakeholders*, particularly customers. Faster in that it does so *as quickly as possible to increase responsiveness*. Cheaper in that it does it to the *highest levels of efficiency*.

The emphasis is on the *elimination* of all non-value-adding activities and the *streamlining* of the core value-adding ones. This approach can be summarized as **ESIA**: Eliminate, Simplify, Integrate and Automate

Eliminate

Elimination is concerned with eliminating 'waste' (sometimes referred to as 'muda') in processes. Waste or non-value-adding activities add cost to processes by consuming resources like time, physical and material resources and money but do not add value and so we need to work to eliminate these wastes. Besides these wastes are Mura or waste of unevenness or inconsistency and Muri or waste of overburden. In fact, Mura and Muri often lead to muda. Any process improvement exercise should eliminate or reduce these mudas There are many categories of waste.

The first seven items in this list of mudas are derived from Ohno within the Toyota production system.

1. Transport, e.g. movement of product between operations and locations which do not change and features of the product;
2. Inventory in terms of work-in-progress items and stocks of finished goods and raw materials that a company holds;
3. Motion or movement of material that does not change form, fit or function of the product;

4. Waiting, e.g. a semi-finished item waiting for the machine to finish,

Simplify

Having eliminated as many of the unnecessary tasks as possible, it is important to simplify those that remain. The search for overly complex areas should include the following items.

1. Forms
2. Procedures
3. Communication
4. Technology
5. Flows
6. Processes
7. Duplication of tasks
8. Reformatting or transferring information
9. Inspection, monitoring and controls
10. Reconciling.

The reduction or elimination of these non-adding value steps is the first target for any structured process improvement initiative.

Integrate

The simplified tasks should now be integrated to affect a smooth flow in delivery of the customer requirement and service task.

1. Jobs
2. Teams
3. Customers
4. Suppliers

Automate

Information technology and relevant process technology can be very powerful tools to speed up processes and deliver higher quality customer service. As well as core processes, the following items should be considered for automation:

1. Dirty, difficult or dangerous jobs;
2. Boring, repetitive tasks;
3. Data capture;
4. Data transfer;
5. Data analysis.

Note that automation should only be applied to processes that are completely specified in all relevant details and are such that the existing resources available, specially the human resource, do not permit convenient implementation of the processes.

9.8 Role of Kaizen

Continuous process improvement encompasses small improvements in process elements involving majority of the people involved as well as interested in a process wholly or partly, both for suggesting improvements and also for implementing the suggested changes that are found feasible by concerned management. In this context, Kaizen—a Japanese Quality Management construct—has been widely accepted and practised for implementing small improvements at many places continuously. Of course, big Kaizens are now quite well known to industry.

Kaizen is a Japanese word that means 'change for the good (better)'. Doing 'little things better' everyday defines Kaizen—slow, gradual, but constant improvement—continuous improvement in any area that will eliminate waste and improve customer (both internal and external) satisfaction. Kaizens continuously implemented along with 'breakthroughs' at intervals together improve process performance. In fact, Kaizens involving many skilled employees sometimes contribute more to process improvement than the major 'breakthroughs' worked out intermittently by seniors and highly rated professionals.

The three structural components of Kaizen are (1) Perceptiveness, (2) Idea Development and (3) Decision, Implementation and Effect. The first component relates to close look at current situations in minute details and admission of problems there. Problems may be obvious and noticed by all, or lying hidden and noticed by very watchful eyes. Awareness about problems and dissatisfaction with the present situation motivate the second component, viz. Idea Development. One has to look for causes, prioritize them if needed, think about the problem from all aspects and try out solutions based on new perspectives. The outcome is a concrete improvement proposal. The third component emphasizes on implementing and following up the results of implementation of the proposal.

A problem is the launching pad for Kaizen. Problems could be

Something that bothers us;
Something that causes inconvenience;
Something that has to be solved;
A discrepancy between the current situation and our objectives;
A discrepancy between our targets and actual results.

Problems could be noticed easily or discovered through vigilant eyes, or could be dug out (in case of Preventive Kaizen) or even could be created when new standards are accepted. Kaizens are important both in process planning and also in process implementation. Even, workplace improvement which contributes to process improvement can be effected through Kaizens on the basis of suggestions given by those who work and also those who frequently visit the workplace.

Kaizens derive strength from simple operator-friendly techniques like the 5-S principles which have yielded visible improvements in process execution. The 5-S principles engage people in standards and discipline. In fact, these promote standard operational practices to ensure efficient, repeatable and safe ways of working in a highly visual workplace. As is pretty well known, the 5-S corresponds to five Japanese words, viz. Seiko, meaning proper arrangement (sometimes replaced by Seiri meaning segregation of what are needed from those not needed), Seiton, implying orderliness or systematic arrangement, Seiso, meaning cleanliness or shining, Seiketsu, implying standardization and Shitsule, mandating discipline to sustain the standard. These have been translated slightly differently to stand as Sort, Set in order, Shine, Standardize and Sustain. In fact, this translation makes the impact of 5-S better understood in practice. In some contexts, safety is taken as the 6th S, though others argue that safety is a direct derivative of the 5-S principles and does not need a separate mention.

Each of these five words goes beyond its dictionary meaning. For example, SEIRI requires sorting things as needed or otherwise, clear unnecessary items and keeping records of things thrown away; SEITON implies straightening and simplification, setting in order, and configuring and includes visual management in terms of colour coding, arrow marks, earmarked danger zones, etc. The basic idea is 'every item has a place and everything is in its place'; SEISO implies laying out of proper drainage, adequate aisles and passageways, besides sweeping, scrubbing, cleaning and checking these; this ensures that everything including machines, tools and jigs, shop floor return to a nearly new status. SEIKETSU takes care of noise, ventilation and illumination, and more importantly implies standardization and conformity to standards, while Shitsuke mandates customization and practice of the standards developed. This, in fact, implies continual improvement. The advantages of 5-S are:

(a) Better utilization of space, equipment and other resources along with
(b) Improvement in safety and work environment.

Some western organizations, unwilling to adopt Japanese nomenclature, have branded these principles as 5C which are Clear, Configure, Clean and Check, Conform and Customize for sustained practice. Some even refer to the acronym CANDO implying Clean up, Arrange, Neatness, Discipline and Ongoing Improvement.

Implementation of 5-S is not to be delegated to shop floor or front-line workers only. Top management has to be involved. Usually, five levels of 5-S are recognized, and Kaizens are introduced in Level 3. A full PDCA cycle of activities has to be taken up at Level 5 to ensure proper results, which can be benchmarked against best practices.

The spirit of Kaizen as doing little things better everyday—slow, gradual but constant improvement—continuous improvement in any area that will eliminate waste and improve customer satisfaction—is brought out by the following poem

For the want of a nail, the shoe was lost

For the want of a shoe, the horse was lost

For the want of a horse, the general was lost

For the want of a general, the army was lost

For the want of an army, the battle was lost

For the want of a battle, the war was lost

For the want of a war, the country was lost

And all for the sake of a nail.

This brings out the possibility of a disastrous consequence of even a small defect by way of omission, or delay or deviation in an early stage of the process. An example of a Kaizen-type improvement is provided by a change in the colour of a welding booth from black to white to improve operator visibility. This change results in only a small improvement in weld quality, but a substantial improvement in operator satisfaction.

9.9 Kaizen Blitz (Event)

A recent format of the Kaizen exercise which is more target-oriented, time-bound (usually taking a week) and concentrated activity by a small group of workers using simple tools, more or less aligned to Lean Six Sigma, Kaizen Blitz or Kaizen Event has come to be widely accepted in the process improvement philosophy. A Kaizen event involves a set of specific actions with clearly defined goals for improvement —may not be very ambitious—along with the results obtained, verified and implemented. Starting with the objective to change the status quo in a process to avoid wastes or remove some glitches, the exercise terminates in an event to celebrate the new standard developed through the (usually) week-long effort. It involves a set of planned, simple and direct actions by small team dedicated to running the event throughout its course without interruption. It proceeds with the assurance that resources needed for the event will be available right now. If the change from the present state to an improved one seems to take longer-than-a-week effort, the problem will be split into two or more elements to be resolved in two or more Kaizen events. It should be noted that a Kaizen event includes implementing the change, assuring that the change can be managed and checking the improvement results after the change has been implemented before the changed state is taken as the new standard. Apart from the simple Five Why's and similar tools to understand the problem and to resolve the same, a value-stream mapping is often needed to eliminate or reduce wastes.

9.10 Quantitative Analysis Leading to Improvement

This is quite often the task taken up in 'off-line quality control'. This will definitely involve some experiments—trying out different values or levels of process parameters in different runs and noting the corresponding output parameters. In some sense, this appears to be a multi-factor, multi-response experiment. And we have to take recourse to Response Surface Analysis to fid out the best combination of process parameter values/levels, best to be defined in terms of values or levels of output parameters along with their respective weights or importance measures. The Analysis of data will involve a multivariate multiple regression which may be assumed to be linear or nonlinear depending on some soft input knowledge about the dependence.

Process analysis (starting with process Monitoring) is essential for all three. Process analysis recognizes (random) variations in output parameters corresponding to variations in input parameters in the presence of some given (uncontrollable) process features.

Usually, a single response variable (may be derived from or may be a combination of several output parameters) is chosen as the dependent variable to be explained in terms of its dependence on several factors (input parameters). Such a dependence study constitutes process analysis in the frequency domain. Given that controllable factor s or input variables are subject to random errors of measurement or can only be controlled within some ranges with some probability, such a relation is really a regression relation.

Let $\mathbf{Z}$ denote the vector of dependent variables or output parameters, $\mathbf{X}$ be the vector of input or independent variables and $\mathbf{Y}$ the vector of given (uncontrolled) variables. We are now required to estimate the regression coefficients in the regression equation

$$\mathbf{Z} = \mathbf{f}(\mathbf{X}; \mathbf{Y}) + \mathbf{e} \tag{9.1}$$

where $\mathbf{e}$ is the vector of errors associated with the regression equation. It will be quite in order if we assume a single dependent variable z and a linear relation between z and $\mathbf{Y}$ in the form

$$z = \alpha + \Sigma \beta_i X_i + e \tag{9.2}$$

where the regression coefficients β_i denote the influence of the process variable X_i on the output variable Z.

Given this relation, ignoring the error component, we can find out the values of components in the vector $\mathbf{X}$ which will maximize the response variables. This optimization exercise should take into account the fact that the given variables $\mathbf{Y}$ lie within certain ranges that are feasible and/or desirable. This estimation can be done by standard methods, viz. method of least squares or of maximum likelihood. Sometimes the regression parameters may have to satisfy some relations or constraints like $\beta_k > 0$ or $\Sigma \beta_i = 1$ or $\beta_k > \beta_l$. Constrained minimization of the sum of squared errors can be taken recourse to in such cases.

To generate data on the observable output and controllable input variables for sets of given variable values, we require to conduct a planned experiment on the process. Such an experiment has to be off-line and should be designed in such a manner that the response surface yielded by the regression equation above can be delineated with minimum amount of experimentation and consequent need for resources. Several such designs for optimum-seeking experiments including sequential designs exist to help experimenters in this regard.

A proper identification of the performance (output) parameter(s) of the process to be improved, followed by identification of factors—controllable as well as uncontrollable—which are known to affect the performance parameter(s) helps us design a screening experiment with several levels of the controllable factors and derive the regression equation of the performance parameter (in the simplest case) on levels of the controllable factors. Applying suitable tests of significance, we can find out the factors which exert significant on performance in terms of main effects of factors and interaction effects between factors and subsequently find out the combination of levels of these factors which yield the best performance, as closely as possible.

Such experiments are greatly facilitated by adopting Taguchi designs based on orthogonal arrays and linear graphs, coupled with the use of signal-to-noise ratio as the performance parameter rather than he classical mean response. This will take care of variability caused by uncontrollable noise factors and even by controllable factors which were left uncontrolled because of reasons like economy, etc.

It has to be remembered, however, that analysis of any experiment to study factor–response (performance) relation will provide a better alternative process design. Unless this alternative is found feasible for implementation within the existing framework and the available resources and is found to be cost-effective, such an alternative will not be accepted as an 'improvement'. In fact, cost-effectiveness has to be established—even by taking account of indirect or remote costs—before the results of any quantitative analysis suggesting a change in the existing process is accepted for implementation.

The SQC and OR Division of the Indian Statistical Institute carried out a large number of investigations for possible process improvement exercises in a wide array of industries throughout India. In most of the cases, results of industrial experimentation were accepted and led to process improvement. We reproduce the findings of one such study where performance of an existing could be improved by appropriate setting of the important factors. However, the change suggested was cost-prohibitive and hence not implemented.

This study was taken up in a company manufacturing Electrical Rotating machines with a view to reduce variability in the field current of an Alternator. The existing design was a proven one, and the company did not experience any field problems. Yet, it was decided to examine whether the design could be made more robust. There are quite a few uncontrollable or noise factors which affect field current variation and losses (stator copper loss, iron loss and rotor copper loss) in an alternator.

An experiment was carried out to evaluate the existing design with respect to field current variation, losses and material cost and to explore the possibility of arriving at a design that is relatively insensitive to variations in the levels of noises.

The designers identified fourteen factors, six of which are uncontrollable, for the purpose of studying their effects on fixed current and losses. Each factor was examined at three levels in this study. A parametric design, as proposed by Taguchi, was selected for the purpose of this investigation. The inner array (design matrix) is an O.A. $L_{36}(3^{13})$. The control factors, 8 in number, were randomly assigned to the three-level columns of this array. For each treatment combination of the design matrix (inner array), there correspond 18 combinations of the levels of uncontrollable factors in the noise matrix (outer array). Thus, there were in all 648 (= 36 × 18) treatment combinations. Obviously, it was not possible to conduct so many physical experiments.

The R&D department had a software package for performance analysis. For given values of the various input parameters, the computer gives a print out of the performance characteristics like field current, losses etc. It was decided to use this package to evaluate the performance characteristics for each of the 648 treatment combinations in the experimental layout. The responses noted were (i) field current, (ii) total loss and (iii) cost.

For analysis of the data, the following concurrent measures (S/N ratios) suggested by G. Taguchi were made use of:

(i) Field current variation.

$$H = -10 \log_{10} s^2$$

$$\text{where} \quad s^2 = \sum_{i=1}^{n} \frac{(y_i - \bar{y})}{17}$$

(ii) Losses

$$= -10 \log \frac{1}{n} \sum_{i=1}^{n} y_i^2$$

where y_i is the response of the ith combination of the noise matrix for a given combination of the design matrix.

The above concurrent measures (S/N ratios) were computed for each combination of the design matrix using a subroutine. Analysis of variance was carried out with the values of relevant concurrent measures for field current variation, total losses and costs.

The analysis showed that three factors—core depth (B), length of stator (C) and Max/Min. gap (F)—had significant effects on field current variation, contributing to as high as 93.5% of the total variation in field current.

The factors B and C were dominant (contribution: 71.1%) for total loss.

The factors A, B and C were significant (contribution: 96.6%) with respect to material cost. Other factors examined did not significantly affect any of the responses.

The optimum combination was identified as $A_2B_2C_3D_2E_2F_1G_2H_2$.

The existing combination $A_2B_2C_2D_2E_1F_1G_2H_2$.

The optimal design substantially reduces the variability in field current. However, material cost per alternator was found to increase approximately by Rs. 1,000. Since there were no problems with this existing design, the company did not opt for any modification of the design.

9.11   Poka Yoke or Mistake-Proofing

Shingo (1986) advocated that that errors in processes are identified as they happen and are corrected right away before serious damage occurs. Shingo proposed his version of zero defects known as 'Poka Yoke' or 'Defect = 0'. Rather than the outcome, viz. zero defect, the procedure to improve a process by recourse to mistake-proofing is to be emphasized.

Mistake-proofing or Poka Yoke, in Japanese, is the provision for design or process features including automatic devices to reduce inadvertent errors from creeping in or to enhance the probability of detecting any such error once introduced. Such errors are quite likely to arise in processes where the human element like operator's attention is important and unintentional errors may arise to create defects.

Mistake-proofing procedures start with a flow chart of the process, in which each step is examined for the possibility of a human error affecting the step. Such an error has to be traced to its source by looking at the entire process. Subsequently, potential ways to make it impossible for an error to arise should be thought of for each error. Some of the suggested actions could be

Elimination—doing away with the step that causes the error, if possible;
Replacement—replacing the error-prone step by an error-proof one;
Facilitation—making the correct action far easier than committing the error. This is often achieved by providing an automatic device like a sensor which will not allow the error to arise.

If it is too costly or inconvenient to rule out the occurrence of an error, we should think of ways to detect the error quickly and minimize its effects. Inspection method, setting function and regulatory functions can also be considered. Successive inspection is done at the next step of the process by the worker there. Self-inspection by the worker concerned to check his own work immediately after doing it and source inspection before the start of a process step to ensure that conditions are proper should also be considered as important actions in mistake-proofing.

Processes which are critical in terms of their impact on the final product or service and which involve some human element at least partly should be subjected to mistake-proofing through appropriate actions and/or devices. Both service production and service delivery processes illustrate this point very effectively.

Table 9.1 Predicted responses: existing and optimal designs

Combination	Field current variation	Field current loss	Total loss	Average material cost (Rs. 1000)
Optimal	1.04	18.7	10.0	16.40
Existing	2.21	25.2	10.6	15.24

Table 9.2 Some fail-safe devices and their functions

Types of fail-safe devices	Device functions
Interlocking sequences	Ensure that the next operation cannot start until the previous operation is successfully completed
Alarms and cut-offs	Activate if there are any abnormalities in the process
All clear signals	Activate when all remedial steps have been taken
Fool-proof work holding devices	Ensure that a part can be located in only one position
Limiting mechanisms	Ensure that a tool cannot exceed a certain position or amount

Source Borror (2010)

Borror (2010) cites the example of a manufacturer who finds that about 1 in 2,000 of its assemblies shipped is missing one of the 165 components assembled. Introducing the principle of Poka Yoke, a bar code is now used for each component and the manufacturer scans serial number and component bar codes as each component is added to the assembly. The software is so prepared that the printer in the shipping department will not print a shipping label if any component is missing.

Poka Yoke is a an important preventive action that ensures that problems or abnormalities in processes will be found out and removed as quickly as possible redundancy is sometimes taken as a safeguard (Table 9.1).

To prevent the occurrence of failures, certain fail-safe devices are used in some situations, and their costs justified in terms of repeated cost of failures averted. Table 9.2 gives out a list of some commonly used fail-safe devices along with their functions.

9.12   Failure Mode and Effects Analysis (FMEA)

FMEA is an analytical technique (to prevent problems) used by a team to identify and eliminate or, at least, reduce the negative effects of foreseeable failure modes of a product or a process and their respective causes before they occur in systems, products, services or practices. In this context, failure means inability to carry out an intended function (documented or otherwise obvious) of interest to some internal or external customer. Non-compliance with a regulatory requirement is also a failure. An exercise involving FMEA can be and is carried out on designs, processes, products (during use) and even systems. Process FMEA is an exercise based on a host of information derived from different sources to prevent failures during process execution. Process FMEA differs from Design FMEA or Product FMEA or System FMEA in some details particularly in working out the risks associated with different modes of failure but with the same basic principles. The focus is on potential failures and not on failures which have occurred, since the objective is to prevent failures from occurring.

This technique requires a sequential, disciplined approach by engineers to assess systems, products or processes in order to identify possible modes of failure and the effects of such failures on performance of the systems, products or processes. The objective is to identify the primary or root causes for any possible mode of failure along with any augmenting or indirect causes, so that failure in this mode can be prevented.

Situations calling for FMEA exercises are illustrated as follows

1. A metallic component meant to support a big structure **breaks down, the** structure collapses, and a **process stops** and/or **some injury is caused to some operator**. A search for causes detects a design problem of using a metal or metallic alloy with inadequate breaking strength, possibly due to inadequate information about possible load to be withstood during use.
2. A conveyor belt with steel cord reinforcement **catches fire** during operation, leading to **stoppage** of the belt and consequent **delay** in materials handling and also a possible **fire hazard**. Analysis may reveal too much friction with the material being conveyed. The belt could snap, because the material conveyed was much heavier than expected or taken into account during the design.

Sometimes, several modes of failure could be detected or were evident, caused by several different mechanisms and leading to different consequences of varying severity, and calling for different tools for analysis and analysis-based actions to reduce or prevent their occurrences. Resources are limited, and we need to prioritize these different modes of failure.

In essence, we have to (1) identify potential modes of failure, (2) identify causes for each mode through a brainstorming exercise, if necessary, and developing a Fishbone Diagram, (3) prioritize causes in terms of their consequences, using a Pareto analysis, and (4) initiate appropriate actions to take care of the more important modes of failure.

There are two primary standards for FMEA, the Military Standard MIL-STD 1629 A and the Society of Automobile Engineers Standard SAE J 1739. Both these are applicable only to design and process FMEA. Some frequently used tools of Quality Management during the FMEA exercise include: Cause-and-Effect Diagram or the Fishbone Diagram, Process Decision Programme Charts, Histograms, Pareto Diagrams, Run charts, force-field analysis, fault tree diagrams and root-cause analysis.

Process FMEA and its extension FMECA with Criticality of each mode of failure taken into consideration are meant to prevent failures during execution of a process. It is a formal, documented procedure that starts with an identification of different anticipated ways in which the given process may fail or deviate from the process plan. The question "how" an anticipated failure of a certain observable type or mode can take place has to be answered on the basis of past experiences or in terms of a detailed analysis of the failure mode. Possible modes of anticipated failures can be identified and documented in terms of a brainstorming exercise.

At this stage, we have to look for a set of conditions or factors which can lead to process to fail or deviate from the process plan in any particular given mode. Possible causes associated with the given mode of failure could include poor or wrong material, poor soldering, inaccurate gauging, inadequate or no lubrication, chip on locator, improper heat treat, inadequate gating/venting, worn or broken tool, improper machine set-up, improper programming, inability of the process equipment to meet specifications or some foreign body affecting the process, besides a host of others. A Cause-and-Effect Diagram can be of help in this task. Only specific errors or malfunctions should be listed and ambiguous phrases like machine malfunction or operator error should not be used.

Next comes the question of estimating the risk associated with each potential failure. Risk assessment is based on occurrence, severity and detection of a potential failure.

This is followed by a detailed exercise to trace a potential failure to its causes to ultimately come up with a plan for preventive action for the most significant risks. This analysis should be repeated until all potential failures pose an acceptable level of risk, acceptability being defined by the user. Then the findings and recommendations have to be documented and reported to appropriate authority for further action.

For each component of risk, a value on a ten-point scale is assigned and these values are multiplied to yield a Risk Priority Number. This estimation will be usually based on knowledge than on data since failures of any given mode are likely to be pretty small and given the too small number of occasions—sometimes zero—in which such a failure could be observed relative to the total number on runs of the process, the classical or Laplace definition of probability will not apply. A fault tree analysis using a Boolean representation of the failure event in terms of Boolean gates to cascade it down to possible causes taken as failures of some entity not functioning properly can provide a reasonable estimate.

This is followed by an exercise to estimate the severity of failure again on a 10-point scale and a similar scale for likelihood of detection of failure. Given these three scales, for a particular mode of failure one can calculate a Risk Priority Number RPN to indicate the relative importance of a given failure mode to attract attention for preventive action. Thus, RPN helps prioritization of different modes of failure. There is generally an accepted strategy for taking action based on RPN which becomes quite important when we have the same RPN for two or more potential failures. This is:

First eliminate the Occurrence, then reduce Severity, then reduce the Occurrence and finally Improve Detection. The following example illustrates the strategy:

Potential failure	Severity	Occurrence	Detection	RPN
1	6	5	8	240
2	8	6	5	240
3	4	10	6	240
4	3	8	10	240

In each case, the RPN is 240. In terms of severity, failures 1 and 2 are more important and of these two, failure 2 has a higher occurrence probability. For the last two failures, failure 3 has a higher occurrence and a higher severity and hence gets priority over failure 4. Thus, preventive actions should be taken against these failures in the following order:

First priority   Potential failure 2
Second priority   Potential failure 1
Third priority   Potential failure 3 and
Fourth priority   Potential failure 4.

Scales for Likelihood of Occurrence, Severity and (Likelihood of) Detection along with the criteria for different scale values as are applicable to process FMEA are more or less standardized, though minor modifications have been incorporated by some agencies or some National Standards Bodies. In fact, the two primary standards mentioned earlier also differ slightly between them in respect these criteria. More importantly, data needed and methods to determine likelihood based on the sparse data usually available are not clearly spelt out in either. Tables 9.3, 9.4 and 9.5 convey the commonly agreed criteria along with their rankings or scores.

9.12.1   Common Source for Tables 9.3, 9.4 and 9.5: BIS Draft Standard on FMEA

From the point of view of process (quality) improvement, identifying different potential modes of failure followed by a search for the causes thereof and initiating appropriate action to reduce the occurrence of failures (and if possible to eliminate some modes of failure) is the more important aspect of FMEA. Computation of risk and of RPN are, no doubt, crucial to comprehend the impact of different modes of failure and to prioritize these modes for remedial action. And here arise some limitations of FMEA in terms of application prerequisites.

It is rather difficult for the FMEA team or the experts there to assign numerical scores to risk factors, due to fuzziness and uncertainty in human thought.

Different experts may have thorough recognition on the scoring target, but their scores based on personal knowledge and experience may reveal some diversity.

The three risk factors carry the same weight, which may not reflect the actual reality in all situations.

Table 9.3 Process FMEA severity criteria

Effect	Severity criteria	Ranking
Hazardous without warning	May endanger machine or assembly operator. Very high severity ranking when a potential failure mode affects safe operation and/or involves non-compliance with regulation. Failure will occur with warning	10
Hazardous with warning	May endanger machine or assembly operator. Very high severity ranking when a potential failure mode affects safe operation and/or involves non-compliance with regulation. Failure will occur with warning	9
Very high	Major disruption to production line. 100% of product may have to be scrapped. Item inoperable, loss of primary function. Customer very dissatisfied	8
High	Minor disruption to production line. A portion of product may have to be sorted and scrapped. Item operable, but at reduced level. Customer dissatisfied	7
Moderate	Minor disruption to production line. A portion of product may have to be scrapped (no sorting). Item operable, but some comfort items inoperable. Customer experiences discomfort	6
Low	Minor disruption to production line. 100% of product may have to be reworked. Item operable, but some comfort items operable at reduced level of performance. Customer experiences some dissatisfaction	5
Very low	Minor disruption to production line. Product may have to be sorted and a portion reworked. Minor adjustments do not conform. Defect noticed by customer	4
Minor	Minor disruption to production line. Product may have to be reworked on-line, but out of station. Minor adjustments do not conform. Defect noticed by average customer.	3
Very minor	Minor disruption to production line. Product may have to be reworked on-line, but out of station. Minor adjustments do not conform. Defect noticed by discriminating customer	2
None	No effect	1

To tackle the first problem and the associated second, attempts have been made to use fuzzy similarity measures and likelihood theory (Mandal and Maiti 2014). Linguistic weighted geometric operator and fuzzy priority have been found to yield better results (Zhou et al. 2016).

9.13 DMAIC Route to Process Improvement

'Six Sigma' signifies the outcome of a highly disciplined, top-down, quantitatively oriented, customer-driven, and project-based approach to organizational performance improvement, usually through improvement in strategic business

Table 9.4 Process FMEA occurrence criteria

Probability of failure	Possible failure rates	Ranking
Very high: failure almost inevitable	>1 in 2	10
	1 in 3	9
High: generally associated with processes similar to previous processes that have often failed	1 in 8	8
	1 in 20	7
Moderate: generally associated with processes similar to previous processes that have experienced occasional failures	1 in 80	6
	1 in 400	5
	1 in 200	4
Low: isolated failures associated with similar processes	1 in 15,000	3
Very low: only isolated failures associated with almost identical processes	1 in 150,000	2
Remote: failure is unlikely. No failures associated with almost identical processes	<1 in 1,500,000	1

Table 9.5 Process FMEA detection criteria

Effect	Detection criteria	Ranking
Absolutely impossible	Not known controls to detect failure mode	10
Very remote	Very remote likelihood current controls will detect failure mode	9
Remote	Remote likelihood current controls will detect failure mode	8
Very low	Very low likelihood current controls will detect failure mode	7
Low	Low likelihood current controls will detect failure mode	6
Moderate	Moderate likelihood current controls will detect failure mode	5
Moderately high	Moderately high likelihood current controls will detect failure mode	4
High	High likelihood current controls will detect failure mode	3
Very high	Very high likelihood current controls will detect failure mode	2
Almost certain	Current controls will almost certainly detect a failure mode. Reliable detection controls are known with similar processes	1

performance solve business problems. Quite often—if not invariably—a business problem is associated with some core or support process failing to completely satisfy the concerned internal or external customer(s), as the case may be. A collection of interrelated processes is also recognized in this context as a process. The problem has been a nagging one, cannot be resolved by a single person in a short time. Hence, the need for setting up a project around the problem process to be taken up by a cross-functional team to work for some six months or so to work out a desired solution which can fix the problem in the process investigated. As against Kaizens which are meant to solve mile-long but inch-deep problems in the process to work out incremental improvements, Six Sigma methodology is applied to solve a mile-deep but inch-long problem, which when solved, results in a breakthrough improvement that can be sustained.

The route to process improvement in this approach has been encapsulated in the five steps involved in applying this methodology, viz. Define (D), Measure (M), Analyse (A), Improve (I) and Control or Stabilize (C or S). This route emphasizes the importance of the Voice of the Customer (VoC), measurements of process quality, use of Normal distribution, Regression relations, and the like. The whole idea is that Improvement as an activity has to be preceded by Definition (of the problem in all relevant details for example baseline situation), Measurement of different process variables and of yield (fraction of non-defective items or units) and Analysis (of yield in relation to process variables). To derive benefits from improvement, improvement has to be stabilized in terms of holding the process at the levels indicated by the Improvement exercise.

DMAIC is essentially an improvement system involving the following steps

DEFINE—project goals and customer deliverables based on Voice of Customer (VOC) and the business problem

MEASURE—process to evaluate current performance in relation to customer requirements

ANALYSE—root causes for poor process performance

IMPROVE—process performance and eliminate defects by devising and evaluating multiple solutions. Pilot solution and compare performance

CONTROL—improvements, through plans to sustain desired performance.

The following are some important concepts used in DMAIC.

- Critical to Quality (CTQ)—Attributes most important to the customer (internal or external)
- Outside In Approach—Looking at internal processes from the customer's perspective and changing them to satisfy the customer
- Defect—Any event/situation that does not meet the specifications of a CTQ attribute
- (Defect Opportunity)—Any event/situation that provides a chance for not meeting customer requirements and which can be recognized and counted
- Defective—A unit of product that contains one or more defects
- Transfer Function—$Y = f(X_1, X_2, X_3, \ldots)$ where Y is the dependent or response variable and Xs are independent predictor variables which control the performance of Ys. The focus of Six Sigma is to control X's to achieve laid down standards for Y's.

Broadly speaking, we need the following types of analysis and hence the corresponding tools and techniques. (1) Dependence Analysis, taken care of through Categorical data Analysis, Regression and Correlation studies. (2) Analysis of factor–response relations in terms of ANOVA /ANCOVA along with their multivariate generalizations and Response Surface Methodology. Optimality Analysis in terms of methods to reach the optimum point on the response surface as closely as possible.

As the very name Six Sigma implies, this approach focuses on reducing variability or increasing consistency in product features so that coupled with a proper setting of the process. The fraction of defectives in the output units (could be small but distinguishable parts of an item that may cause customer dissatisfaction) is extremely small like 0.00135 or even smaller.

In the step 'Improve', we have to (1) generate alternative solutions to the problem by identifying suitable combinations of levels of contributory factors or by looking at requirements or specifications for some qualitative process parameter as given in the process plan (2) evaluate the solutions in terms of the output levels and select the best one, (3) assess risks in implementing this solution and develop the pilot solution in all relevant details, (4) develop implementation plan, (5) plan and test actions that should eliminate or reduce impact of identified root causes and (6) plan how to evaluate results in control.

9.14 Benchmarking and Beyond

Benchmarking has been used as a general business practice since around 1980, starting with IBM and Xerox in USA. The urge to perform at least at the same level as the competitors got crystallized in the form of benchmarking. And benchmarking has benefitted quite a few manufacturing and also service organizations in improving their business processes and also performance. Usually, process benchmarking and performance or results benchmarking are considered as the two important areas where this practice has been adopted; some people talk of benchmarking competitive advantages and of strategies. Four philosophical steps in process benchmarking have been mentioned by several authors. These are (1) know your operations, (2) know the industry leaders or competitors, (3) incorporate the best practices and (4) gain superiority. Camp (1995) refers to four types of benchmarking, viz. internal, competitive, functional and generic. Internal benchmarking focuses on best practices within an organization across different areas or divisions or departments. Of course, benchmarking for improving process quality mandates the same or similar processes to be carried out in the best-practice situation. Competitive benchmarking provides a comparison among direct competitors, most of whom may not like to share information regarding distinctive key features of their processes. Functional Benchmarking is across units within the particular industry or even going outside that generic benchmarking focuses on innovative work processes in general.

Whichever formal definition for benchmarking like any of the following

The continuous process of measuring products, services and practices against the toughest competitors or those companies recognized as industry leaders or
The search for and implementation of best practices or

A systematic approach by which organizations can compare themselves in certain selected areas or dimensions of their operations or results against the best-in-class organizations can measure and analyse gaps and initiate necessary improvement actions

is accepted, the common strands are

Gathering various types of business information—derived from primary data collected through direct interactions allowed by the benchmarking partners as well as from secondary data gathered from published documents or put in the public domain.

Creating new business knowledge by analysing specifics of various business factors of competing companies, specially those known to perform better, and comparing those with corresponding factors in the present organization.

Using this knowledge to come up with new decisions for controlling such factors and to develop actions to improve processes and results.

The goal of process benchmarking is to gain knowledge about the characteristics of planning, designing, executing and controlling and even evaluating various business processes and activities involved in the processes selected for improvement by which competitors successfully implement their strategies. The goal is to improve the process(es) in the given organization.

Since benchmarking involves comparison with other competing companies, an important problem is to choose the right companies against which to benchmark the selected process(es). The general advice is to consider

Other companies within the same group—internal benchmarking

Competitors within the industry

Other companies in the industry which are not direct competitors and

Other companies in other Industries.

The last one is more applicable in the case of process benchmarking where a particular process like space utilization or interacting with customers in an industry that offers a completely different product or service profile but is known to perform excellently well in the selected process. Of course, such a choice of the benchmarking partner requires a deeper knowledge by decision-makers in the company to use benchmarking about the given process and of the organizations which are required to perform this process and which have performed with a high level of efficiency.

Xerox in 1980s studied their direct competitors among Japanese companies to discover that

Unit manufacturing cost equalled the Japanese selling price in USA

Number of vendors was nine times that of the best companies

Assembly line rejects were 10 times higher, product lead times were twice as long

Defects per hundred machines were seven times higher

Xerox started process benchmarking seriously and effected the following changes:

Number of vendors was reduced from 5000 to 300 only.

Concurrent engineering practice was introduced.

Commonality of parts was increased from 20 to 60–70%.
Cross-functional teams were put in place.
And the results were surprisingly huge in terms of
Quality problems cut by two-thirds
Manufacturing cost cut by 50%;
Development time reduced by two-thirds;
Direct labour cost dropped by 50% and corporate staff by 30%.

In essence, a benchmarking exercise provides inputs for process improvement, and in most cases for breakthrough improvement. These inputs along with others like findings from FMEA have to be converted into feasible corrective and preventive actions to improve process quality. Actions being used in the best-in-class organization may not apply mutatis mutandis to our organization, and the task is to work out suitable modifications so that such actions are suitable for our organization. And evaluation of results on completion of one round of benchmarking is a must to ensure that the organization remains where it was at that point in time.

Creative elements have to be injected within and beyond current practices prevailing in the concerned industry to achieve improvement in a situation where most of the units or organizations have already reached their limits of technological and managerial competence. These extensions of benchmarking require serious creative thinking or breakthrough followed by appropriate actions. This implies the need for thinking out of the box, as opposed to thinking in the box (finding an existing best practice and adopting that with or without adaptation). In the first case, we need creative thinking to move beyond current best practices.

In this direction DeBono (1992) refers to six hats in creative thinking and to six action shoes. While the need for breakthrough improvement was emphasized earlier by Juran, the approach espoused by DeBono avoids usual arguments and counter-arguments during never-ending discussions often resulting in nothing concrete to push the organization or the processes towards improvement.

The white hat has to do with data and information and corresponds to questions like

What information do we have now about a process to be improved?
What relevant information is missing currently?
What information would we like to have for our improvement effort?
How are going to get the additional information?

The red hat has to do with feelings, intuitions, hunches and emotions. In fact, intuitions have on occasions helped the origin of creative ideas. Even misgivings about possible failures if a suggested idea is implemented can be expressed by wearing the red hat.

The black hat is a 'caution' hat that prevents people from making mistakes from working on ideas that may cause violation of regulations inviting penalty, causing decrease in profits, increase in costs and fall in markets.

The yellow hat provides boost to positive thinking and to optimism about the outcome by looking at possible ways to take care of adverse consequences indicated by the black hat and working out feasible directions to put a promising idea to practice.

The green hat provokes participants to explore new, beyond exploring feasibility of ideas already suggested and criticized objectively. The green hat raises questions like

Can we identify some additional alternatives?
Can we do this process in a way different from what has been suggested earlier?
Can we have another explanation for consequences of an idea that came up earlier?

DeBono argues that the green hat provides space and time for creative thinking to flourish.

The blue hat takes us to the next step in our thinking process and, in some sense, controlling that process to come up with subsequent actions to be thought out for implementing the finally agreed upon idea for improvement.

Thinking has to be followed by appropriate actions. We wear shoes to reach some destination and DeBono speaks of six colours of shoes for six types of action. The navy shoe stands for formal, routine action. It covers formality, routines and procedures. The grey shoe represents Investigative Action and includes exploration, investigation and collection of evidences. The purpose of this action is to gather useful information. The brown shoe corresponds to Enterprise Action. Practicality and pragmatism mark this action which implies 'Do what is sensible and practical'. The orange shoe stands for Emergency or Crisis Action. In case of danger and emergency, such an action is to be designed to ensure safety and security as the prime concern. The pink shoe defines human values action and involves care, compassion, and attention to human feelings and sensitivities. The last is the purple shoe that marks leadership action. It clarifies the need for leadership, authority role and command chain. The person here is not acting in his or her personal capacity, but in an official role.

9.15 Concluding Remarks

Process improvement holds the key to quality improvement and eventually to improvement in organizational performance, when applied to all business processes—core as well as support. And the key to process improvement lies these days with improved process planning, given that competence in implementing a process plan is already in place. Process planning has to be target-oriented and, at the same time, resource-based. It should take due advantage of relevant tools—mostly quantitative—and of relevant information about processes to be improved from sources wherein such processes are carried out. During process execution, appropriate preventive actions including devices have to be introduced to prevent the process failing to meet plan requirements and, that way, to meet requirements of internal and/or external customers.

Such requirements have often to be found out from the voice of the customer and not just visible or directly available. Improved process monitoring to detect deviations from an improved process plan as soon after such a deviation takes place also plays an important role.

Improvement in process design contributes the most to improved process quality. And this improvement preceded process implementation, followed by process monitoring and control, and process changes for the better. Improving process design can take advantage of many inputs and also various techniques. Thus, Quality Function Development (QFD) holds out a promise for process improvement as part of an exercise to meet customer requirements for products. One can conveniently use QFD in a particular process to satisfy requirements of internal customers.

References

Automotive Industry Action Group. (2001). *Potential failure modes and effects analysis*. MI: Detroit.

Bogan, C. E. (1994). *Benchmarking for best practices: Winning through innovative adaptation*. New York: McGraw Hill.

Borror, C. M. (2010). *The certified quality engineer handbook* (pp. 342–343). Milwaukee, Wisconsin: ASQ Press.

Bureau of Indian Standards. Draft Standard on FMEA

Camp, R. C. (1989). *Benchmarking: the search for best Practices that lead to superior performance*. Milwaukee, WI: Quality Press.

Camp, R. C. (1995). *Business process benchmarking*. Milwaukee, USA: ASQ Quality Press.

Chang, R. Y. (1993). *Continuous process improvement*. USA: Kogan page.

DeBono, E. (1992). *Serious creativity: Using the power of lateral thinking to create new ideas*. New York: Harper Collins.

Geber, B. (1990). Benchmarking: Measuring yourself against the best. *Training, 27*(11), 36–44.

Harrington, H. J. (1995). *High performance benchmarking: 20 steps to success*. New York: McGraw Hill.

Imai, M. (1986). *Kaizen*. New York: McGraw Hill.

Indian Statistical Institute. (1997). *Quality case studies* (pp. 281–283). Indian Statistical Institute: SQC & OR division.

Kolarik, W. J. (1999). *Creating quality: Process design for results*. New York: McGraw Hill.

Mandal, S., & Maiti, J. (2014). Risk analysis using FMEA: Fuzzy similarity value and possibility theory based approach. *Expert Systems with Applications, 41*, 3527–3537.

MIL-STD 1629A. (1980). *Procedure for performing a failure mode, effects and criticality analysis*. Washington DC: Department of Defence.

Nadler, G., & Hibino, S. (1994). *Breakthrough thinking*. Rocklin, CA: Prima Publishing.

Ohno, T. (2008). *Toyota production system*. Productivity Press.

Ohno. (2008). *Workplace management*. Productivity Press.

Pryor, L. S. (1989). Benchmarking: A self-improvement strategy. *The Journal of Business Strategy., 10*(6), 28–32.

Pyzdek, T. (2001). *The Six-Sigma handbook*. New York: McGraw Hill.

Shingo, S. (1986). *Zero quality control: Source inspection and the Poka Yoke system*. Portland, OR: Productivity Press.

Walleck, A. S., O'Halloran, J. D., & Leader, C. A. (1991). Benchmarking world class performance. *The Mckinsey Quarterly, 1*, 3–24.

Womack, J. P., & Jones, D. T. (1996). *Lean thinking: Banish waste and create wealth in your cororation*. New York: Simon and Shuster.

Zhou, Y., Xia, J., Zhong, Y., & Jihong, P. (2016). An improved FMEA method based on the linguistic weighted geometric operator and fuzzy priority. *Quality Engineering, 28*(4), 491–498.

Chapter 10
Quality in Higher Education

10.1 Introduction

Discussing quality in the field of education is treading on a slippery ground. The absence of a consensus definition of education coupled with the enigma of quality creates serious problems in this context. The inspiring idea of Swami Vivekananda that 'education is the manifestation of the perfection already in man' is too philosophical to allow assurance of quality in education imparted by and in an institution. Similarly, a statement like 'quality is a way of life' maybe a laudable one, but may not be of any help in the context of quality assurance in education. Education corresponds to a wide spectrum—covering formal, non-formal and informal education on one dimension, primary, secondary and tertiary or higher along a second dimension, liberal versus professional education along possibly a third dimension, besides similar other differentiations.

  Two disturbing and noticeable features of the prevailing education system in many developing countries and even in some developed countries have caused worries to their educational planners and administrators. These are a deterioration in the standard of education (particularly at the primary and secondary levels) and the growing unemployability of many pass-outs from the tertiary-level institutions. The growing mismatch between the world of education and the world of work has its contribution recognized in the social fabric of a country. Of course, these disturbing trends may not be visible to the same extent in different types of education (such as liberal vs. professional) at different levels (such as elementary, secondary and tertiary) in all countries (or even in different parts of the same country characterized by

This chapter draws upon some material from the author's previous work published in Total Quality Management & Business Excellence, Vol. 6, 571–578. (1995) Thanks are due to the publishers Taylor & Francis.

different socio-economic-politico factors. At the micro-levels, problems maybe more numerous as well as more serious in some educational institutions than in others.

Engelkemeyer categorized the shortcomings of present higher education systems as (1) lack of competent and committed teachers and consequent fall in quality of teaching, (2) curricula for different not updated from time to time to reflect recent developments in the subjects, (3) curricula in a programme not balanced against curricula in previous stages, (4) excessive cost causing difficult access even to the deserving and (5) growing and inefficient administrative bureaucracies, coupled, in some cases, with undesired interference by political and social leaders. One could easily add to this list in specific contexts.

There is a strong belief that higher education or post-secondary education has entered a new environment in which quality plays an increasingly important role. Fiegenbaum believes that 'quality of education' is the key factor in 'invisible competition between countries' since the quality of products and services is determined by the way that 'managers, teachers, workers, engineers and economists think, act and make decisions about quality'. Many exponents of quality with special interest in education strongly feel that Total Quality Management with appropriate orientation has significant potential to address and even resolve to some extent challenging problems plaguing the world of higher education in a globalised and yet highly competitive world.

With recent advances in methods and practices to improve quality in services and the growing emphasis on use of methods and techniques for this purpose, there has been a feeling in some quarters that quality in higher education can be discussed within the broad framework of quality management. There are others who strongly resent this idea, on the plea that quality in education has got much to do with human behaviour, while Quality Management has been occupied primarily with behaviours of machines, materials and methods. Further, identification of stakeholders and their needs and expectations and subsequent incorporation of such elements in a quality policy guided by a vision or a mission statement are being attempted only very recently and on a very limited scale. These and many more such considerations render quality in higher education a complex entity to comprehend, or control or improve.

In the present chapter, some thoughts on the topic expressed by the author as also by several other exponents have been presented briefly to focus on the problems involved, rather than on methods or procedures claimed to solve such problems.

10.2 Quality in Education

The Summary of Declarations made in the World Conference on Higher Education (Paris, 1998) states—among other items—that 'quality in higher education is a multidimensional concept, which should embrace all its functions and activities, teaching and academic programmes, research and scholarship, staffing, students, buildings, facilities, equipment, services to the community and the academic environment'. Particular attention should be paid to advancement of knowledge through research. Higher education institutions should be committed to transparent and external evaluation, conducted openly by external specialists. However, due attention should also be paid to specific institutional, national and regional contexts in order to take into account diversity and to avoid undesired uniformity. There is a perceived need for a new vision and paradigm for higher education, which should be student oriented. 'To achieve this goal, curricula need to be recast so as to go beyond simple cognitive mastery of disciplines and include the acquisition of skills, competencies and abilities for communication, creative and critical analysis, independent thinking and teamwork in multicultural contexts.'

According to Gola (2003), the definition of quality, as applied to higher education by the International Organization for Standardization (ISO), could be 'specifying worthwhile learning goals and enabling students to achieve them'. The basic goal should be to enrich the student in terms of mental and moral development and enable the student to accept and discharge appropriate responsibilities at home, at workplace and in the society at large.

Specifying worthwhile learning goals would involve articulating academic standards to meet:

  i. Expectations of society through acquisition of moral and ethical values
 ii. Aspirations of students about roles they like to play in the society and the economy
iii. Demands of the government, business and industry; in terms of competence to man different tasks and
 iv. Requirements of professional institutions along with institutions for higher education.

Enabling students to achieve these goals would require good course design, an effective teaching/learning strategy, competent teachers and an environment that enables learning.

The quality of higher education is determined by the relevance (fitness of purpose) of its mission and objectives for the stakeholder(s) and the extent to which the institution/programme/course is also judged by the extent to which it satisfies the minimum standard set for inputs, processes and outcomes, which is called the standard-based approach to quality. It must be remembered that standards relating to different processes or activities carried out by an institution in relation to the various educational programmes offered by it—whether these are externally set, e.g. by an affiliating body or are internally developed to meet its own vision and mission requirements—are dynamic in nature. Based on contemporary changes in the field of education, the availability of new educational resources and the changing demands for knowledge, skills and attitudes of the participants in the process of education as well as more and more stringent criteria put forth by accrediting authorities, Standards have to be continually revised.

Relevance in higher education should be assessed in terms of the fit between what society expects of institutions and what they actually do. For this, institutions and systems, in particular in their re-enforced relation to the world of work, should base their long-term orientations on societal aims and needs, including the respect for cultures and environmental protection. Developing entrepreneurial skills and initiatives should become major concerns in higher education. Special attention should be paid to higher education's role of service to society, especially in activities aimed at eliminating poverty, intolerance, violence, illiteracy, hunger, environmental degradation and disease and to activities aiming at the development of peace, through an interdisciplinary approach. To make matters complicated, societal aims and needs have been changing—swiftly in recent times—and responsiveness of the higher education system to such changes becomes important in judging relevance of higher education as one major aspect of its quality. Quality has thus become a dynamic concept that has to constantly adapt to a world whose societies are undergoing profound social and economic transformation.

Harvey and Knight (1996) mention five different but inter-related ways of thinking about quality in higher education based on the project taken up by the University of Birmingham. These can be thought of as several connotations of 'quality' in the context of higher education, viz. excellence, consistent conformity (to Standards), fitness for purpose, value with a price tag, and self-development of the learner. The following provides some amplification of these connotations.

(1) Quality is associated with the attribute 'exceptional', and that way with the recognition as 'distinctive', by way of exceeding high standards and as passing a set of required standards, where 'high' standards and 'required' standards being somewhat context-specific and not generic.

(2) Quality is consistency as has been espoused even for manufactured items, specially of the processes involved in relation to corresponding specifications, to be achieved through a 'zero defects approach and a quality culture'. Given that consistency of conformance to some of these specifications which relate to student actions and achievements can rarely be predictable, this consistency is a questionable proposition.

(3) Quality is relevance and adequacy for the purpose. The point that this purpose maybe the achievement of an abstract all-time universal set of attributes or could be set forth by different stakeholders in their respective perceptions makes it difficult to spell out this relevance and adequacy or fitness for purpose (Crawford and Shutler 1999).

(4) Quality is 'value for money through efficacy and effectiveness', clearly linked to accountability and calling for credible indicators of performance. Available indicators are neither unique nor comprehensive to beget confidence.

(5) Quality is recognised through the transformative ability of education to promote learners' cognitive development as also to enhance their creative and emotional development.

Quality, particularly in the context of higher education, is a quandary—difficult to define but appreciated by all, not directly amenable to measurement but quite often subjected to discussions and initiatives for control, assurance and improvement. Different perceptions about quality are quite expected in relation to diverse goals and objectives meant to be achieved in different socio-economic and cultural backgrounds. Recently, attempts are being made to incorporate ideas about quality and Quality Management as are embodied in the generic ISO 9000 standards. While there are good reasons for this, there are some fallibilities also. It maybe better if we draw support from the more flexible Total Quality Management approach to analyse, assure and improve quality in the field of education maybe better if we draw support from the more flexible Total Quality Management approach to analyse, assure and improve quality in the field of education.

It will be too mundane and too simplistic to look upon an education system (or any segment thereof) as an industry, even if one takes a broad view of the term 'industry' to imply an enterprise which procures (from vendors), processes (by itself) and provides (to customers). A crude analogy could consider student entrants as raw materials, teaching and evaluation as processes and graduates or pass-outs as finished products.

Quality in education in terms of excellence is more a consequence of creativity fostered in an environment of flexibility, rather than an expected pay-off from the religious (and not necessarily imaginative) implementation of some rule-based decisions.

10.3   Components of Quality

In the established field of Quality Management relating to products and processes initially and subsequently to services and systems, quality is comprehended in terms of three important components, viz. quality of design, quality of conformance and quality of performance. Again, we speak of quality due to product support as supplementing quality of performance. It maybe worthwhile to identify these components in the field of education, even if one argues—and not without reasons—that quality in the context of education need not invite a standard definition or a standard analysis. Diversity of approaches and analyses is just expected. It should also be borne in mind that 'quality in education' as is imparted in an educational institution or as is received by the students there maybe different from 'quality of education' as is perceived by the participants or assessed by the concerned stakeholders including the society and the national need for persons with requisite domain knowledge, skills and proper attitude towards work.

Speaking about quality in education, imparted by an educational institution, the vision–mission statement along with concretely spelt out goals and objectives for different functions carried out by the institution maybe taken to define the quality of design. In fact, this design must first recognize the different processes under its control and then set up standards or norms for each process, segmented into distinguishable sub-processes whenever necessary. Important processes relate to admission of students, recruitment of teachers and non-teaching support staff, development of curricula for different courses offered, periodical as also final assessment of students' performance, adoption of appropriate pedagogy, interaction with parents/guardians, as also with the state administration and employment market, review of standards and procedures at suitable intervals, etc. Some of these processes may not be relevant in situations where the corresponding decisions are taken by an external authority and passed on to the particular institution. The fundamental idea is to work out appropriate decisions whenever this is left to the institution in such a way that the goals and objectives can be achieved duly.

The process approach which is basic to the generic Quality Management System Standards IS/ISO 9001 can be implemented imaginatively with necessary modifications with some benefits in institutions of higher education. Guidelines for this purpose prepared by the Bureau of Indian Standards mention the following processes, not all of which will be present in all the institutions.

Strategic Management Process for determining and establishing aims and policies of the institution, keeping in view the existing societal needs

Recruitment and promotion of faculty
Development and updating of courses and curricula
Production and distribution of learning materials
Creation and maintenance of facilities and conducive work environment
Selection and admission of students
Follow-up and assessment of education provided to students
Final assessment for grant of degree/diploma/certificate of competencies
Providing support services to students
Organisation of co-curricular activities
Monitoring and review of programmes
Measurement of relevant features of educational processes
Review, corrective and preventive actions to improve the overall performance
Career counselling and placement activities
Training and professional development of faculty and support staff
Planning and executing internally funded as also externally sponsored research

Quality of conformance relates to the nature and extent of concordance of practices and procedure actually followed with the relevant norms and standards. For each process to be planned and implemented by the institution, there has to be a detailed and documented plan developed by involving top management as also the concerned functionaries. Quality of conformance can be found out through periodic checks and audits or reviews findings whereof should be properly documented and discussed with the concerned personnel to take suitable corrective actions as also to initiate some preventive measures to avoid some more deviations surfacing in the near future. A real problem is that in the arena of higher education, conformity cannot be strictly enforced, since conformity to set standards is regarded as less desirable than excellence achieved possibly through a creative effort that does not fall in line with the prescribed standards.

Quality of performance is generally judged by the performance of students in public examinations, in the employment market and in the whole society. Results in public examinations by students completing their courses, acceptability by employers and entrepreneurship of pass-outs and acceptance and discharge of responsibilities by them as competent persons on their respective jobs and as good citizens in the society are some indicators of quality of performance. This is also revealed through the performance of teachers in terms of their contributions to knowledge creation and dissemination. Collectively speaking, this quality determines the effectiveness, efficiency and productivity of units engaged in production of goods and services—both in public and private sectors.

The way opportunities for refreshing and augmenting domain knowledge for pass-outs can determine quality due to product support. This quality can be indirectly assessed by the upgradation of skills, enhancement of (domain) knowledge and orientation of attitudes towards discharge of responsibilities as revealed in terms of performance of pass-outs before and after participation in continuing education programmes.

10.4 A Word About Professional Education

While the idea of dichotomizing education as liberal versus professional may not be acceptable in educational philosophy, the distinction has taken deep roots in reality. In terms of the background and the career options of the entrants, the infrastructure existing or desired on the campus, the method of teaching and evaluation, fees to be paid by students and perquisites extended to faculty, placement interviews for students on campus, and the like mark out professional education as distinct from liberal education. Of late, we find a spurt in activities that need professionally trained people, causing in its wake a rush among school-leavers to join such professional courses.

Competition is the buzzword and we notice cut-throat competition—sometimes unscrupulous—for admission, for private tuition, for positions in the list of successful examinees, for placement in jobs, for promotion to better jobs and higher positions, etc. This is quite evident in the fields of engineering and management education in many developing countries, meant to equip the participants with not only knowledge and skills but also with a frame of mind that can enable them to provide leadership in their respective workplaces.

Professional education has to be concerned with cross-societal demands and challenges. Values, in the context of professional education do involve some context-specific virtues like patience and perseverance, sensitivity and empathy, commitment and diligence. Some of these virtues maybe branded as leadership skills. All these values can best be instilled during education and training by teachers who realise the importance of these values and have imbibed these, at least to some extent.

Currently, emphasis on ethics and values has been incorporated in the curricula of professional education in many institutions. However, the ambience in the concerned institutions, the process through which students enter these institutions, sometimes bending rules and cutting corners, the role models put up before them during interactions with captains of industry and other-related arrangements tend to place the students away from the core values. We really need a wholesome change in our education system to inculcate values among our students so that they can squarely face challenges in life and achieve the success they deserve. Values and ethics constitute an important component of quality in education, in general.

In the next section, we take up two issues, viz. customer orientation, commonly considered in industry as a factor promoting quality, and involvement of teachers which is universally regarded as the cornerstone of quality in an education system.

10.5   Customer Orientation

Education is definitely a crucial service that benefits students directly and the entire society, economy and polity indirectly. Institutions of higher education which impart education to those who enter their portals along with organizations which are in place to support them in terms of finance, expert advice and manpower as also to regulate and control their activities plan and deliver one of the most important public service, viz. education. These service providers in the public or the private sector have to deal with a wide array of individuals and institutions who expect and receive some services directly or otherwise from them. Some may abhor the use of terms like providers and customers—generally applied in the world of business and industry—as somewhat derogatory in the arena of education. It must be remembered, however, these terms are no longer repugnant even in the sphere of education. And institutions engaged in the noble task of providing good quality education to those who deserve it have to deal with suppliers of goods and services in the usual business on industry world. Unless the quality of such goods and services is assured, institutions will not be enabled to offer quality education, which is becoming more and more demanding on physical and material resources of high quality.

Customer focus or customer orientation in all its activities—as a major element of TQM—is acclaimed as having led to improved quality (of products and services) in many manufacturing as well as service industries. In this context, identification of customers, both internal and external, characterization of customers' stated needs and implied expectations, orientation of organizational activities to satisfy such needs and expectations and assessment of customer satisfaction or otherwise have to be comprehensively understood. And, in the context of education, all these pose significant conceptual and operational problems.

If the system under consideration is, say, a secondary school accommodating several classes or grades, one may look upon a higher grade/class (represented by the latter's faculty, say) as the (internal) customer for a lower grade/class, while the university (one to which the school is affiliated or some other university which admits pass-outs from this school) or even the employment market maybe considered as external customers. One may also view parents or guardians or even donors/patrons as customers. For the University Grants Commission, customers could possibly be the beneficiary or affected university or college. For a university, the employment market (encompassing entrepreneurs) is the most important customer.

Are such characterizations of customers (of the education system) absolute or unique? Can we not regard students taking a public examination as customers looking for some service from the examining body? Is it not true that—as in the context of industrial quality assurance—every individual or organization is simultaneously a provider and a customer of educational services? A teacher, as one who provides education, is also a parent who, as a customer, expects some tangible results from his/her children's education. A teacher as one engaged by an educational institution can also rightfully expect and receive some compensation and some respect from the management of the institution. The latter in this case is the provider and the teacher is a customer. A university procures resources and services from the state/national educational bursars and authorities, on the one hand, and provides education to its students, on the other. In the complex process of teaching and learning, both students and teachers are providers as well as customers. Teachers stimulate and students respond, and vice versa.

For a given system like a university (the provider of service), students as customers maybe denied the role of specifying their needs by themselves (of course, they have their expectations about benefits to be derived from university education in the labour market or the sphere of self-development). The university may feel that even parents and guardians as customers need information, education and assistance to formulate their needs, which the university should satisfy. The university may not even appreciate the reasons for satisfying the requirements of skill formation as needed by the organized sector of the employment market. In fact, the university may well argue that its activities should be oriented to the 'creation, conservation and communication of knowledge' as enshrined in its charter, and customers (whoever they are) should orient their needs and expectations accordingly. Incidentally, there exist other institutions which are meant to enhance skills. Of course, the university should accept the need for changing its goals and objectives (as incorporated in its charter) in harmony with changing national needs and aspirations.

Given that there are various customer groups for an education system, are the needs and expectations of these different groups mutually consist? Not always. Even different requirements of the same customer group may appear to be mutually contradictory for a system, unless resources at the disposal of the system are heavily augmented or the nature of service (rendered by the system) and the corresponding procedures are drastically simplified. This is the case with large-scale public examinations, where students, guardians and society as a whole expect efficient, credible assessment and certification by the examining body within a reasonable time.

This does not mean that an education system can continue to be serenely oblivious of the needs and expectations of different segments of society—clear or vague, just or otherwise, modest or ambitious. The imperative necessity is to review these needs and expectations and revise the system's goals and objectives suitably, from time to time. The cryptic motto (of a university, say) should be amplified to formulate a quality policy that embodies goals and objectives, on the one hand, and strategies and plans, on the other. Through the hierarchical structure of the university system, this policy should be deployed right down to the task of each teacher, each student, each support staff member, and so on.

10.6   Involvement of Teachers

It is well appreciated that teaching–learning constitutes the most important process that ultimately determines the quality of education. It should be understood that just teaching by the teacher(s) in the classroom or the laboratory is not enough and one cannot look upon learning by students as a second distinct process. These two are inextricably linked one to the other. And if we accept the idea that the process of teaching is meant for the students, quality of teaching has to take due account of the academic and mental make-up of the students. However, in most discussions on quality in higher education, it is presumed that students have the right background and the right attitude towards learning and what matters in terms students' performance in public examinations as well as in their work lives is 'quality of teaching'.

'Quality of teaching' is not amenable to dissection into elements followed by integration—as could be the case with materials, products and services. This is particularly so, given that teaching concerns humans with diverse backgrounds, current needs and future aspirations. The 'consistency' implication of quality cannot be expected to hold water in the context of the teaching–learning process. There has been a general impression that institutions of higher education are not delivering the 'right' type of education to the large majority of their students. It must be admitted, in the same breath, that there exist institutions which continue to impart quality education to their students. There could be a rider here. Students who enter into these latter institutions succeed in an entrance examination and are mostly keen to learn. Given the ever-increasing number of students and correspondingly of institutions as well as of teachers and support staff members, concerns about quality of teaching have been voiced in many quarters.

In this context, it maybe wise to attempt an explanation of this unsatisfactory performance of a majority of our teaching organizations in terms of the famous Ishikawa model for 'quality of working life'. (Readers will please bear with me if I leave the research components in the activities of our institutions for higher education beyond the purview of the present portrayal.) According to this eminent exponent of Japanese quality management, collective performance of an organization depends on individual performance besides physical and material resources at the disposal of the organization for deployment to carry-out different activities in the manner desired or planned. Individual performance (Ishikawa continues to argue) is determined by the individual's ability on the task to be performed, coupled with his/her motivation to perform. Probing further into the matter, Ishikawa proposes that an individual's ability to perform is the composite of his/her knowledge (in the relevant field) and skill (to communicate knowledge to others and/or translate knowledge into action), while the two components of motivation are attitude (towards the organization, towards the profession and towards job being performed) and work environment (in terms of organizational culture, peer relations, access to requisite facilities, etc.). Knowledge without the requisite skill to transfer or use it, or no self-knowledge but some skill to translate or use another's knowledge, will not make for an individual's ability. 'The gift of the gab' alone does not make for an able teacher, nor does introverted scholasticism. The model clearly indicates that only a person with adequate knowledge in the subject(s) to be taught along with appropriate communication and interaction skills to facilitate learning by the students, having a positive attitude towards teaching as a profession, and possessed of pulling on well with peers and administrators can qualify to be a well-performing teacher. Of course, facilities like laboratories and library, aids for projection and demonstration, etc. should be available to enable the teacher to perform to his/her capability.

Thus, the basic determinants of the performance of an academic institution are the knowledge of the teachers (in the subjects/topics they teach rather than the ones in which they specialized), their communicative and interactive skills, their attitude towards the teaching profession per se and the teaching they do, the work environment that characterizes the institution as well as others which influence or affect the former, and lastly the physical, material and technological resources available to the institution.

Many thinkers may not contribute to the views of Ishikawa (who links components with a product symbol to signify that an absence of one component cannot be compensated for by excellence in another), and come up with claims that a band of dedicated teachers makes up for scant resources, that the profound knowledge of a teacher creates the desired impact on students despite the former's unimpressive delivery of lessons, or that keeping its own house in order even in the midst of poor surroundings (not physical) is not an impossible task for a college or university. One must remember that such cases are exceptions, which exist only to prove the rule or the prevailing but unfortunate situation.

We sometimes forget that training programmes, meant to update or refresh or enrich knowledge as well as enhance skill, may also lead to a change in one's attitude towards a given job. A job that seems to be challenging or which matches one's level of knowledge and skill at present may subsequently appear as routine or monotonous or as drawing upon a lower level of knowledge and skill, once the individual reaches a higher ability level. A favourable or positive attitude in respect of a job may turn into a negative one and may thus lead to lower motivation resulting in poorer performance.

The performance of an individual teacher or research worker or administrator on an assigned job or task in a given organization depends not merely on his/her ability to do the job but also on the extent to which he/she feels motivated to do the job. We assume that physiological needs like job security are taken care of. Ego or self-actualization needs have to be satisfied through such motivation. We can recognize attitude towards the job as being challenging versus routine, easy versus difficult, rewarding versus demotivating. Also important as a factor is the attitude towards the organization/environment. And this attitude is moulded through a better and clearer perception of organizational constraints, of links—existing and potential—between the organization and society, between organizational performance and societal development.

Mobility in jobs from easy to difficult, from usual to unique, from routine to challenging and from underrated to fully rated (in terms of remuneration, rank and reward) must be ensured, wherever possible. Otherwise, investment in training goes waste and more significantly negative attitudes develop.

Environment, spoken of by Ishikawa, essentially meant work environment prevailing within the institution or agency. However, this environment has to take into account external influences exerted by different stakeholder or interest groups on teachers, support staff members and academic administrators. In fact, such influences, in some cases, imply regulation and control over different processes planned and carried out within an institution of higher education by some external authority or even by some political or social groups.

We need dwell on Ishikawa model to emphasise the fact the performance of a teacher in the teaching–learning process is a complex outcome of several factors—some manifest, others latent—which should be identified, maybe uniquely in a given context, and steps have to be initiated to orient these factors in a way that ensures their positive contributions to this performance.

10.7   Role of Evaluation

Evaluation of performance is an integral part of education, providing desired assurance about quality of human and material inputs as also a basis for improvement in the processes involved. For admission of entrants to the institution—and that way to the system—also for recruitment of faculty and non-faculty support staff, we carry-out various evaluation exercises. We have to carry-out an evaluation of students passing out of the institution and entering into the world of work. The purposes for such evaluations may differ and obviously the forms of evaluation. However, tests are common instruments for evaluation and are often augmented by interviews. Criticisms of such tests are not that rare and not in all cases such criticisms are unwarranted. And the quality of these evaluations does affect the quality of education, whatever be the implication of this 'quality' accepted by us.

An important evaluation exercise in a situation characterized by more aspirants for admission than the number that can be accommodated in a course or a programme is the admission test, carried out by the institution concerned or by any related agency that conducts tests for admission to various institutions. Once admitted to a programme, we have periodic evaluation to assess how the participants have been progressing. Subsequently, we have the end-of-course evaluation. Sometimes, this evaluation takes into account performance of an individual in previous periodic tests besides the test at the end.

For entrance or admission purposes, evaluation should involve prognostic tests with a high predictive validity, where validity can be established in terms of the correlation (based on past data) between grades in the admission or entrance test and grades in the final end-of-course evaluation. Such a test need not possess a high discriminative ability, all that is needed being to discriminate those who are fir for entry from those who are not. This fitness depends on special skills and interests, e.g. in drawing, in experimentation, in exploration, in histrionics, besides a favourable attitude towards any distinctive features of the intended programme or course. In fact, given that tests of knowledge acquired in concerned areas at the previous level of education are dependable, entrance test need not duplicate the knowledge component of fitness. Unfortunately, this duplication is quite common in many countries.

Periodic evaluation should be diagnostic in nature, with the primary intention to identify inherent deficiencies in students and in pedagogy or the relevant teaching–learning process in enabling students to receive and absorb lessons properly. Validity of such a test can be indicated by differences in results of two consecutive tests (between which some remedial action is presumed to have bee initiated), confirming the persistence of some deficiencies or reduction or elimination of some. Discriminative ability is not much expected for such a test.

The end-of-course test should be assessing levels of knowledge and skills acquired by the assesses. Validity is usually visible in terms of future performance of the pass-outs. Of course, face validity is crucial for the image of the institution and concurrent validity against some current reference norm maybe attempted. And, in many countries, such tests are desired to have high discriminative ability.

It should be evident that construction, and even administration, of these three types of tests should be different one from the other. Contents must be relevant to the purposes and scoring or grading must also follow different principles and have different ranges. Unfortunately, what happens in usual practice—with a few exceptions that stand out differently—is that all these three become just assessment tests, with little ability to prognosticate or to diagnose. Those involved in planning and preparing the tests should devote more time to ensure appropriate content of the tests along with appropriate grading manuals.

Apart from evaluation of performance by current students, we also need to evaluate performance of teachers as facilitators in the process of learning. In some cases, feedback from students in prescribed forms is used as an input in this process. Just like admission tests for students, some institutions require an applicant for a teaching position to make a presentation on a topic of his/her choice before a group of existing teachers in the domain to judge the potential ability of the individual.

There is also a need to evaluate courses and programmes to be offered by an institution in terms of the contribution by a course or a programme to acquisition of knowledge per se or to employment generation or to economic, social and cultural development. Evaluation of performance of institutions will be covered broadly under the section on accreditation. However, such evaluation leading the award of ranks at national or international level globally across all types of institutions or within specified classes of institutions are also being carried out by several agencies.

10.8   Internal Quality Assurance

There are two types of quality assurance programmes in the context of higher education—internal and external. Internal quality assurance ensures that an institution or programme has policies and mechanisms in place to make sure that it is meeting its own objectives and standards. External quality assurance is performed by an organization external to the institution. The organization assesses the operation of the institution or its programmes in order to determine whether it meets the agreed upon or predetermined standards. Obviously, internal quality assurance is the key.

Each institution should develop its mission statement after adequate deliberations among its stakeholders to spell out its long-range objectives in the context in which it currently operates as also the context in which it is likely operating in foreseeable future. Consistent with its documented mission, it should develop objectives and targets for each of its processes like admission of students, recruitment of teaching and support staff, teaching–learning, evaluation of students' performance, planning and execution of research projects, interactions with the society at large, interactions with stakeholders including regulatory bodies, collaboration with other institutions including potential employers, prospecting for sponsors to support innovative programmes and practices and similar other activities. In some cases, it maybe desirable to break down a process into several sub-processes to be considered individually. Thus, admission of students to liberal education courses and to professional courses or to distance education programmes usually demand different procedures and hence each of these should be regarded as a distinct process by itself.

For each process, the need for a process plan with a focus on quality of output should be appreciated as the basis for internal quality assurance. All concerned people should be involved and their commitment secured to put these plans into practice. Management must review conformity with the plans at regular intervals and take appropriate corrective or preventive actions should be taken at the earliest opportunity. Sometimes, based on actual results achieved through conformity to process plans relative to the desired or expected results, the appropriate action maybe to modify the process plan itself.

Quality of any such process is to be understood and controlled only in relation to requirements specified in the corresponding process plan. The plan for any process, to be monitored during execution and controlled to ensure conformity with the plan, must give out clearly the criteria for any decision, the procedures to be followed in executing an activity, the time required to complete an activity, etc. It will be necessary to have numerical targets for any parameter associated with the input or the process or the outcome. We have to remember that in the field of education, the inputs are human beings with all their diversity in attitudes and aptitudes, as also their entry-level knowledge and skills reflected somewhat inadequately in the results of examinations passed by them before an try into the educational institution under consideration. The same is true for the outcome, since 'value addition' in terms of augmentation of knowledge and skills during the process of teaching–learning is beset with many conceptual and measurement problems.

Coming to process planning, the problem of specificity of the situation under consideration is quite important. Consider the process of selecting students for admission to a particular course in an institution with a number of seats prescribed exogenously by a regulatory body. Only in a few specialized courses, this number maybe decided on internally in keeping with aims and objectives of the institution and the resources—physical, material and human—available. In the case of a reputed institution, the number of applicants for admission will exceed the stipulated number to be admitted and we have to specify selection criteria and procedures. And the plan may specify if simply scores obtained in the previous public examination—total and/or concerned or relevant subject(s)—will be considered against a threshold set up in the plan or if a test will be administered and scores in that test alone or along with scores in the previous public examination will be taken into account. A vexing question in some contexts could arise when applicants come through several board/university examinations and the score distributions are not identical. One can possible take equi-percentile scores in such a case, provided score distributions for all candidates clearing each board/university examination are available. The last date for admission has also to be specified in the plan. The number selected should be somewhat larger than the number to be eventually admitted. Here also a problem may arise if the better candidates selected turn up later than the mediocre ones and all the seats are filled up to exclude admission of the better candidates. Whenever discretionary decisions have to be taken, the plan must specify who, how and when can such decisions be taken. For example, if some duly selected candidate pleads inability to pay admission fees or comply with other admission formalities, the case should be referred to the appropriate authority. In fact, the process plan may even mandate an admission committee to recommend modes of selection and to clear names for admission. For admission to specific courses, applicants may need counselling to advise them on appropriate choices keeping in mind their aptitudes and attitudes. If necessary, a differential aptitude test maybe administered to identify the course best suited to an individual applicant.

Equally important is the process of identifying and appointing competent and committed teachers and support staff members. Faculty selection is an important and a difficult process. Attitude towards teaching and research, empathy for students and colleagues, adherence to norms, sincerity of purpose and commitment to values and ethics should be given due importance in selecting staff members. No doubt competence in terms of domain knowledge and communication skill will be basic requirements. Beyond, selection, staff members should be involved in some orientation programme with two distinct components. The first and a generic one should cover topics like educational planning, evaluation methods, communication skills, emotional intelligence. While the second should be a specific component to convey to the participants' distinctive features of the particular institution, its strengths and weaknesses, its vision and mission and even its problems. The orientation programme should focus on strong cohesiveness among the faculty members.

By far the most important process to be planned for is the teaching-learning process which should spell out how to involve students in this process actively, by encouraging discussions and debates, literary and social service activities, group learning and similar other practices. An important exercise is the development of a curriculum in consonance with a prescribed syllabus, wherein the number of lecture hours, of tutorials of practical classes (wherever relevant), of periodical tests and of homework would be specified. Teachers should encourage the habit of reading textbooks as also reference books among students. They should not provide complete class notes or similar materials that would send a wrong signal to the majority of students that a perusal of those notes will suffice for the purpose of learning the subject, beyond passing the corresponding examination. Teachers should devote differential times to explain concepts, derivation of results, applications and potential uses. Students should be encouraged to address their classmates on some topic in the presence of the teacher who should act as a resource person to help the speaker. Teacher–student interaction in academic matters beyond classrooms to resolve doubts and to facilitate assimilation of what has been discussed within the classroom or even to augment the latter is an important means to improve the quality of teaching–learning. Teachers should take upon themselves the task of acquiring new knowledge if that is warranted to satisfy some inquisitive and bright students willing to know beyond what is prescribed in the syllabus or contained in the books referred to or discussed by the teachers.

Coming to another important process, viz. assessment of students' performance, it is worth remembering three different types of tests to be developed and administered to capture three different facets of performance during three different time periods. The first is a prognostic test to determine the ability of a candidate to undergo and complete successfully a given course to which admission has been sought. Such a prognostic evaluation should be focused on entry-level knowledge about the subject(s) concerned, attitude towards the course and the institution as also specific abilities and interests required by the course. Thus, laboratory work or outdoor experimental work may not be liked by some candidates. Then, there are the periodic evaluation exercises which, by design, should reveal or diagnose shortcomings among some students to receive and assimilate lessons or conduct experiments. These problems could be traced in some cases to deficiencies in teaching. Maybe the teachers concerned are quite competent to deal with the subjects or topics taught, but they may not always orient their teaching to the interests and aptitudes of their students. In fact, periodic tests are rarely designed and delivered to be diagnostic evaluation. Finally comes the end-of-the-course assessment evaluation of the gain in knowledge and skill acquired during the course. Generally, these assessment tests are more or less properly developed to serve the purpose.

Here, the process plan should specify the nature, content and frequency of periodic evaluation exercises as well as the actions to be taken by all concerned on the basis of their findings. Diagnostic evaluation should become a part of the pre-admission selection of students.

The process related to student support services has to get an input from each student about his/her physical and mental health through a proper mechanism. A mechanism to provide counselling and such other services as can provide some relief to the students concerned must also be specified. In situations where campus interviews are not carried out, some guidance about possible placements has to be in place.

Identifying learning resources needed to maintain a reasonably good quality of education, acquiring those in right time and in right quantities as also maintaining such resources to ensure ease of access, readiness for use (operational readiness) and proper and timely upkeep and updation. At the same time, efforts should be made to encourage regular and adequate use of such resources by students and teachers alike. The library should be well-stocked with print and digital books and should allow users on-line access to journals. Necessary reprographic facilities should also exist to facilitate authorized copies of reading materials. Similarly, laboratories for different science subjects should have all requisite equipments and materials. Equipments have to be calibrated at regular intervals. And compliance with good laboratory practice requirements should also be maintained. A language laboratory may also be built up to help students and even teachers develop their communication skills.

A plan for the general administration process should spell out mechanisms involving teachers and support staff to ensure all-round discipline. We have to ensure that classes are held regularly and on time and that all concerned students do attend classes.

Once plans have been developed for all the important processes, our next task is to ensure conformity to the respective plan requirements and initiate appropriate actions whenever deviations are noted through internal quality control checks. The nature and frequency of such checks and the people to do this should have been specified under the process of general administration.

As is true generally, quality of the process plans in terms of comprehensiveness and expected effectiveness, followed by quality of conformance to the plans together constitute quality of

Assessment of conformity to process plans has been generally left to independent teams within the institution who are not involved in the processes being assessed. Such independent assessments are referred to as Quality Audits in the language of quality management. It must be borne in mind that audits are checks of conformity to plans already in existence and not checks of absolute quality in the processes. There are international standards on audit procedures and related issues. Audit is not a fault-finding exercise, rather it is a confidence-building activity. And the stress in internal quality assurance is on maintaining conformity to process plan by the concerned people in the first go.

To speak of process quality to be reflected in a process plan, one may note the following seven generic ways in which quality is measured inversely in a process and the fact that each of these refers to a certain standard or norm stated in the process plan. These are

Defect or deviation from the specified requirements, e.g. attendance of students not noted on a particular occasion as a cumulative number

Rework involved in removing the deviation, once detected, e.g. a periodic test scheduled on a given day skipped and conducted sometime later, maybe clubbed with the next test

Rejection in case rework is infeasible or unable to remove the defect satisfactorily, e.g. some part of the prescribed syllabus not covered in time and cannot be covered with the time available without creating additional class load on students

Delayed start of an activity behind schedule, e.g. a class test beginning later than the scheduled time

Late delivery of the output likely to cause work behind schedule in the next step, e.g. submission of mark sheets for the last periodic test not submitted in time for the final mark-sheet to be prepared

Lost item of work which cannot be recovered in time for its use, e.g. an answer book-marked duly but somehow misplaced and not found before submission of mark sheets in due time

Items of work not required or redundant steps in the process, e.g. making multiple copies of mark sheets manually.

The following presents a schematic diagram (Diagram 10.1) connecting inputs, processes and outputs in an institution of higher education.

10.8.1 Dimensions of Quality Assurance in Higher Education

Emphasising on equity, efficiency, relevance and effectiveness of education imparted by an institution, we can develop some structural relations among the different elements involved in a given institution, looked upon as the system, as under (Diagram 10.2).

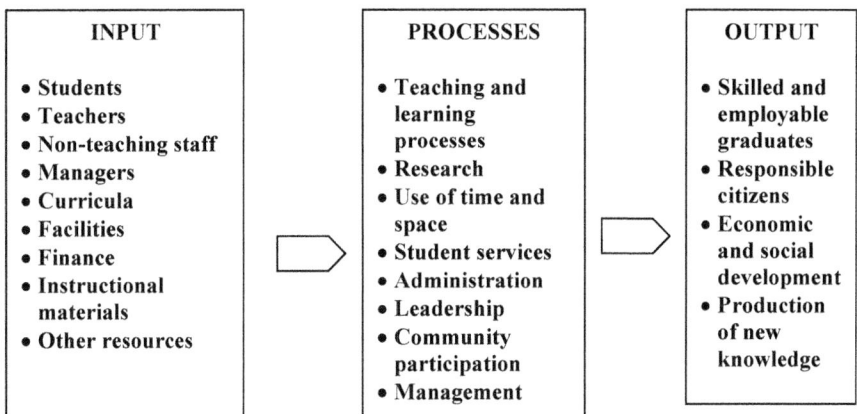

Diagram 10.1   Input, processes and output in an education system. *Source* Author's own construct

10.9   Accreditation

Accreditation has become a major issue for higher education over the past few years. The development of new technologies, progress in distance and virtual education, multiplication of new providers, attempts to generalize higher education into a commodity, internationalization of higher education and, as a consequence of all these factors, the need for trustworthy systems to ensure quality of relevance of institutions and programmes. Measure at the national and international levels is being justified in order to ensure quality and protect countries and students from unacceptable products and from fraudulent providers of educational services.

Accreditation is the most widely used method of external quality assurance in the field of higher education. In fact, it goes beyond an external quality audit to identify non-conformities in the institution or programme in relation the stated mission and objectives; it usually assigns a credit rating to the institution or the programme seeking accreditation. Accreditation ensures a specific level of quality according to the institution's mission, the objectives of the programme(s) and the expectations of the different stakeholders, including students and employers.

Sanyal and Martin (2007) point out that accreditation will ensure (1) quality control (at least at minimum standards internally in higher education institutions), (2) accountability and transparency, (3) quality enhancement and (4) facilitation of student mobility. In addition, this exercise will provide confidence to government and potential users of the pass-outs for national development objectives. Accreditation is commissioned by a suitable and recognized agency and encouraged by stakeholders to ensure, 'value for the money'. They also identify seven types of accreditation being carried out in different countries. These are

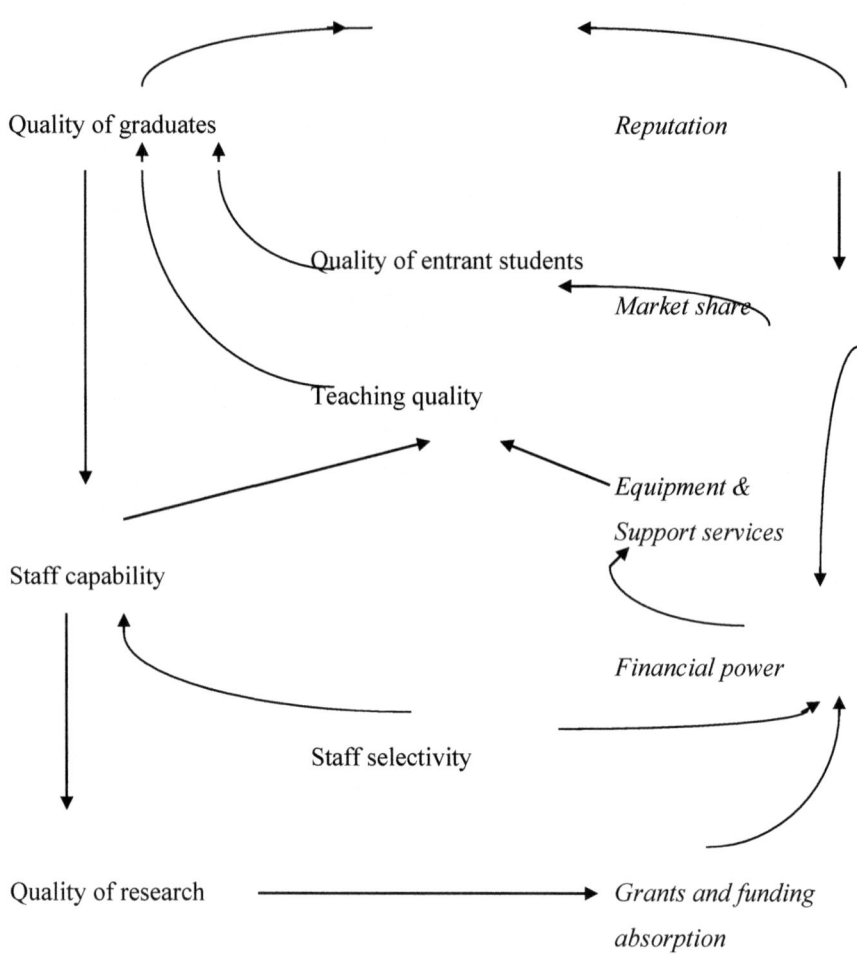

Diagram 10.2 Industrial effectiveness of graduates. *Source* Author's lecture in NASI symposium, Lucknow

(a) Fitness for purpose versus standard-based approach
(b) Voluntary versus compulsory accreditation
(c) Accreditation by geographical coverage

    (i)   Accreditation at the sub-national level
    (ii) Accreditation at the national level
    (iii) Accreditation at the regional level
    (iv) Accreditation at the international level

(d) Accreditation by control of higher education.
(e) Accreditation by type of higher education (University and Non-University Institutions)

(f) Accreditation by unit of analysis like individual programmes offered
(g) Accreditation for distance learning higher education

The process of accreditation is carried out in different ways in different countries, following different criteria, involving different agencies and making different uses of the credit rating. Some of these problems have pointed out by van Ginkel and Dias (2006) in a GUNI publication. However, there is a broad consensus that has been gradually unfolding with regional, continental and even international organizations engaged in developing similar accreditation practices for their constituents. The following accreditation criteria are quite widely accepted across countries:

Mission, governance and administration, human resources, educational programmes, academic standards, quality of learning opportunities, quality management and enhancement, research and other scholastic activities, community involvement, and consolidated development plans. Specific criteria for accrediting special categories of institutions and of programmes are also used.

Accreditation attracts many criticisms—particularly by those who look upon the process as an infringement on the autonomy of HE institutions and a reflection of the system's failure to tackle problems of inadequacies and resulting poor-performing institutions. In fact, each country should have a strategy to ensure that accreditation is accepted by the stakeholders as a transparent, credible and meaningful exercise to improve the status of higher education.

Distance Education institutions pose serious problems in their accreditation many of the measures used in traditional accreditation do not apply to online institutions, among these are the institution's full-time faculty strength, the number of volumes in the library, and the amount of time that students are in class. How can the equivalence between a course offered on a distant mode and a traditional classroom-based course on the same subject be established remains a big question. What happens with a university that has no campus poses a genuine problem about the type of accreditation based on geographical coverage? Even if a campus does exist, a student located at a great distance may never set his/her foot on the campus.

A Delphi exercise to pool opinions of concerned functionaries across countries on merits and limitations of accreditation was attempted by GUNI (Global University Network for Innovation) and the findings were reported for different regions. The commonly agreed merits and limitations pointed out by respondents from the Asia-Pacific region, along with their recommendations regarding accreditation are reproduced hereinafter in Table 10.1

Somewhat related to accreditation, though not conveying the same idea about quality of education imparted in an institution of higher education is the concept and use of ranking of such institutions. Different rankings are based on different frameworks, with criteria or parameters and their weights in the overall rank. Unfortunately, such rankings are mostly derived from data or documents provided by the institutions concerned, and in several situations data called for are not available. Even some of the parameters on which data are required do not have unique operational definitions.

Table 10.1 Benefits and problems of accreditation pointed out by respondents from the Asia-Pacific Region

Regions	Benefits (mean over 4.00)	Problems (means over 3.70)	Recommendations (means over 4.50)
Asia and the Pacific	1. It protects students from fraudulent programmes and institutions (4.57)	1. The connection between accreditation and the financing of universities might discriminate against universities that are growing and universities in developing countries (4.07)	1. Make the outcomes of the accreditation process transparent and information accessible to society (4.86)
	2. It is a tool for decision-making in institutional strategic planning as it identifies strengths and weaknesses (4.29)	2. There is a need for financial support for the accreditation process (4.00)	2. Accreditation processes should be periodically reviewed (4.85)
	3. It facilitates the international recognitions of national and cross-border degrees (4.21)	3. Quality is not understood in the same way by agencies (4.00)	3. Public and social needs should be reflected in the accreditation process (4.79)
	4. It assists students in choosing HEIs (4.21)	4. The process maybe bureaucratic and complex (3.93)	4. Cooperation between accrediting bodies, HEIs and society should be strengthened (4.79)
	5. It facilitates comparability and equivalence between local, regional and international HE programmes (4.14)	5. The process is highly time consuming (3.93)	5. There should be continuous quality improvement as a result of the accreditation process (4.69)
	6. It facilitates cooperation, the sharing of good practices and benchmarking (4.14)	6. It involves heavy workloads for participating institutions in preparing documentation, databases and so on (3.92)	6. Accreditation agencies should be periodically assessed by external auditors (4.69)
	7. It creates a quality culture in HE worldwide and enhances the educational system as a whole (4.08)	7. There is a lack of understanding in society of accreditation's strengths and weaknesses (3.86)	7. The autonomy of HEIs should be respected (4.62)
	8. It facilitates the exercise of continuous improvement of quality and the introduction of a quality culture (4.07)	8. It is difficult to develop acceptable and pertinent performance indicators (3.86)	8. Qualified personnel who are knowledgeable in the fields to be assessed should be trained to undertake accreditation (4.62)
	9. It enhances transparency, provides information and makes institutions accountable to society and stakeholders (4.00)	9. The concept of accreditation is not clear enough, as it is understood in different ways in different national and regional settings (3.86)	9. Accreditation bodies should follow good practices (4.46)
		10. There is a shortage of qualified personnel to undertake accreditation (3.71)	

Source Global University Network for Innovation, Higher Education in the World 2007

10.10 Concluding Remarks

It has been a known fact that several developing countries do have well-documented quality systems for their education systems, particularly at the tertiary level, with inputs for experts in their respective countries as also from international agencies and foreign experts. However, such systems could not be implemented by the concerned institutions, primarily because of inadequate resources. In many countries, educational institutions at the tertiary level are funded by both the governments as also by the corporates and trusts. Government funding for any one institution has dwindled over the years with the increasing number of institutions and privately funded institutions have their own priorities, and sometimes vagaries in allocation and utilization of funds. The net result id quality of conformance to recommended processes and procedures suffers and in turn quality of performance falls short of expectations. One can possibly argue that the commitment of management, the involvement of all concerned, customer orientation in all activities, focus on continuous improvement and other positive steps can hardly make-up for deficiencies in resources—space, teaching aids, libraries and laboratories, museums, observatory, facilities and accessories for sports, exercises and other extra-curricular activities. Lack of journals, unkempt museums, out-of-date aids, dysfunctional observatory, unused equipment, inadequate space and similar other hindrances sap at the quality of education imparted in an institution of higher education.

In an egalitarian society, we require to strike a delicate balance between fewer institutions with adequate resource support including competent and committed staff and equipped to impart high-quality education and many more institutions accommodating more students, employing more people and spreading a bigger umbrella over society, but with limited support, insufficient to meet the requirements of high quality.

Distance education is a widely accepted proposition now. And in terms of the reach and the resources, the distance education system cannot carry-out much of a control exercise over the most important teaching–learning process. The focus should be on quality improvement, with the kaizen approach or the six sigma methodology.

References

Barnalee, C. (1995). *Introduction to total quality in education*. Montreal: Inter Universities Press.

Crawford, L. E. D., & Shutler, P. (1999). Total quality management in education: Problems and issues for the class-room teachers. *The International Journal of Educational Management, 13*, 67–72.

van Ginkel, H. J. A., & Dias, M. A. R. (2006). *Institutional and political challenges of accreditation at the international level (GUNI)*. Palgrave Macmillan.

Gola, M. M. (2003). Premises to accreditation. ENQA workshops report 3, 25–31. European Network for Quality Assurance in Higher Education, Helsinki.

Harvey, L., & Knight, P. T. (1996). *Transforming higher education*. Buckingham: SRHF and Open University Press.

Montano, C., & Utter, G. (1999). Total quality management in higher education. *Quality Progress, 32*(8)

Mortimore, P., & Stone, C. (1990). Measuring educational quality. *British Journal of Educational Studies. 39*(1)

Mukhopadhyaya, M. (2001). *Total quality management in education*. New Delhi: Sage.

Mukherjee, S. P. (1991). Human resource management and national development. *Everyman's Science, 26*, 19–20.

Mukherjee, S. P. (1992). Quality systems—A philosophical outlook. *EOQ Quality, 3*, 14–16.

Mukherjee, S. P. (1995). Quality assurance in an education system. *Total Quality Management, 6* (5), 571–578.

Mukherjee, S. P. (2005). Value-based leadership and professional education. Philosophy and science of value education in the context of modern India (pp. 114–122). Calcutta, India: The Ramakrishna Mission Institute of Culture.

Owlia, M. S., & Aspinwall, E. M. (1996). Quality in higher education—A survey. *Total Quality Management, 7*(2), 161

Padhi, N. (2010). *Total quality management of distance education*. New Delhi: Atlantic Publishers and Distributors.

Sallis, E. (1996). *Total quality management in education*. New Jersey: Prentice Hall.

Sanyal, B. C., & Martin, M. (2007). Quality assurance and the role of accreditation: An overview. In *Higher education in the world. 3—19. Global university network for innovation*. New York, USA: Palgrave Macmillan.

Schragel, F. (1993) Total quality in education. *Quality Progress*, October.

Stella, A. (2004). Quality assurance in distance education: The challenges to be addressed. *Higher Education, 47*.

Chapter 11
Quality in Research and Development

11.1 Introduction

While the need for quality is ubiquitous and there has been a growing emphasis to expand the scope of quality assurance and quality improvement to a wide array of human activities and outcomes thereof, Research and Development with its diverse network of processes and people with specialized knowledge in varying domains working to achieve different objectives—not all clearly or uniquely understood or stated—justifies an approach to quality which is bound to differ from the approach adopted with success in manufacturing and even in service enterprises.

In the manufacturing or the service sector, organizational performance depending on value addition by a producing unit is judged by the quality of its products or services or in terms of process efficiency and similar other criteria. And, it is quite appropriate to speak of quality of a product or a service. R & D as a sector embraces a wide range of processes corresponding to an equally wide range of R & D projects, differentiated in terms of organizations involved and purposes envisaged. Most of the core processes would be project specific, though some support processes may be somewhat generic. Even broad requirements for quality in processes and outcomes will differ from experimental R & D to those which involve 'soft' laboratories. Research and development activities carried out in independent research organizations or in R & D set-ups within universities or manufacturing or service industries were often found to fall short of targets in terms of concept-to-market time or in quality of the output—which could be a technology or a process or a control mechanism or a product or a service—or in-process efficiency and similar other criteria. People engaged in such activities often hold the opinion that R & D efforts cannot be judged by objective and quantifiable criteria which could validly apply to manufacturing or even service units. In fact, the very concept of quality as could be defined in a manner acceptable to many, if not all, has been often considered as not even consonant with R & D culture.

© Springer Nature Singapore Pte Ltd. 2019
S. P. Mukherjee, *Quality*, India Studies in Business and Economics,
https://doi.org/10.1007/978-981-13-1271-7_11

In this context, it may not be out of place to start with some largely agreed definition of Research, though it is easier to describe what a researcher does or has done than to define 'Research'. The definition that will not attract much of a criticism states that *Research is search for new knowledge or for new applications of existing knowledge or for both.* Knowledge, here, connotes knowledge about different phenomena which take place (often repeatedly) in the Environment or the Economy or the Society—together constituting the perceptual world—or even in the conceptual world of human imagination or thought. To ampliate, knowledge encompasses concepts, models, methods, techniques, results, algorithms, softwares, etc.

Linking Research with Development efforts, we talk essentially about knowledge acquisition in the perceptual world dealing with concrete entities, processes involving such entities and outputs of such processes. And in many situations, we attempt to find out new applications, through development of appropriate technology or mechanism, of existing knowledge. And this application does not necessarily mean that we are engaged in Applied Research. Such applications may not always be data based or empirical in nature and can become Theoretical Research as well.

It must be admitted that existing concepts and measures of quality along with methods for quality control, assurance and improvement practised by manufacturing industries do not apply mutatis mutandis to the service sector. Further, such concepts, measures and methods as have been oriented to meet the needs of the service industry will not be straightaway applicable to Research and development activities. In fact, development activities need less modifications while research activities mandate major modifications to existing methods and practices.

Research and Development are two distinct activities, differing on many counts. Usually development follows research, being primarily an activity or a process to transform the findings of some research into some concrete entities like a technology or a process or a device or a product ready for consumption. A development activity, that way, leads evidently to enhance national income or to provide social benefit to a large segment of the population or both. It is also possible to that 'development activity' motivates and/or facilitates further research on the subject. Development of the first type, which accounts for the bulk of development activity, is strongly related to Technology.

The ultimate output of Research is to throw up new knowledge by processing various relevant and interrelated pieces of information generated through observations or experiments or experiences. Such information items have to be reliable and reproducible (from one location or institution to another), and the knowledge they generate has to be accessible by all interested parties. A major distinction between research and development could be the relatively larger uncertainty characteristic in a search for new knowledge, compared to that in an attempt to apply that knowledge and come up with a concrete entity in a development activity. At the same time, both during research into a phenomenon and in development of a concrete entity by suitably applying some newly found or existing knowledge, we are quite likely to come across steps and consequences which could have been better avoided since those were not putting us on the right track. However, such avoidable steps and undesirable consequences were only to be found out, since those were not just obvious. And such findings constitute valuable information that can be used to development better procedures or algorithms and should not be dubbed as simply 'negative' and a wastage of efforts. In fact, such information is as important a research output as a 'positive' one. This is applicable to an equal measure in activities to make use of existing knowledge to come up with some concrete entity.

We restrict ourselves to research activities in the perceptual world and, there again, to research areas where some outputs are targeted to benefit not merely academics but also a large cross section of the people in terms of coming up with a process or a product or a service or a method or a practice or even a concrete and comprehensive idea that could lead to any of these entities. We try to bring out some aspects of quality in development activities in a limited way. Talking of Research and Development together, we remain confined to scientific and industrial research carried out in institutions rather than by single individuals.

11.2 Research: Process and Output

Research can be comprehended both as a process as also as its outcome or output. And 'quality' as a concept and a measure could apply to both the Research Process as also to Research Output. Of course, the latter is most often regarded as the more important aspect of quality, it has to be admitted that the latter depends on the former and, that way, the first aspect of quality in the Research process should be accorded due priority in any discussion on quality in research.

Although research activities in different subject areas and taken up by different individuals or teams under different conditions are quite different among themselves, we could draw upon the basic features common to all of them. Thus, for example, we agree that Research is a process with many inputs of which knowledge in the relevant domain along with a spirit of enquiry is possibly the most important and that the output or outcome of Research is something not very concrete always.

Just as the quality of a process is comprehended in terms of pertinent features of inputs, influences operating on the conversion of inputs into output(s) and controls exercised during this conversion to ensure that the process is on the right track, the same exercises have to be carried out to comprehend quality in a Research Process and to ensure that we can achieve the desired quality. Quality also relates to the output of the 'research' process. While the tangible outputs of research are easier identified than quantified, their assessment in terms of values or benefits is a big problem. And the problem is aggravated by the fact that we may not be able to begin with a clearly formulated statement of objectives. Sometimes, these objectives unfold themselves as research progresses. For example, we take up a research into the behaviour of a metal or metal compound under high temperature and, at the end of our research, we establish some results relating to the behaviour. But then which aspects of behaviour, e.g. mechanical properties, or electromagnetic features or bonding with other metallic compounds or metals, could not be specified at the beginning.

Outputs of Research are not always easy to identify. In fact, there remains a considerable gap between the outputs as derived or obtained by the researcher(s) and the outputs as communicated in a documented or a comparable form. The manner in which the output(s) is (are) communicated can lead to a big difference in assessed quality of research. While the communications emanating from research follow quickly the derivation of output, the real worth of the output may be revealed after and over a long period of time. Thus, judged by contemporaneous impact either in terms of creating new knowledge or providing solution—fully or partly—to some nagging real-life problem, the worth of some path-breaking research could be pretty small, though it may so happen that the output of such a research is appreciated better by some research worker(s) in future who come up with some substantial result(s).

Quantification of the value of research output defies any completely objective and universally accepted procedure. In fact, many alternative ways to assess the worth of a published communication by way of an article or a note or a full-length paper or thesis, taking account of the journal or book or edited volume of articles from different contributors wherein it was published, the number of investigators or authors who have cited this communication in their work, etc. For the same research project or programme, results may be communicated in a single lengthy paper or in several short communications or notes. This difference may contribute to a great extent in the number of citations for the work as a whole.

While it is true that the process implication should be regarded as more important than the output or outcome connotation, we quite often judge the process only in terms of the output value or worth. Should this be true for Research also? A research programme or process taken up with a mission that has been as clearly stated as is possible at that point in time may be executed in reasonable conformity with a well-developed process plan and unfortunately fails to meet the mission—at least in some aspects. How do we judge the quality of this research programme? Possibly such a programme will reveal reasons for not achieving the mission requirements and will also hint at possible alternative ways of executing the research process which can meet the requirements better. In the context of scientific or technological researches, a programme that comes up with some concrete results that were stated in the objectives or mission achieved without developing a detailed work plan and/or a high degree of compliance with such a plan will not be rated as of high quality. Indeed, one can debate the relative importance of the procedural part and the substantive part of a research and, probably, end up by admitting that both are important and that the assignment of importance measures to these two parts is unwarranted.

11.3   The Research Process

In recent times when researches are taken up more by teams rather than single individuals, more in terms of costly and sophisticated equipments and facilities being deployed rather than using makeshift devices, more through long-drawn experiments than by using a few observations or trials, the research process has to be well-planned.

Let us look at the broad steps in research that covered in the following list offered by many writers on the subject.

(a) Concrete formulation of the research problem, indicating the phenomena to be studied along with their aspects and concerns noted about the problem
(b) Review of relevant literature and available experiences, if any
(c) Formulation of the expected outcome or the working hypothesis in terms of removal of concerns about the problem or of limitations in earlier investigations and results thereof
(d) Research Design which provides the essential framework for Research and guides all subsequent steps to achieve the research objectives as are reflected in the problem formulation and the working hypothesis
(e) Collection of evidences in terms of primary and/or secondary data throwing light on the research problem and of all supporting information, methods, tools and results
(f) Analysis of evidences through use of logical reasoning, augmented by methods and techniques, both qualitative and quantitative, to come up with a solution to the problem
(g) Establishing validity of the research findings to ensure face validity, concurrent validity and predictive validity.
(h) Preparation of the Research report and dissemination of research findings to all concerned.

To distinguish Research Process from Research Output, one may possibly consider the first five steps as corresponding to the process, steps (f) and (g) to the output and the last step to communication/dissemination of the output. It may not be wrong to combine the last three steps to discuss the quality of research output.

Research, in general, may not even allow a formulation of its scope and objective(s) right at the beginning and may have to grope in the dark in many cases. The problem area may be, of course, identified as one that is currently associated with some doubt or confusion or ignorance or uncertainty and may even be one which has been investigated by earlier research workers. The only delineation of the research problem in such cases is provided by an outline of the remaining doubt or confusion or ignorance.

In this sequence of steps, Research design is the most crucial and plays the role of a Process Plan in the context of a Manufacturing or Service Industry. And quality in the research process combines the concepts of quality of design and quality of conformance (to design requirements). In modern Quality Management, quality of design outweighs quality of conformance and this is verily true of the Research process.

Incidentally, there are strong advocates of complete flexibility in the Research Process with no initial directions or guidelines. To them, the Research Design cannot be pre-specified at the beginning: it evolves gradually as the investigation proceeds, with steps at the next stage not known until some implicit or explicit outcome of the earlier stage is noted. Using the relevant language, the Research design is sequential and, even if pre-specified, calls for modifications and even reversions as and when needed. However, this flexibility may be desired in research meant just to acquire new knowledge in the conceptual world than in the case of sponsored research or applied research in the perceptual world.

11.4 Quality in Research Process

The quality of research, in such a context, should be assessed in terms of quality of process planning. And one can reasonably expect that a comprehensive process plan that is developed keeping in view the research objectives or the mission requirements, if properly and honestly executed, should enable or facilitate the research team to achieve the targets or, at least, reach close to the targets.

Inputs to the Research process include, among other elements, (1) motivating factors, expectations of the organization as also of the research community or even the concerned segment of the society (2) documents, patent files, standards for processes and procedures likely to be followed in executing the process besides the most important soft input, viz. (3) intellectual and creative ability of the research worker(s) and (4) softwares for simulation, computation and control. Quality of inputs into the main process as outlined in the steps stated earlier as also of support processes like equipments, materials and utilities, work environment and the like turns out to be important in affecting the quality of the Research Process. And quality of some of the 'hard' inputs can be examined in terms of calibration of equipments and the resulting uncertainty in measurements to be generated, process and procedure standards being up-to-date, available softwares having requisite efficiency, laboratories having control over ambient conditions, reference materials being duly certified, etc. Knowledge of the subject domain, of relevant models and methods, of algorithms and softwares and the like can also be assessed in broad categories if not in terms of exact numbers.

In Research, conversion of the so-called inputs into what ultimately will be treated as the output is so complicated and subject to so much uncertainty that relating quality of output to quality of inputs in an acceptable form may be ruled out. There could be cases where some relations—if not quantitative—can be established and made use of to assure quality in all conceivable inputs.

Formulation of the research problem reveals a wide spectrum. In sponsored research, the problem has already been identified by the sponsoring people and has also been formulated in some details which may require further amplification, concretization and focus on what is meant by the 'intended solution'. The intended solution may be too ambitious in terms of its cost, scalability and utility features and may require some pilot experiments followed by a relook at the initial characterization of the intended solution. In some rare situations, the sponsoring agency may be quite modest and the pilot results may even suggest some improvement over the initially expected result or solution. Sometimes research problems are identified out of a box of problems suggested by some national agency to meet some planned targets for improving socio-economic conditions in the country or to preserve or enhance the quality of environment. A comprehensive perusal of relevant literature in some chosen problem area and an assimilation of related experiences may help finding out the deficiencies in knowledge to describe, analyse, control and optimize some phenomenon so that the research worker/group can come up with a clear formulation of the research problem. There exist some situations where, of course, research workers have full freedom in identifying their research problems based on various inputs including their own interests in knowledge acquisition and in service to humanity.

Problem formulation must take due account of resources—material as well as human—currently available or likely to be augmented. Of course, resource sharing among institutions or among research groups should not be a problem necessarily.

Validation of research results is a crucial step and is sometimes compromised in terms of a limited dissemination of the results to some selected stakeholders with a view to getting their honest feedback.

Thus, we come to the crucial problem of understanding the components of a research plan and of proposing suitable measures of quality for each of these components. A Research Plan is guided by the objectives of research to be formulated as a concrete and clear statement of what is expected to come out of the research project. The Research Design should spell out the nature of evidences to be gathered—either by planning and conducting some experiment(s) and/or by compiling evidences generated by others, after checking whether such evidences an be accepted for the research in hand or not. A Research Design incorporates quite a few components like sampling design (whenever required), measurement design and operational design. Sampling is involved more often than not, in selecting specimens or experimental units, in choosing factors affecting some response(s) to some treatments or operations carried out on the specimens, in considering only some out of all possible responses that can arise from the same experimental unit and so on. Measurement of factor levels or responses along with co-variate poses problems associated with the measuring equipments and the methods of measurement. We have to collect measurements by ourselves or from some other source and assure ourselves of acceptable levels of uncertainty in them. A research worker has also to decide on the manner in which experiments or surveys will be carried out.

The Research Design follows from the scope and objective(s) of research, takes into account resources available and constraints on their deployment, and rolls out a comprehensive guideline for various activities to be performed. It should specify the variables on which data are to be collected including operational definitions to be used, the choice of proxies or substitutes for variables which cannot be conveniently measured, the quantifiable considerations that accommodate latent variables like user-friendliness or convenience in inspection and maintenance of the research output in the form of a new process or product, the manner in which evidences gathered should be analysed so that inferences related to the research objective(s) can be reached, and even the way results of research should be validated before being used as inputs to some 'development' effort.

While research has always involved some trial and error—possibly more in earlier days than at present—resource limitations make it almost mandatory that even such trials be so planned that they can lead to the desired result(s) as quickly as possible. For example, in an experiment to find out the composition of a metallic alloy in terms of proportions of basic metals and non-metallic substances in such a manner that a structure out of the alloy will have the largest time to rupture under some specified stress situation should use a sequential design for experimentation, rather any fractional factorial design or any other design attempting to minimize the number of design points or of experimental runs. An experiment in this case is quite time-consuming even if we make use of accelerated life testing methods and therefore, the need for 'optimization' is much stronger than in some other situations. The objective of research could even require the experiment to achieve the largest time to rupture subject to a minimum desired ductility. This will complicate the analysis of experimental results.

The role of planning or designing a research is of great significance in influencing the research outcome. It is true that unlike in the case of a manufacturing set-up, conformance to design or plan is not that essential in the context of research. In fact, as research proceeds it may be sometimes desirable or even compelling to deviate from some details specified by the initial design. Research, that way, involves a lot of learning through wilful deviation from the path chalked out in the Research Design. It may even come out that following an initial plan, we end up in a failure and learning from such failures is a must. In such cases, the need to leave aside the initial plan and to tread on some new and uncharted ground can well provide inputs for design modification. However, arbitrary or subjective deviations from the plan will definitely affect the quality of output and, that way, of research itself.

Good laboratory practice is a mandatory requirement to carry out experiments, to record measurements of factors and responses, to carry out checks for repeatability and reproducibility of test methods, to estimate uncertainty in measurements, to get back to original records in case of any doubt, and thus to provide a credible basis for analysis of results. Results of such an analysis should be duly interpreted in the context of the experiment and the objective behind the same.

Ethical considerations have assumed great significance over the last fifty/sixty years as R & D activities were being taken up by many agencies—both private and public—over and above by a fast-increasing number of individual research workers in academic and professional bodies. Some in the burgeoning research community were not scrupulous and did not maintain adequate honesty or transparency or commitment to contracts and the like. This created development of standards for ethics in Research—both in the Research Process as also in the Research Result. In this context, the term 'ethics' takes care of legal or regulatory requirements along with moral and ethical norms and practices.

In fact, the Research process right from the Research Design to documentation and Dissemination of research results should incorporate ethical considerations to the desired extent. In case of biological, medical and even sociological research or in any research where humans are used as subjects or units in experiments or observations, it is mandatory to seek informed consent of the participants in the investigation, to maintain strict confidentiality of information gathered or response observed, to avoid any encroachment on privacy during experimentation or observation, and to debrief such participants about the research findings before these are brought to the public domain.

In experimental or observational research, it is essential to maintain properly all raw data and associated notes about how such data had been generated and how had those been used to generate the final or smoothed or modified data to be used in the analysis stage.

Whenever some 'material' is borrowed from some source, it must be duly acknowledged. Further, in cases where permission is need to borrow such material, it is mandatory to seek such permission from the appropriate authority.

11.5   Quality of Research Output

To assess what can be construed as quality of Research Output is an extremely difficult task. The level of difficulty is more in the case of theoretical research meant to augment the existing stock of knowledge in the domain of interest and the task is to be handled carefully and with sensitivity by experts in the domain. Once the findings of such research appear in the public domain, follow-up by other investigators and experts may throw some light on the quality we are talking of. Quality in terms of richness of the findings as well as in terms of the methods followed to arrive at the findings and the manner in which these findings have been documented and disseminated should be considered. In case of a sponsored research, we can possibly take the extent to which research output satisfies the sponsoring body as an indirect reflection of quality. In cases where research was taken up to find out a solution to a vexing and nagging problem in real life by way of a new device or material or process, quality could be judged in terms of features of the concrete output entity. In fact, positive or beneficial aspects like economy, convenience, stability or longevity, ease in maintenance, even ease in disposal at the end of useful life can be considered in a quality assessment, along with possible negative or harmful features like adverse impact on environment, hazard in use.

For their own survival and growth, industries particularly those who are quality conscious and who operate within highly competitive market environments have taken up research from time to time. Sometimes, basic research has already been carried out in industries. E.G. pharmaceutical industries who wanted to remain ahead of others in coming up with new formulations were not merely playing with known molecules of chemicals having medicinal properties but also trying to develop new molecules. Of course, quality of research output would be looked upon differently by industries based on impact on business as implied by both the offering of a new product with novel or augmented features as also the reduction in costs of producing some existing products and services, without comprising on quality.

In fact, quality of output in industrial R & D exercises is usually judged by certain desirable features of the prototype or model of the end product developed. The product has to be such that potential customers can acquire it at an affordable, still leaving a reasonable profit margin for the organization. It must admit of being scaled up from the bench scale or pilot plant scale to full production scale to meet the projected market demand. It must be sustainable in the sense that the end product should have a good length of life. And it must be distinctive to create a niche segment for it in the existing market or create a new market for itself. General electric considered several dimensions of research quality: technical quality of research (assessing the research process), impact of research (in terms of research output) being 'game-changer' or 'disruptive' versus incremental, business relevance and timeliness (early or late relative to the target market requirement.) At DuPont, R & D quality is defined as 'creating, anticipating, and meeting customer requirements' which required 'continual improvement of knowledge, application and alignment with business objectives'.

Thus, customer satisfaction by effectiveness of features added through R & D resulting in value addition and efficiency resulting in cost reduction together provide some idea about quality of industrial research (research intended to benefit Industry). From the customers' point of view, satisfaction with the features of the information provided and the absence of deficiencies of the information in actual use environments are of great importance. Features of research information include timeliness, utility, accuracy and costs. Deficiencies can occur either during the research process or be reflected in the end products. Possible deficiencies in research products may be that the knowledge on which the product is based is late, inaccurate, irrelevant or of relatively poor value for the investment. Deficiencies in a research process may be ascribed to process 'rework' or 'scrap', e.g. having to repeat an experiment or an analysis of results of an experiment or a wrong interpretation of results of analysis. Repetition of an experiment with consequential increase in cost and time may be due to use of substandard reference sample, or poor control over experimental environment or poor choice of experimental units.

Thus, falling in line with the generic definition of quality given by ISO, one can define research quality as the degree to which features of information and knowledge provided by the research function meet users' requirements.

The primary outputs of exploratory and applied research are information, knowledge and technology. Research quality can thus be judged.

A critical dimension of quality of research output prior to its communication to others is validity or 'ability to measure or reveal or analyse or optimize what was intended to be measured or revealed or analysed or optimized'. Before research results are accepted for documentation and/or dissemination, it is quite important to check the validity of the results. Often, this validation will require that operational definitions and instruments for measurement as well as instruments for analysis of such measurements are valid. And, whenever it is possible to repeat the study and the results thereof, we should check the reliability of the results. Of the many different forms of validity which apply in the case of research results, the more important ones are face validity, concurrent validity, predictive validity and construct validity. In some cases, the results seem to be absurd or very surprising or quite unexpected or even fallible, right at first sight. Thus, some theoretical research on inventory management may yield a formula or a function to work out the optimal purchase quantity or the optimal size of a production run. Once some numerical values of the parameters involved are fed to the model, the optimal order quantity comes out to be far different from the expected demand. Here, the research result lacks face validity. Similarly, if the research result grossly differs from that of a related research on a related subject or aspect of the phenomenon, we may suspect concurrent validity. And in some cases, we fail to use results of research study to predict what happens in something not yet observed or even observed but not yet analysed. E.G. if we fail to predict the outcome of using our research result based on a part of the total sample that we had to the rest of the sample, then predictive validity will be lacking. And construct validity is said to be there if we can build up a knowledge base about the entire population or phenomenon from what we have observed and analysed.

Quality of research output is generally spoken of in the case of published research findings and, there again, mostly those published in journals as distinct from many web-based dissemination. Post-publication assessment of quality is often judged in terms of Peer Review. Such a review could be double-blind, single-blind or even open. Peer Review properly conducted attracts deep and strong support across the research community. However, Peer Review has been criticized on several counts, viz. it causes delay, it is not always effective in detecting misconduct and malpractice, the judgements made are subjective and inconsistent, it tends towards conservatism and stifles innovation, it discourages interdisciplinary research, etc. With an enormous increase in research activities and consequent increase in submissions for publications, it becomes pretty difficult to identify and engage knowledgeable reviewers to forward their reviews within prescribed time limits.

Quality of Research Output—understood somewhat crudely as an analogue of quality of Performance of a manufactured product—can be assessed in terms of parameters like impact on the research community or, better, the academic community that includes potential research workers, impact outside the academic community and impact on human life. Each one of these parameters is multidimensional and some of the available indicators like impact factor of the journal wherein the research has been reported based on several features of the journal including the average number of citations of articles published in the journal are not beyond well-argued criticism. While originality and novelty of findings distinct quality features that admit of agreed assertion or otherwise, contributions to the extant body of relevant knowledge denies consensus judgement, except in those few researches which stand out as exceptional or path-breaking.

It is also an established fact that absolutely new concepts or methods or results emerging from sustained research by an individual or a group who did not at all consider or foresee any possible future applications of their contributions did not find takers in the research community for further follow-up of such findings until decades or even centuries later those are found by some enthusiasts to pave the way for some highly useful research. Convincing illustrations are provided by the contributions to the theory of numbers by G. H. Hardy or to the theory of knots by. The former had a profound impact on coding theory developed much later and the latter found extensive modern applications in topics ranging from the molecular structure of DNA to string theory.

Impact on human thoughts to be considered as a reflection of quality in the output of some research, especially one in the world of abstract entities, may be well-nigh impossible in some cases, e.g. identified four distinct levels of parallel universes. In Level I, there are universes with the same law of physics but different initial conditions. In Level II, there are universes with the same equations of physics but perhaps different constants of nature. Level III employs 'the many worlds' interpretation of quantum mechanics and in Level IV, there are different mathematical structures. How do we propose to assess the 'impact' of Tegmark's findings on further research?

High impact is generally followed by individual or professional recognition—accolades and/or awards or fellowships of highly rated scientific societies and academies. While some of these recognition really reflect on quality of research report, recent times have seen the emergence of a whole host of such recognition which are based more on extraneous considerations than on impact of research output. Moreover, a research worker who has carried out several researches in related subjects or even in a diversified field with impact above average may well qualify for some such award as an individual. A single research study deserving such an award usually corresponds to a much higher impact and throws new problems and opportunities for research. One may justifiably argue that the second situation is characterized by a higher quality of research than the first.

In the case of sponsored research, particularly when the sponsoring organization is an industry, one may argue that customer satisfaction with the output of research is a reflection—at least indirectly—of performance quality of research output. Measuring (institutional) customer satisfaction and related measures in the case of R & D institutions engaged in sponsored research invites some special features. In particular, customer loyalty should be comprehended in terms of a customer's (a) acquisition of another product/service offered by the R & D institution, (b) approach to the R & D institution for possible design and development of a new product/service exclusively for the customer, (c) recommendation to others about products/services offered by the institution as meriting acquisition and (d) request/suggestion to the R & D institution for a possible expansion of the latter's product/service profile.

Rounding up a discussion on quality of (or 'in') Research, one may not like the idea of dissecting quality into two components, one relating to the research process and the other corresponding to the research output. It is not an easy task to combine these two into an overall indicator of quality. There have been attempts to consider both the process and the output together and to characterize 'high quality' research in terms of some features which will not, hopefully, raise any debate. Such a characterization is summarized in what follows:

A High-Quality Research

* addresses an interesting research question—which could be posed for the first time or could reflect on the findings of some earlier research, need be too complex or could be quite challenging, could be oriented directly or indirectly to some real-life problem (being faced) or potential problem (likely to be faced) or could just relate to some abstract entity
* is based on some logical rationale and is tied to some tested theory
* can be replicated and generalized to other settings
* states clearly the variables and the constructs to be examined
* ensures validity and verifiability of its findings
* may or may not come up with a complete or even a partial answer to the research question addressed. Even the impossibility of answering a research question could be a stimulating and useful finding
* has the potential—in some way—to suggest directions for further research by offering more research questions and motivating investigations into those questions
* makes some contribution—may be even incremental—to the existing body of knowledge.

11.6   Quality in Development Process and Output

Closely linked with research is the equally challenging task of developing research infrastructure, specifically equipments, measuring instruments, gadgets and control devices required in experimental research. In fact, some of the path-breaking researches over the years have been rendered possible through successful development of the research infrastructure. Such development activities were taking place sometimes after some significant research results came out or even synchronously with the progress of research in certain other cases. Such development activities take due advantage of scientific research outputs and depend quite heavily on technological developments. In the process, some of these development activities result in new technologies altogether.

Many of the equipments and instruments developed through science-based findings and technology-based processes have not only facilitated scientific and industrial research, but have also impacted human life in many significant ways. Sprinklers in targeted agriculture, plasma cutters for extremely hard metallic substances, gene-sequencing machine in the study of plant and genomes to bring about genetic improvements, pacemakers and other implants including ceramic eyeballs or hip joints, atomic clocks and particle accelerators in study of atomic physics, satellites and satellite launch vehicles for space and atmospheric research including their components like transponders, and a whole host of other concrete entities illustrate the contribution of the Development process. We can even think of parallel developments in the abstract world like expert systems and search engines, optimization algorithms to deal with complex non-regular objective functions like simulated annealing or ant colony optimization, cloud computing, mobile telephony and the list goes on.

The process of Development in this context is more often streamlined with, of course, some amount of flexibility in-built in it. Different stages in this process including core processes of developing the system design (noting that the output is generally a system rather than a single product or process or service), validating the design against its expected performance (which may be to offer a technical support for some experimentation or some analysis) and verifying the design elements at each stage of assembling or building up. The process of procurement complying with design requirements is also quite important, as are the various engineering processes through which the inputs will eventually come up with the system as designed.

In the above context, it is at least possible to think in terms of quality of a design and quality of conformance during different stages of production and to relate the ultimate quality of the output system to these two quality dimensions. And quality of performance of the output can be objectively assessed provided the requirements for developing the output system have been clearly spelt out in the mission plan behind the development programme. In fact, many situations permit determination of the rate of failure or success of the output system during actual use or deployment.

Speaking of R & D as somewhat distinct from Development of a new product or service, we can illustrate the linkage between quality improvement effected through new product development and can assess the quality of such a development activity in terms of the nature and extent of quality improvement (in a broad sense) it brings about.

New product development (NPD) is one of the most important activities in business and engineering today. Whatever is designed, developed and delivered—including a service—can be generically treated as a 'product'. In fact, services are not necessarily product-free. There are services in the hospitality industry, for example, which are heavily loaded with products, while professional services like rendering legal advice illustrate services with almost no product involvement.

NPD may involve

- Function augmentation in an existing product
- Core product revision
- Changes to an augmented product
- Development of a completely new product

A new product may imply a product* new to the market

- New to the company
- Completely new to create new markets.

Industry leaders look upon NPD as a proactive process, in contrast to a reactive strategy in which nothing is done until problems occur or some competitor introduces an innovation.

The NPD process involves

Idea generation—Ideas for new products may be obtained from basic research using a SWOT analysis, competitors, focus groups, employees, market and consumer trends, salespeople, corporate spies, trade shows, etc.

- Idea generation involves Brainstorming.
- The second step is idea screening to eliminate unsound concepts prior to investment of resources on them, raising questions like
- Will the customers in the target market benefit from the product?
- What is the size and growth forecasts of the market segment/target market?
- What is the current or expected competitive pressure for the product idea?
- Is it technically feasible to manufacture the product?
- Will the product be profitable when manufactured and delivered to the customer at the target price?
- Concept Development and Testing become the third step to develop marketing and engineering details. Costs of manufacturing and reactions of some potential customers to the Idea are tested.
- Business Analysis is then carried out to estimate likely selling price, sales volume and break-even point.
- The next step is BETA Testing of a physical prototype or mock-up (as also its packaging) in typical usage condition. Market Testing should follow in terms of focus group customer interviews and trade shows. Necessary adjustments are made, an initial run of the product is made and the product sold in a test market.
- The next step is Technical Implementation including resource estimation, logistics plan, supplier collaboration, engineering operations planning, etc.
- The last step is, of course, Commercialization involving product launch, promotional efforts and filling the distribution pipeline.
  NPD—as a process—is closely linked to quality improvement, not always apparently.
  This requires a broad view of product quality revealed through performance during usage by the customer.
- If NPD targets a completely new product, quality improvement does not become an apparent corollary. In other cases, NPD is linked up with quality improvement.
- Quality improvement—in a very broad sense—may imply improved or better levels of some existing product features which make it more user-friendly, more robust against use environment fluctuations and cheaper in terms of life cycle

cost (if not in terms of acquisition cost alone), easier to maintain and even to dispose of, less harmful to general environment and even more appealing to the customer.

Thus, quality in new product development process meant to improve upon an existing product can be analysed in terms of the usual quality of design and the quality of conformance, resulting in quality of performance. The last can be judged in terms of the added beneficial features along with the reduced harmful features of the original product.

In case NPT targets a new product based on some existing technologies, quality should be judged by examining the realized performance of the new product against the requirements incorporated in the Product Design. This design has to implicitly recognize merits and deficiencies in an existing product in the same category where the new product is to be developed and to specify how the deficiencies can be removed and/or the merits can be enhanced.

We also have development activities taken up by a research organization which intends to come up with a concrete entity that is based on findings of this research. This entity could be a pilot plant-level process or a prototype which can be passed on to customers for being examined and scaled up—with or without modifications —to enter the market. In such situations, feedback from customers can provide inputs for assessing the quality of the product.

A very well accepted approach to NPD and/or quality improvement is provided by TRIZ (in English TIPS—theory of inventive problem-solving) developed by Altshuller and tried out successfully by several leading manufacturing and service organisations across countries during the last three decades.

The quality of a development process can be defined—again on the lines of ISO —as the extent to which the development process efficiently provides product and process features capable of meeting their targeted design goals, e.g. for costs, safety and performance on a sustained basis.

11.7   Concluding Remarks

Quality and Research (along with Development) complement each other and are strongly interdependent, R & D is a bad necessity for improvement in quality. Research on new materials, processes, procedures, control mechanisms, evaluation and assessment or accreditation principles and practices are needed to ensure a quantum jump (as against an incremental change) in quality of goods and services. Quality—in a broad sense—is crucial in the R & D sector, especially in organized R & D activities carried out by S & T institutions and more so in Industrial R & D and in Industry-sponsored R & D effort.

Quality in R & D activities is a virgin field, awaiting pioneering contributions where self-innovation coupled with a 'learn-as-you-go' approach is critical. Quality Management will definitely gain ground in industrial research where the concept of

Return on Investment continues to remain valid, though with appropriate conno-tations of 'return' And quality of research in such cases is judged in terms of quality of performance revealed in some measure of return against the planned investment in research, allowing for some flexibility in amount corresponding to on-the-way hits that miss the target.

Quality in Research and even in Development (of new technologies, processes, products and services) can better be appreciated than assessed or compared.

References

Clarence, M. E. Jr. (1992). Applying quality to R & D means 'Learn-As-You-Go'. *Research Technology Management*, 24–31.

Davidson, J. M., & Pruden, A. L. (1996). Quality deployment in R & D. *Rersearch-Technology Management, 39*(1), 49–55.

Endres, A. I. C. (1999). Quality in research and development. In Juran, J. M., & Godfrey, A. B. (Eds.), *Juran's quality handbook* (pp. 19.1–19.34). New York: Mcgraw Hill.

Francis, P. H. (1992). Putting quality into the R & D process. Research Technology Management, July–August, pp. 16–23.

Hardy, G. H. (1940). *A mathematician's apology*. Cambridge, England: Cambridge University Press.

Miller, R. (1995). Applying quality practices to R & D. Research Technology Management, March–April, pp. 47–54.

Rantanen, K., & Domb, E. (2002). *Simplified TRIZ*. Boca Raton: St. Lucie Press.

Research Information Network. (2010). Peer review: A guide for researchers

Szakonyl, R. (1994a). Measuring R & D effectiveness—I. *Research-Technology Management, 37*(3), 27–32.

Szakonyl, R. (1994b). Measuring R & D effectiveness—II. *Research-Technology Management, 37*(3), 44–55.

Smolin, L. (2006). *The trouble with physics: The rise of string theory, the fall of science, and what comes next*. Boston: Houghton Mifflin

Tenner, A. R. (1991). Quality management beyond manufacturing. *Research Technology Management*, September–October, pp. 27–32.

Tegmark, M. (2004). In Barrow, J. D., Davies, P. C. W., & Harper, C. L. Jr. (Eds.), *Science and ultimate reality*. Cambridge: Cambridge University Press.

Wooding, S., & Grant, J. (2003). *Assessing research: The researchers' view*. Europe, Cambridge, England: RAND.

Chapter 12
Quality of Information and Quality System in an Information Industry

12.1 Introduction

Individuals and institutions need information, irrespective of their activities and engagements, as producers or users of goods and services in the public or the private sector. Types and amounts (quantity) of information needed will vary from one individual or institution to another. Information is necessary for manufacturing industries and service organizations; for the state administration; for planners and policy-makers; for representatives of people wanting to know about present states of development in different regions and on different dimensions; for the armed forces to meet their operational, tactical and strategic needs; for regional and international organizations attempting to assess relative positions of different countries in regard to economic growth, poverty and hunger, education and health care, violation of human rights; and so on. Besides, there are organizations trying to improve the quality of our environment and the quality of our lives which also need information.

Information is possibly the most important input in making decisions and even in forming opinions. Informed decisions are expected to result in effective actions. This must be true in all spheres of human activity, including Quality Management taken in its widest sense. Of course, quality of decisions and of opinions will depend not only on the Quality of Information inputs but also on the way information has been processed to provide knowledge about the underlying event or phenomenon and the way this knowledge has been used. Quality of decisions or of opinions, Quality of Information and of information-processing and information-using activities are concepts which defy straightforward and operationalization. Notwithstanding this serious problem (that is common to many latent variable situations), these constructs have been gaining wider and wider implications. And it is worthwhile to attempt reasonably good operational definitions and measures of

S. P. Mukherjee, *Quality*, India Studies in Business and Economics, https://doi.org/10.1007/978-981-13-1271-7_12

these three entities, since the ultimate objective is to improve the quality of decisions and opinions and since without appropriate measures, improvement cannot be ascertained.

National Statistics Offices are facing new and challenging expectations, demands and requirements from (a) statistics users who want faster, easier and less expensive access to statistical data—through media and routines that are better adapted to their own processing needs; (b) data providers like individuals, households, establishments and the like who demand less invasive, less burdensome reporting—through media and routines that better suit their own information systems; (c) governments and taxpayers who want more value for less money; and (d) international organizations, requesting member countries to provide timely, comparable, good quality statistics which comply with international standards.

In this chapter, the author first considers the variety of information needed by a whole range of users and then proceeds to address the issue of objectively assessing Quality of Information. Important and widely used measures of quality of both quantitative and qualitative information have been reviewed. Subsequently, some issues in implementing a Quality System on the lines of some standard like ISO 9001 in an Information Industry have been discussed briefly by offering an outline of some changes needed in this connection. Based on his personal interactions with public sector Information Industry, the author offers some operational definitions of concepts and procedures required to implement a Quality System there. It is hoped that the ideas suggested in this particular context will apply in similar situations.

12.2   Nature and Diversity of Information

Information—usually derived from data generated or collected through properly designed experiments and planned surveys or compilation of records maintained by concerned agencies may be purely qualitative or quantitative. Most often these are quantitative in nature and could be presented in terms of a single index for a country or a state or a suitably defined region, sometimes along with its component-wise break-up or its values at disaggregated levels. Thus, figures relating to Human Development Index or Gender Parity Index or Global Innovation Index or Happy Planet Index or any similar figure provided by some international bodies are often used by different countries to launch new initiatives for improving their current ranking. Within a country, figures like Gross Domestic Product per capita or doctor–population ratio or percentage of literates in the population or proportion of the population below the poverty line presented for different states or regions reveal important disparities which call for remedial measures. Quite important are figures relating to stocks and concentrations of pollutants of different types and with different implications on human and animal health in different water bodies or in air. While such items of information are important to administration, industry and business need lots of information about supply and demand, prices of goods and services and purchasing power of the people, lifestyles and purchasing habits of people in different socio-economic segments, production and storage as also transportation facilities, government rules and regulations, taxes and subsidies as well as many other items.

Quantitative information may be provided in terms of distributions of some parameter(s) of interest, e.g. distribution of household consumption expenditure, the distribution being summarized to give quintiles or deciles which are of importance by themselves as also for planning other surveys with this expenditure being an auxiliary or a stratification variable. Similarly, information in the context of environmental management may be presented through multiple time series on concentrations of several suspended pollutants, separately for several locations. Instead of the figures comprising the time series data, even the corresponding trend curves could be available for information.

In some contexts, qualitative information may be crucial; thus, names of individuals or organizations involved in some specified activities, locations where some deposits of minerals or metals are possibly available, some undesired activities being planned by some individuals or groups, some new materials recently developed and used for some purposes of interest to some party, some new products or services entering the market and likely to disturb the overall market scenario, and a host of similar other items of information.

Currently a lot more of information is generated and gathered on a wide array of subjects across the globe than be processed even with the contemporary capacity available for this task. Not all processed information is used as bases for decisions and not many decisions are transformed into actions. Often un-informed decisions lead to actions whose outcomes are not always the desired ones. Informed opinions on crucial policy matters are not always available from the concerned groups or agencies. Little evaluation is taken up for many information items that involve non-negligible—if not considerable—resources.

12.3   Quality of Information

Quality of Information (sometimes referred to as Information Quality and abbreviated as IQ) has become an important subject for discussion, so that institutions meant to provide information inputs in decision-making and subsequent actions to be initiated, specially those affecting public interest, can put in place mechanisms to come up with high levels of IQ. Here one can come across two distinctions between related entities which merit our attention. The first is between Quality of Data and Quality of Information and the second is between Quality of Information as is inherent in the message or document containing it and Quality of Information as is received and comprehended by the user or recipient. There is the general understanding that data duly processed yield information. This processing could involve summarization and visual presentation, computation of some derived quantities and interpretation of such displays and computations. Thus, the manner in which this processing activity has been carried out will differentiate between Quality of Data—which depends on a whole host of controllable and some uncontrollable factors—and Quality of Information which will depend on the Quality of Data and the quality of data processing. And the latter is contained in the message or document, while the quality that eventually affects the information-based decisions of the user depends on quality of communication of the information—verbal or written. Consider the case of information relating to incidence of inequality in income or of poverty or of malnourishment among children aged 6 years or less derived from data on household consumption expenditure on items of food, clothing, housing, fuel and light and miscellaneous items obtained in a large-scale survey. Data quality will depend on various non-sampling errors and biases apart from sampling errors. Quality of Information will depend on checks on data consistency or absurdity or incompleteness as also on operational definitions adopted for inequality or poverty line or malnourishment. Even if data quality is reasonably good, with unacceptable or misleading or inconsistent definitions, Information Quality may suffer.

There are many organizations—essentially in the public as also private service sector—which are primarily involved in gathering and processing of data pertaining to different fields of enquiry or public interest and providing information usually in terms of serial publications released in the public domain. With a broad definition of Industry, any such organization can be branded as an Information Industry. And to ensure requisite quality in their outputs, viz. information items, and to take up initiatives to improve the quality of such information items, such Information Industries can also implement Quality Systems covering their processes and procedures. Quality Systems applicable to an Information Industry may need some modifications over systems that apply to manufacturing or other service industries.

In this chapter, the author first addresses the issue of objectively assessing Quality of Information and then proceeds to outline some changes needed by an Information Industry to implement a Quality System on the lines of ISO 9001. Based on his personal interactions with a public sector Information Industry, the author offers some operational definitions of concepts and procedures required to implement a Quality System there. It is hoped that the ideas suggested in this particular context will apply in similar situations.

Information can be treated as the output of a process carried out on data as inputs. The process could vary from a manual and simple display and/or summarization of the data to a very elaborate and complex analytic process using software on modern computing systems. Obviously, Quality of Information will depend on quality of processing, given the data inputs. It should be borne in mind that Quality of Information, unlike quality of a manufactured item, is more aligned to requirements of use and user.

Information Quality, taken broadly as quality of content and presentation of information pertaining to a subject or to the needs of a particular user or user group like consultants or research scholars, has engaged the attention of many investigators in the domain of Information Science and Technology as also of discriminative users of information. Quality of Information is, beyond doubt, a multi-dimensional concept requiring assessment of each dimension separately. The relative importance of the different dimensions may differ from one user to a second as also from one context to another. Further, Quality of Information is a dynamic concept, starting with only a few dimensions and recognizing new ones as advances in Science and Technology take place leading to enhanced user expectations and demands.

12.4   Assessing Quality of Information

Three basic quality parameters to comprehend Quality of Information can be identified as accuracy, adequacy or completeness and contemporaneity or timeliness (besides being up-to-date or current). Relevance to the need or purpose (as envisaged by the user) is taken for granted as a parameter. Even high-quality information may be irrelevant to a particular purpose or need, and it should be the responsibility of the user to look out for and obtain relevant information. Before proceeding further to discuss other aspects or dimensions of Information Quality, let us consider the implications of these three basic parameters. First, noting that such implications for each parameter may vary from one user or use to another.

To be accurate, information provided must be free from bias or computational errors or systematic errors caused by a wrong operational definition of an information item or error due to deficiencies in data which have been processed without necessary cleansing or checking or errors due to improper processing or even errors in presentation. Sometimes, internal consistency among different items of information is separately quoted as a quality parameter. Internal consistency is a crucial parameter to be checked before use of the information. The level of accuracy may depend on the use, and insistence on a high level of accuracy, especially in terms of data processing quality, may become cost-prohibitive. In the context of decision-making, accuracy is somewhat like assurance for robustness of decision/ action based on the information. The need for accuracy should be judged in relation to the sensitivity of the decision (based on the information to likely changes in information accuracy and to the level of accuracy achievable in a given framework, characterized by the information-gathering mechanism along with its resource boundaries).

Measuring accuracy of an information item or a collection of information items when information involves some number(s) attempts has been made to define accuracy in terms of the difference between the value contained in the information and some True (or Reference) value obtainable from some other (more credible) source. This is taken as the error. Sometimes, the error ratio (error/true value) is considered. Considering a group of information items, one can speak of the error rate as the percentage of items where the difference exceeds some acceptable limit.

Because of the inherent problem in dealing with (absolute) accuracy, internal consistency has been stressed as an important IQ. To establish this quality in an information, presumably containing multiple information items on a smaller set of intrinsic transactions of features of individuals about whom information is being sought, we should check whether logical relations which should connect some information items are satisfied in the information set presented to the user.

The next quality parameter, i.e. adequacy, is relative to purpose and requires that complete information as is needed should be provided, no facts or figures should be missing or hidden, no necessary computational result should be omitted, and all unit-level records should be appended or annexed with the document containing the information, whenever needed. Inadequate information bearing on some aspect of a phenomenon or a state may lead to a wrong assessment of the entire situation, in the absence of information related to the other relevant aspects.

While information about the past may be flowing in now, what is needed is to secure and deploy information revealing the current situation, may be with as little a delay as possible. The current profile is what is needed, and information about the past may help comparing this profile, e.g. of air at a certain height above the ground level over time. Information production from data has a cycle time, and similarly, there is a cycle time for valid use of an information. As timeliness or currency is insisted upon, cycle time reduction in these processes through adoption of new processing technology becomes a real necessity.

As more and more organizations surfaced to provide information on diverse items to a vast number of users, questions about authority, authenticity and resultant credibility became important IQ parameters. Several authors have provided somewhat long lists of such parameters. A look at Wikipedia reveals a list provided by Wang and Strong (1996) which classifies these parameters into four groups as follows:

Intrinsic IQ: accuracy, objectivity, believability (credibility), reputation;
Contextual IQ: relevance, value addition, completeness, amount of information (adequacy);
Representational IQ: interpretability, format, coherence, compatibility; and
Accessibility IQ: accessibility, access security.

Some other authors offer a slightly different list of eight characteristics of IQ, of which security and conciseness were influenced strongest by employees' general satisfaction levels:
Accessible, Accurate, Believable, Complete, Concise, Consistently Represented, Secure and Timely. There could exist trade-off relations between some of these characteristics, e.g. accessibility and security.

Some agencies in the private sector and even some government agencies with inadequate facilities and competence in planning and conducting field surveys to collect credible data either because they fail to work out a good rapport with the respondents or because they cannot engage experienced field staff are also providing information on issues of public interest. Information by way of summary figures released by such agencies sometimes do not agree with corresponding figures released by a duly authorized government agency equipped well to plan and conduct background surveys, and they create confusion among users. Thus, the question of credibility of information linked up with credibility of the organization providing the information becomes quite important. Credibility of the organization ensures believability of the information it provides. And in this context, compatibility with comparable information provided by other authorized agencies—presumed to be credible—becomes an important parameter of IQ. Another important parameter is verifiability in terms of all supporting data from which the information has been derived being in place along with the procedure adopted to process the data and derive the information.

In recent times, institutional users of information (generated within or outside the institution) are more concerned about information security. And quite often this Quality of Information is explained by the popular triad CIA (confidentiality, integrity and accessibility). Wikipedia defines Information Security as 'the practice of preventing unauthorized access, use, disclosure, modification, inspection, recording or destruction of information'. However, Information Security with all its ramifications and connotations and their impact on business, defence, research, administration, banking and finance, etc., should be regarded as a critical quality parameter of an information or of an Information System. Loss of Information Security poses a grave risk and risk management is done through various methods like administrative, logical and physical controls, access control in terms of identification, authentication and authorization processes and cryptography.

The triplet of confidentiality, integrity and availability has been sometimes referred to as security attributes, security goals, critical information characteristics and basic building blocks of information security. Accountability has sometimes been added to the triad. The OECD, in its 1992 Guidelines for the Security of Information Systems and Networks (revised in 2002), proposed nine principles, viz. awareness, responsibility, response, ethics, democracy, risk assessment, security design and implementation, security management and reassessment. There have been other modes and sets of key principles. Important would be the ISO 27000 standards.

Confidentiality is the property that information is not made available or disclosed to unauthorized individuals, entities or processes. Information integrity (or rather data integrity) means maintaining and assuring the accuracy and completeness of data over its entire life cycle. This implies that data cannot be modified in an unauthorised manner. Availability implies that the information must be available when it is needed. Ensuring availability also involves preventing denial-of-service attacks, such as a flood of incoming messages to the target system essentially forcing the same to shut down.

12.5 QMS in an Information Industry

The service sector will play an increasingly important role in the economic development of a country, not merely in terms of a growing contribution by various services to the gross (net) domestic product, but also through enrichment of the manufacturing sector by more effective and efficient support services. In particular, information support—technological, financial and marketing—for better planning and coordination of production, distribution, pricing, import and export and even of product use in core sectors like iron and steel has gained paramount importance for the survival and growth of such a sector in the fiercely competitive global market today.

Earlier, such information services, especially those relating to items under state regulation, were highly bureaucratic, cost-ineffective, provider-oriented and not-so-transparent. Today, these administrative setups have to become information industries which should compete with manufacturing industries in customer orientation, profit motivation, process management, people involvement, etc. Such industries should embrace requirements of ISO 9000 standards for Quality Systems and should further implement harmonized systems and procedures for quality, cost and environment management.

The aim of this section is to consider some basic conceptual issues which arise in the context of Quality Management in an Information Industry. Usual interpretations of various terms and phrases that suit the manufacturing and the service sectors require appropriate modification before an attempt is made to implement a Quality Management System on the lines of the ISO 9001 standard. And this is the limited task of this author. No attempt has been made to discuss models and methods for implementing a Quality Management System on the lines of ISO 9001 or any other comparable Quality System standard in an Information Industry.

One can define an Information Industry as an organization that procures data (facts, figures, opinions, complaints, etc.), processes such data into information and provides such information (by way of responses to queries and requests and also in terms of regular or ad hoc reports to those who explicitly ask for as well as those who indirectly use the information).

Today every country has a National Statistical System, elaborate in some cases and concise in some others, which provides a variety of information relating to economic, social and political conditions of a country over time. Besides the official agencies engaged in information services (generation, authentication and dissemination of information), there exist non-official organizations which also cater to needs of information by specific user groups. In India, we have a separate Ministry in the Union Government as well as in some state governments branded as Ministry of Statistics and Programme Implementation. In some countries, including India, a National Statistics Commission functions as a coordinating body. All these organizations illustrate what we have named as information industries.

Let us consider the case of one such Information Industry which, in the traditional perception, is an autonomous organization (attached to a federal ministry dealing with a basic industry) created to provide the industry with a variety of information related to production, price, demand, dispatch (from each unit of the industry), technological advances, export markets, etc. In fact, representatives of the more important units of this basic industry happen to be members of this organization which also includes several bureaucrats in the government. The organization is manned by sufficiently educated and skilled personnel having adequate physical and material resources at their disposal and enjoying a comfortable work environment. It has several units spread throughout the country to collect raw data from industrial units, local markets, customs offices and other relevant sources and to transmit these to the headquarters for subsequent processing. For the present, information is provided free of cost though the industry occasionally accepts requests to collect, collate and provide information of a special nature on payment.

The very concept of an 'Information Industry' was not initially acceptable to the semi-government organization that was keen to preserve and project its superior non-industrial character. There was some appreciation that a service organization is a better characterization and that whatever applied or oriented research was being carried out by one of the units of this organization is not amenable to standardization and systematization. The wide implications of terms like process, product, design, test and inspection status, inspection and measuring equipment, etc., were not properly understood. And, of course, terms like customers, contracts, non-conforming products, traceability and identification were unfamiliar to the staff.

Such a scenario is possibly not unique and not typical of a particular country. There are genuine problems in perceiving an information agency as an industry and in translating the commonly accepted quality vocabulary into operational language relevant to such an agency. Thus, the very first step in implementing a Quality System in such an organization would be to develop reasonable operational definitions of terms and phrases that pervade such a system.

Some Quality-related Terms and Their Explanations

a. Product:

A product is a piece of information (resulting from processed data or lower level information pieces) provided in the form of a statement (as a matter of course or in response to an enquiry/a request/an advice) or a report—released regularly or ad hoc. A printout or a publication containing the information is an augmented product and coupled with necessary guidelines, explanations, notes, etc., that facilitate appreciation and use of the information we have an extended product.

The piece of information may have minimal content, e.g. a single figure relating to the burden of a specific disease in a specified geographical area or the percentage of people below the 'poverty line' in a state or a district. On the other hand, the term 'product' in this context may refer to a comprehensive report of an Employment–Unemployment survey.

b. Quality (of a Product):

Quality of Information (as a product) is not easily defined. However, some of the important quality characteristics—not all of which admit of absolute measurements—are (i) accuracy, (ii) adequacy and (iii) currency or contemporaneity. Of course, authenticity and acceptability are also quite important.

Accuracy in terms of accepted standard operational definitions of various concepts like 'disposable income' or 'expenditure on account of a domestic tour' or 'consumption of a certain item during a reference period' and the like being used and response errors and biases being absent is important in all information products arising from a census or a sample survey. In the case of sample surveys, relative standard errors of estimates thrown up by the survey for the parameters of interest do provide some idea about accuracy.

The information product should be adequate for use as inputs to planning, in programme implementation as also in monitoring and evaluation of programmes and projects. This, in turn, may require disaggregated details up to a certain level, instead of macro- or meta-level aggregates only. Thus, state-level figures may not suffice for block-level planning and monitoring.

Socio-economic scenarios and related features are undergoing fast changes, and we need data regarding the current levels of different indicators as also regarding changes in such levels over time. Sometimes, we provide information based partly on some features relating to a previous period of time. e.g. when we offer Consumer Price Index Numbers for a particular segment of the population and the base year is quite in the past to the effect that data from a family living survey carried out in the base year fails to reflect a vast change in the consumption pattern of the people in the particular segment, the credibility of the Index Number suffers.

Another aspect or dimension of quality concerns the compatibility or comparability of the information item(s) with similar information dished out by other organizations within the country—in the public or private domain—including international agencies.

An assessment of Quality of Information given out by a particular organization or agency is reflected by the demand for information currently provided as well as currently unavailable from the organization by existing or potential users.

As applicable generically to analysis of quality of any product, here also quality of the final product released to the customer by an Information Industry—on-line or off-line—will have three determinants, viz. Quality of Design, Quality of Conformance (to design requirements) and Quality of Performance. Further, quality due to Product Support provided to the customers is also quite relevant.

Quality of Design is understood in terms of the clarity and specificity of the scope and coverage of the information (in a particular project like a particular survey or a particular round of an on-going survey), the sampling design, the clarity, comprehensiveness and user-friendliness of the Instruction Manual if primary data have to be collected from the field, the scrutiny of data collected and actions to be taken on detection of deviations from the specifications, the tables to be prepared, the way the report is to be rendered as also the manner of dissemination of the information.

c. Customer:

For an Information Industry, differences between seen and unseen users and that way between current and prospective customers are worth an examination. While the Information Industry caters—on or without payment—to the needs of those users who approach it or even direct it, its products are all too easily accessible to many remotely placed and indirectly connected users—mostly without any payment.

Some customers—in terms of a direction or a request or an advice—communicate their information needs to the industry, and the industry is required to take such needs into account. These could mean some statutory bodies or some industry or Business Association or some Chambers of Commerce. Beyond these visible and directly approachable customers, there exist many users, from whom information needs can be sought.

In fact, the concept and practice of Contract Review have to be appreciated in a manner that differs markedly from their counterparts in the manufacturing and similar industries. There is really nothing analogous to a contract, except in specific sponsored surveys, though an Information Industry has to plan its activities keeping in view Potential Needs and Expectations of possible users of its products. These can and should be analysed among its own people, may be in meetings wherein representatives of different sectors of potential users are invited.

d. Inspection:

Speaking about inspection of incoming materials to an Information Industry, inspection really refers to the checks carried out on data flowing in or collected regarding internal consistency, apparent accuracy, trend compatibility, input datum (that is not many received by way of computer input devices) and development of appropriate software to apply these checks on data entry to computers meant to process the incoming data.

In terms of test and inspection status, the software mentioned earlier should also 'hold' any data set not meeting the stipulated checks. Thus, the products are the inspection equipment used and their maintenance has to be carefully done.

e. Quality Policy

Once these terms and phrases were discussed among senior executives in a particular organization and agreed definitions/explanations were finalized, a Quality Policy statement was documented after a comprehensive and prolonged debate on its various implications—organizational and functional. This runs as follows:

To organize all its quality-related activities within the framework of a Quality System, geared to provide complete satisfaction to all the users and beneficiaries, through error-free and regular publication of Reports/Periodicals and rendering services as per the need and requirements of the national and international segments.

Our strategy is

  a. To ensure quality in all acquired data, related to performance of iron and steel industries including import and export from all major Indian ports;

b(i). To use knowledge and skills of the people at all levels of the organization and to enhance their motivation and participation;

b(ii). To integrate efforts of all our employees to provide satisfaction to our end-users by ensuring quality products and services;

b(iii). To impart necessary training to the personnel and providing them necessary resources for achieving quality standards;

c. To emphasize computerization in all sections/departments/regional offices;
d. To improve communication and interactions within the organization; and
e. To develop a system/mechanism of getting feedback from our end-users/
   beneficiaries for effecting improvement in quality of our products/services.

f. The Procurement Process

While materials and equipment as also accessories and consumables are to be procured through purchase from approved vendors more or less in the same manner as in any other industry, the most important material unique to an Information Industry is the body of data which have to be processed to yield the desired information. And quality of the incoming material in an Information Industry needs appropriate definition and, even, measurement.

In a situation where the industry collects data of its own by engaging field investigators, developing appropriate questionnaires or schedules, preparing the Instruction Manual, training the investigators and supervisors, carrying out adequate checks during field operation, and scrutinizing the filled-in questionnaires or schedules for possible gaps or mistakes or inconsistencies, the task of procurement is more onerous than the task faced by an Information Industry that collects secondary data from relevant sources.

The most important step in data collection through a sample survey is to develop a proper sampling design. Problems of defining the sampling units, obtaining or developing a sampling frame (a complete list of ultimate sampling units), of noting the parameters to be estimated, of deciding on the relative standard errors of the different estimators which would be acceptable, of determining the total sample size, of identifying the need to stratify the population of interest and of suitable stratification variable(s), of allocating the total sample across different strata and of using appropriate estimators have to be adequately addressed. Validation of data is also of great importance and provides an effective tool for inspection at most levels.

g. Preventive and Corrective Actions

In the case of an Information Industry, it may be quite important to remember the distinction between a non-conformity that simply makes the product appear less attractive or one that the user can easily correct on the one hand and a non-conformity that renders the product unfit for use, e.g. figures given out in a document which are not mutually consistent and cannot be confidently corrected by the user. This implies the need on the part of the organization to do careful examination of results before these are disseminated and even after the release. Some deficiencies will be reported by users subsequently.

Reasons for all types of non-conformities internally detected or pointed out by external sources should be investigated and traced to some activities which were missed or not carried out properly. Inadequate training of investigators, inadequate supervision over data collection, inadequate scrutiny of data collected, inadequate instructions for tabulation of data, improper interpretation of results and the like. Accordingly corrective actions in regard to non-conformities detected should be first eliminated to the extent possible. Beyond this, the organization has to hold discussions, reach decisions and implement actions to exempt future deficiencies.

h. Document Control

In an industry that collects or compiles information items of interest to many individuals and groups on a continuing basis, the types of information to be collected during one round or period of time, the way those are to collected or compiled, the manner in which those should be disseminated and similar such decisions are often taken during meetings of concerned bodies. The proceedings of such meetings along with the decisions reached there become important documents which should be duly authenticated by the concerned authorities and preserved even beyond the completion of work during the relevant round or period for checking the practices followed with the ones decided on and documented as also matching the information items actually collected or compiled with the expected items. In addition, document required to meet the provisions of ISO 9001 also falls under the process of Document Control.

i. Management Commitment

This is quite often a serious problem in an Information Industry and more so in a public sector organization or a government agency. Such organizations are characterized by transfer of executives from one organization to another, sometimes totally unrelated, organization. Whatever commitment or involvement they exhibit or whatever leadership they can provide personally to the Quality Management System within an organization may not necessarily be picked up and displayed by those who succeed them. More than that, those who join may not like the Quality System to operate in the same manner as with their predecessors. This is particularly found to be true for the most activities like preventive and corrective action as also with improvement efforts.

12.6 Concluding Remarks

A strong and effective information system is a critical-to-success element in any organization. This could be a self-sufficient part of the organization or could draw upon the strength of selected agencies to gather adequate, credible and timely information about different activities or tasks carried out by the organization. In most situations, information to be procured has to be secure. In the present age of information, we should obviously expect growing investigations on how to get hold of information that we need in the manner that we desire. And as competition grows, Information Quality and security become top priorities.

Information Industry is the backbone for National Development Planning, and units in public or private sectors in this Industry owe a lot of responsibility to their users. Hence, they should try their best to implement Quality Systems in their operations.

References

Dhillon, G. (2007). *Principles of information systems security: Text and cases*. Hoboken, NJ: John Wiley and Sons.

English, L. P. (2009). *Information quality applied*. Indianapolis: Wiley Publishing.

Fehrenbacher, D. D. (2016). Perception of information quality dimensions from the perspective of commodity theory. *Behaviour and Information Technology, 35*(4), 254–267.

Layton, T. P. (2007). *Information security: Design, implementation, measurement and compliance*. Boca Baton, FL: Auerbch Publications.

Miller, H. (1996). The multiple dimensions of information quality. *Information Systems Management, 13*(2), 79–82.

Wang, R., & Strong, D. (1996). Beyond accuracy: What data quality means to data consumers. *Journal of Management Information Systems, 12*(4), 5–34.

Chapter 13
Quality of Life

13.1 Introduction

Starting with quality of manufactured products, we began discussing quality of processes and went on to analyse quality of services. We also characterized the quality of our environment in terms of qualitative and quantitative aspects of its abiotic components, viz. soil, water and air as also the diversity in its biotic components, viz. plants, animals and micro-organism communities. More recently, we experience many different norms and regulations influencing the way we behave and work. As a natural extension of our concerns for quality, we should talk about Quality of Life. And, in this context, we come across wide disparities among different segments of the human population—in different countries, within the same country and within the same social or religious group—in respect of the situations in which they live and work and the extent to which they can satisfy needs for their self-fulfilment. Looking from a different angle and going back to our earlier engagement with quality of products, processes and services, we find instances where we appreciate the quality of some produce and where the people behind are —within or beyond our knowledge—denied of many basic necessities of life. We also come across situations where people who have access to and who actually possess adequate opportunities for satisfying their needs offer goods and services that fail to satisfy the users. Do we then find a link between Quality of Life available to a producer and the quality of the produce? And does such a link require a definition of Quality of Life that is somewhat different from quality of living?

Dwelling on this subject, questions that plague our minds are like the following:

Is it at all relevant to discuss Quality of Life in the context of quality and more so in the context of Quality Management?
Can we come up with at least a largely agreed definition of Quality of Life?
What are the possible dimensions of Quality of Life?
Should Quality of Life be the same for all within the existing framework of our existence?

© Springer Nature Singapore Pte Ltd. 2019
S. P. Mukherjee, *Quality*, India Studies in Business and Economics,
https://doi.org/10.1007/978-981-13-1271-7_13

Can we identify the determinants of Quality of Life?

How does quality of our environment including that of 'built' environment impact Quality of Life? And, can the latter assess quantitatively to allow comparisons to be made?

Can the concept of Quality of Life be free from socio-politico-economic set-ups? Can we change the level or position of any such determinant to improve Quality of Life?

How do we quantify Quality of Life in terms of measurable aspects reflecting on the different dimensions of Quality of Life?

Do we have credible data to build up measures for each such dimension?

How do we combine measures reflecting on different dimensions into a single index?

Does Quality of Working Life of an individual or a group of workers within an organization affect the overall Quality of Life for the extended group of people depending on these workers?

How is any such index related to an indicator of development—social, economic and political—of any segment of the population?

Quality of Life as a concept as also in terms of a quantified indicator has been a matter of interest and even concern by all sorts of people including those not directly dealing with this subject in their core activities. Just to mention the Central Mining Research Institute which, of course, has a stake in the Quality of Life of miners and people living in mining areas, has put in some materials on the concept in their website. The concept in particular has been examined discreetly by various learned men and women across the globe over the centuries whose precepts about life and living tend to impact upon Quality of Life. Attempts have been made over the years by individuals as well as institutions to answer these and related questions through the development of different models and measures. In fact, people tend to compare Quality of Life in one country with that in another, or between two different cities or towns within the same country in terms of some such measure or index. Even international comparisons are being attempted using some index or the other.

It must be admitted that any discussion on Quality of Life, howsoever comprehensive, cannot be conclusive. It can only highlight the problems involved in treading this slippery ground and indicating the available solutions along with their limitations. The present chapter is one contribution in that direction.

13.2 Quality of Life 'Multidimensional'

Quality of Life is essentially a multidimensional concept, and one can identify these dimensions from different perspectives. It is important to remember, in this connection, that Quality of Life has two distinct dimensions, one personal and the other institutional. The first reflects the attitude of an individual towards life, determines his/her needs and desires and delineates his/her adjustment to the environment that encompasses physical resources and facilities available to the individual. The second is provided to the individual in terms of goods and services made available to the individual by the state (taken in a broad sense). The latter includes availability of and accessibility to facilities for housing, food and clothing, education, health care, sanitation, recreation and a host of other earthly entities. Quite often, it has been argued that the second dimension subsumes the first in the sense that without the second the first makes little meaning to the majority of people. Thus, economics of production and of distribution get the upper hand. The big assumption here is that if an individual does have access to the required level of consumption of goods and services, irrespective of individual attitude and aspiration, Quality of Life enjoyed by the person is high. Contrary to this view, some others hold that Quality of Life is in terms of a comparison between what one gets and what one expects in life. And the latter is moulded strongly by the influence of culture, and traditions and mores, of social and economic norms. Though the second dimension may provide facilities and even impetus for improving Quality of Life to a certain extent, any striking improvement in the Quality of Life enjoyed by a population depends primarily on individuals, the gifted and the aspiring individuals.

Attention has been generally paid to the institutional or, in some sense, the collective dimension of Quality of Life and to the development and use of standardized measures or indices that relate to a specified population group like a country or a region within it, a state, a community or a class of people or the like. These take into account availability of, access to and consumption of goods and services that are considered as determining or reflecting on the Quality of Life. As can be expected, such measures are based on available data on such determinants and correlates of Quality of Life and worked out on a completely unwarranted assumption that the parameters bearing on the determinants and correlates depict a uniform distribution across individuals. In fact, any parameter that reflects the Quality of Life as enjoyed by one individual or as some features that determines the former is known to exhibit variation from individual to individual, even when relatively homogeneous groups of people are considered. In fact, differences among individuals in any group in respect of attitudes and aptitudes, values and ethics, as well traditions and cultures are likely to be quite considerable. And hence, any collective measure is likely to have an altogether different composition and connotation compared to the perceived Quality of Life enjoyed by an individual.

13.3 A Philosophical Note

It is difficult, but also perhaps unnecessary, to seek a very precise definition of Quality of Life. It is partly a question of 'to have' and partly one of 'to be'. We need not also go by simplistic formulations such as 'the less you have, the more you are', or Gandhiji's concept of voluntary poverty or the Biblical saying 'blessed are the poor'. Each of them may have some validity in a larger moral contest, but one cannot really preach ethics to hungry people nor extol the virtues of non-acquisition of those who have nothing. Besides, a minimum, or should we say optimum, level of living is essential for ensuring happiness which is the ultimate objective of development, self-fulfilment, self-realization and the various other attributes of the Quality of Life. Here again, there could be debates about the 'optimum' level (the more you have, the more you want, etc.) and even the parameters that determine the Quality of Life. We have struggled with human development indicators, tried to refine data and work out alternative frames for analysis. Different cultures in different historical times and as on today have different notions of what determines the Quality of Life. But venturing a generalization one may say that it is basically a balance between 'to have' and 'to be', between material possessions for satisfaction of immediate physical wants (food, clothing, housing, health, etc.) and improvements of the self, more knowledge, more emotive satisfaction, more of the sense of fulfilment, a better and brighter 'self'—by itself and in relation to others.

True and sustainable Quality of Life should take care of man in entirety. To maintain the Quality of Life in a society and to upgrade it further, it is necessary to provide proper education to harmonize the external life with the internal life of individuals so as to attain an optimum considered beneficial for individuals and the society. Vices like greed goad man for more conveniences, facilities and privileges. But greed is insatiable. It endangers the Quality of Life of an individual who pursues his unbridled greed. This greed invites corruption and other dishonest practices which are bound to adversely affect the Quality of Life of others in the society. To check such tendencies, a society has to put adequate restrictions on such practices that threaten the Quality of Life of the society. Apart from restrictive and remedial measures, the society must place some high ideal before its members, the pursuit of which will help individuals and the community improve their Quality of Life. Every society is influenced by some role models who uphold such ideal in life.

It is said that unless a person enjoys good health, he or she cannot have a good Quality of Life. For this, individuals have to take proper care of themselves in terms personal hygiene, physical exercises and controlled habits. Of course, the state has to provide adequate healthcare services to all its members.

Quality of Life depends a lot on quality of environment looked upon as our surroundings like soil, water, air besides plants and animals. A healthy environment is a bad necessity for a good Quality of Life, and this implies a pollution-free environment.

13.4 Quality of Working Life

A discussion on Quality of Working Life is surely of great significance in an attempt to capture Quality of Life comprehensively, since this affects the Quality of Life of not just the workers but also of those who depend on them and are affected by them. Most people spend a large fraction of their life at work and some of them enjoy this situation while some others are forced to accept the situation. Studies have revealed that a low level of well-being at work leads to a loss of productivity and, thus, a drop in national income.

In this context, one can consider a relatively small group of people working within and for a particular organization—a manufacturing unit, a service provider, an educational institution, a research laboratory, a social welfare organization, a government department or a similar establishment. A model offered by a Japanese exponent of Quality Management K. Ishikawa to explain Quality of Working Life is worth consideration in view of its originality and general applicability. The model is in terms of a set of equations, each equating the product of two characteristics as one determinant of Quality of Working Life. One can easily take the 'product' as 'the composite of'. And this model eventually speaks of Quality of Life experienced by an individual worker. One can possibly aggregate or combine these measures for individuals to come up with a collective measure for the organization.

$$Knowledge \times Skill = Ability$$

$$Attitude \times Environment = Motivation$$

$$Ability \times Motivation = Individual\ Performance$$

$$Individual\ Performance \times Organisational\ Resources = Organizational\ Performance$$

$$Organisational\ Performance \times Society = Quality\ of\ Working\ Life$$

Thus, the determinants of Quality of Life as is being enjoyed by an individual worker depends on his/her knowledge in the domain of work and his/her skill to act, to react and to interact in order to translate the knowledge into performance on a task. Also, contributing to performance will be attitude towards the task (s), the organization and its management. These alone cannot explain the individual's performance. As one determinant of motivation, we should take into account the environment or the organizational climate in terms of relations among peers, between peers and superiors, and between self and supporting members. This climate is partly created by the organization in terms of certain facilities and procedures determined by the management as also by the worker as an individual as influencing the given environment. This concept of Quality of Working life is no doubt a collective concept that is closely linked up with the performance of the organization where the people work as also on the impact created by the organization—its values and ethics, principles and practices, concerns for and compliance with societal needs and norms—on the society within which it operates.

Even for a single individual, who has to work for or within an organization, this concept is quite relevant. This model links up Quality of Working life with the level of performance and the resulting satisfaction and gratification. It may be noted that this model connects Quality of Working Life to attitudes and aptitudes of individual workers as also to resources available to the organization and, quite importantly, the image of the organization and its impact on society. Going a bit deeper, one may note that an individual joining to work in an organization carries with him/her a bag of knowledge and skill and may be placed on a job that matches—more or less, if not perfectly—this bag. And the individual is satisfied and tries to perform to the best of his/her ability (as the composite of knowledge and skill). However, the individual subsequently undergoes programmes—in-house or otherwise, sponsored by self or by the organization—to augment the initial stock of knowledge and enhance the initial level of skill. Unfortunately, however, the individual is more often than not put back on the same job, may be with a higher designation and/or better compensation. The individual now finds the old job to be less satisfying, demanding a much lower level of knowledge or skill than what is currently possessed by him/her. This affects the individual's attitude towards the organization and, more so, about the job. And according to this model, his/her motivation diminishes, eventually causing less-than-expected performance and leading to a poorer Quality of Working Life.

Several other models for Quality of Working Life including a wide range of factors have been offered by researchers over the years. Some of them drew attention to what they described as psychological growth needs as relevant to Quality of Working Life. It can be argued that acquisition and application of varied skills to perform diverse jobs within a broad framework are looked upon by a majority of workers as providing a better work environment. Every sensitive worker likes his/her job to be rated as significant to the organization by peers as also superiors. Barring a few nagging individuals who tend to shirk responsibilities, workers like to be identified with the tasks they are on. They also expect some autonomy to decide how best to carry out the tasks assigned to them. This does not imply any disregard for instructions given to them by superiors. Feedback on their performance rather than a simple reprimand for some job not properly done or a simple silence when a job has been completed well is also expected by many workers.

While models focusing on needs for psychological or ego satisfaction have been around as experience-based theoretical constructs, Taylor (1978) more pragmatically identified the essential components of Quality of Working Life as basic extrinsic job factors of wages, hours and working conditions as also intrinsic job notions of the nature of the work itself. He suggested that some of the following factors could also be added, including individual power, employee participation in management, fairness and equity, social support, use of one's present skills, self-development, a meaningful future at work, social relevance of the work or the product and effect on extra work activities. Taylor rightly pointed out that concepts and components of Quality of Working Life may vary according to organization and employee group (within the same organization).

Commenting on the present Quality of Working Life in Australia, Watson et al. (2003) in their book entitled Fragmented Futures: New Challenges in Working Life remark that working life has become more fragmented as a result of significant social and economic changes in the last quarter of the twentieth century. Such fragmentation has been the result of unemployment and underemployment, structural changes in the industrial sector of the economy, emergence of automation and consequent disappearance of certain types of manual jobs and occupations, growth of non-standard forms of employment often requiring non-traditional skill sets, longer hours of work causing greater physical and mental stress, retirement and superannuation at fixed ages and the like. Prosperity resulting from greater and better inputs from employees is not shared among all stakeholders evenly. Diversity, choice and opportunity have diminished, and exclusion and inequality have increased.

In fine, Quality of Working Life is apparently analogous to overall well-being of employees, but differs from job satisfaction which represents the workplace domain only. Quality of Working Life incorporates work-based factors such as job satisfaction, satisfaction with pay and relationships with work colleagues, but also factors that broadly reflect life satisfaction and general feelings of well-being which are affected by happenings in the non-work domain. Time disposition in family and in society along with the satisfaction derived therefrom also plays an important role in building up the experience about Quality of Life.

13.5   Quality of Environment

It is just natural to expect that Quality of Life at the individual or the collective level will be influenced by the environment in which we live, move and work. Environment is in terms of all that surrounds us—not just in our immediate neighbourhood, but anywhere outside that can exert some influence on our lives. Quality of our environment is understood basically in terms of quality of the abiotic compartments of this environment, viz. land, water and air. Of these three, quality of land does not directly affect quality of our lives and we remain concerned about quality of water that we drink and use for other purposes and air that we breathe. Water and air quality have been being adversely affected by many anthropogenic activities including those which tend to enhance the quality of our living, providing more of conveniences and comforts, options and openings.

Quality of Life depends a lot on quality of water and of air. This is more easily realized in terms of the consequences of poor air and water quality on human health and longevity. In fact, an increasing proportion of people has been getting exposed to risks of various fatal diseases and even succumbing to such diseases. This fact is established beyond doubt and need not be substantiated by numbers. This appreciation has led to the development of air quality indices and water quality indices and their relations to human health, introduction of regulations to keep water and air clean and free from pollutants.

Air quality is assessed in terms of concentrations of several pollutants suspended on ground or in the atmosphere and then combined into some unit-free indices. For each index, a few categories of air quality are spelt out and their impacts on human health are identified. Different countries come up with different indices based on varying numbers of parameters monitored. In many countries, five parameters are measured, viz. concentrations of ground-level ozone, along with particulate matter, carbon monoxide, sulphur dioxide and nitrous oxide suspended in air. An index ranging from 0 to 500 is obtained. A value of the Air Quality Index (AQI) less than 50 is good, while a value exceeding 300 is hazardous. In India, eight parameters are monitored including particulate matters PM_{10} and PM_{25} (instead of simply PM), ammonia and lead. Six quality categories are spoken of to cover the range 0–500 for the index, viz. good (–50), satisfactory (51–100), moderately polluted (101–200), poor (201–300), very poor (301–400) and severe (401–500).

Water Quality Index is obtained similarly with separate indices available for quality of drinking water and river water quality as also groundwater quality. The minimal index is based on only three parameters, viz. dissolved oxygen, turbidity and total dissolved solids. Other indices take into consideration parameters like pH, biological oxygen demand (BOD), faecal coliform count.

There are genuine problems of monitoring concentrations, stocks and flows of the pollutants at different locations and times and relating these to human health hazards, because of many aggravating or ameliorating factors which get confounded with the impact of air and water quality.

13.6 Assessing Quality of Life

Considering Quality of Working Life as an important component of Quality of Life —of course not for the entire population and not taking care of factors outside the work domain that influence Quality of Life. There exist a few measures of Quality of Working Life that have been found to have reliability and validity. These are **the Brief Index of Affective Job Satisfaction** and the **Work-Related Quality of Life Scale** along with its sub-scales. In fact, the Work-Related Quality of Life (WRQOL) scale does include the job and career satisfaction scale, the general well-being scale, the stress at work sub-scale, the control of work sub-scale, the home–work interface scale and the working condition scale. Thus, we get a detailed set of metrics to gauge the Quality of Life of an employee or a worker that is slightly more comprehensive than the usual measures of Quality of Working Life.

There is little doubt that people need adequate food of right quality, adequate clothing, proper shelter, basic education, primary health care, gainful employment, etc. It is right and proper for our planners to concentrate their attention on provision of such necessities for the entire population and improvement of the available basic services. But they also need to recognize that 'man does not live by bread alone' and that good life is more than eating and drinking. Economic development and adequate provision of material needs do not ensure good life or human well-being. The ultimate aim of all economic activities is to contribute to the general 'well-being 'of the people. In this context, development is ultimately about people and not about 'things'; it is about 'being' rather than 'having'; it is about Quality of Life and not about quantity of goods and services procured and consumed. This universal truth is sometimes forgotten by our planners. The fact is evident in the ways our cities are allowed to grow into slums, our crime rate is climbing rapidly, our cultural heritages are neglected, and our religious precepts are cast aside.

There have been two perspectives of human development and, hence, of Quality of Life in recent literature on the subject and correspondingly two different approaches to its measurement. One is called the output or constituent perspective and the other the input or determinant perspective (Dasgupta 2001). In the first, we focus on different aspects or dimensions of Quality of Life in terms of which the present state prevailing in a country or a region or a population segment can be understood. The second is more important in drawing our attention to the determinants of the present state, their adequacies or shortfalls. To initiate steps for improving Quality of Life, it is necessary to identify the current shortfalls in these determinants or factors along with measures which can mitigate these shortfalls. Of course, data availability questions may sometimes require us to accept only the first option.

Five major goals for the development process (going beyond the much-hyped economic development) have been generally agreed upon. Development should meet basic needs of food, shelter, clothing, sanitation, education and health for all people, should ensure social justice, should enable people's participation in the development process, should enhance the level of self-reliance among the people and should enrich the Quality of Life enjoyed by the people.

Quality of Life is a broad concept that encompasses a number of different dimensions (by which we can comprehend the factors or elements that make up for the whole entity and each of which can be measured through a set of sub-dimensions with an associated number of indicators for each). In this context, we have several objective factors (like employment status, living conditions, access to resources needed for productive work, capacity to afford recreation and entertainment, access to public services) where we can assess the nature and level as are available currently with reasonable confidence. At the same time, we have to involve a number of subjective factors also in terms of the perceived need, the facility enjoyed and the satisfaction or otherwise with the currently available resources and facilities. The latter significantly depends on the citizens' needs and priorities and these in turn depend on their realization of the limitations imposed on the availability of resources and opportunities. It is clear that in dealing with perceptions or realizations and experiences, we have to involve latent variables for which proxies have to be obtained from responses to carefully crafted questions and, subsequently, scaled in an appropriate manner.

Assessing Quality of Life in measurable terms that can enable fair comparison across different populations and countries is a complex task. A single measure by way of an index may serve the limited purpose of ranking different situations (countries, population groups, periods of time) and motivating the respective parties to look closely at the determinants or factors and their current adequacies or otherwise.

A big hurdle is posed by the non-availability of credible data reflecting on the states of the dimensions during a reference period. The National Statistical Systems in quite a few developing and underdeveloped countries are hardly equipped to compile data bearing on the different factors or determinants on a regular basis and with reasonable precision. The choice of indicators to be considered for each determinant has not been uniformly laid down and accepted. Specially designed surveys and non-conventional methods of analysis may have to be adopted to throw up the required indicators.

Nine dimensions have been recognized by OECD based on academic research and several initiatives. These can be formulated as follows

Material living conditions (income, consumption and material conditions);
Productive or main activity (both market and non-market activities to be included);
Health (a multidimensional concept realized through several facets like morbidity, mortality at different ages);
Education (including adult literacy, transfer ratio among school-leavers);
Leisure and social interactions (facilities and times spent);
Economic and physical safety (looked upon as cornerstone of development);
Governance and basic rights (as protected through governance);
Natural and living environment (determined by governance as also people's participation) besides; and
Overall experience of life (as perceived by the people).

For material living conditions, several different indicators for which data exist or can be collected have been proposed for each of the three sub-dimensions, taken from both national accounts and household surveys (net national income, household disposable income). [In some countries, household disposable income figures are not directly collected. Instead, we get data on household consumption expenditure.]

For the second dimension, both the quantity and the quality of jobs available (working hours, balancing work and non-working life, safety and ethics of employment) are some of the indicators used in Europe.

Health conditions can be measured in terms of objective health outcome indicators like life expectancy, infant mortality, number of healthy (disease-adjusted) life years as also more subjective indicators such as access to healthcare facilities and self-evaluation of one's health.

Currently available indicators of education that are relevant for Quality of Life in Europe include number of early school-leavers or dropouts, self-assessed and externally assessed skills and participation in lifelong learning.

In Europe, the dimension 'leisure and social interactions' is measured in terms of answers to such questions as how often citizens spend time with people at sporting or cultural events, do they volunteer for different types of organizations outside work life, etc. In addition, the potential to receive social support and the frequency of social contacts are also taken as indicators, data for which can be compiled through household time-use surveys.

Safety is measured in terms of physical safety (e.g. number of homicides as an inverse) and economic safety (inversely measured by extent of indebtedness and similar other measures). For the latter, ability to face unexpected expenses, having or not having arrears, is used in OECD framework as proxy variables to build up indicators of economic safety.

The right to get involved in public debates and to influence the shaping of public policies is an important aspect of Quality of Life. Good governance depends on the participation of citizens in public and political parties, trade unions, rights groups and advocacy groups. This is reflected in the level of trust of citizens in the country's institutions for security and justice, satisfaction with public services and the lack of discrimination. Gender discrimination measured in terms of the unadjusted pay gap is currently taken as the relevant variable.

Exposure to air, water and noise pollution has a direct impact on the health of individuals and even the economic prosperity of the country. Both objective indicators like concentration of pollutants in air or water as also subjective indicators like individual's own perception about the living environment are included in the Eurostat framework.

Overall experience of one's life is measured using three sub-dimensions which fall in line with the three psychological attributes, viz. cognition, affection and emotion followed by action. These three have been branded as life satisfaction (cognitive appreciation), affect (a person's feelings or emotional states, both positive and negative, typically measured with reference to a particular point in time) and eudaemonics (a sense of having a meaning and purpose in one's life, or good psychological functioning).

The collection of micro-data on well-being is a serious concern for any National Statistical System. It should be admitted that collection of some of the data needed to provide a comprehensive picture about Quality of Life is fraught with many problems, particularly in developing and less-developed countries characterized by lower levels of public awareness. It may be noted that most of the dimensions considered above relate to outcomes rather than factors or determinants. These suffice for getting an idea about the prevailing state of quality, while a consideration of the determinants is a necessity to explain that state and to bring about an improvement.

13.6.1 Gross National Happiness

Departing completely from the emphasis on national income as the fundamental or the most important measure of national development and recognizing the fact that social and cultural well-being contributes more to Quality of Life, the tiny state of Bhutan came up with a big idea that a reflection of Quality of Life is to be revealed in happiness of all people and that a sustainable development cannot be achieved without caring for nature and environment.

The Gross National Happiness concept and the index-based thereon have evolved through the contribution of international and local scholars and researchers to become an initiative beyond the borders of Bhutan. It is true, however, that the concept was first accepted by the fourth king of Bhutan Jigme Singye Wangchuck in the early 1970s. In 2006, the International Institute of Management published a policy white paper calling for the implementation of GNH philosophy in the USA.

The four pillars of GNH are sustainable and equitable social and economic development, environmental conservation, cultural preservation and promotion and good governance. The component of environmental conservation is recognized in the Happy Planet Index, but preservation and promotion of culture and good governance have not figured explicitly in any other concept of Quality of Life and not incorporated in any other measure of the same. GNH also focuses on nine domains, viz. psychological well-being, health, time use, education, cultural diversity and resilience, good governance, community vitality, ecological diversity and resilience, and living standard. It is quite clear that GNH does not consider economic growth as the paramount determinant of Quality of Life. In fact, it takes into account adequacy of time available for leisure and recreation as an indicator of Quality of Life. Needless to add is that with the recent focus on economic activity only, work–life stress has increased a lot and psychological well-being is at stake. This is why the GNH-motivated approach emphasizes on non-economic activities that lead to spiritual and moral development like meditation.

Implementing the GNH approach, Bhutan has over the last 20 years almost doubled life expectancy at birth, achieved almost universal enrolment of its children in primary schools and enhanced its agricultural produce considerably.

The real problem with GNH and the Gross Happiness Index is that several of its constituents are subjective in nature, eluding objective measures.

13.6.2 Physical Quality of Life Index

In appreciation of the fact that Gross Domestic Product or Gross National Product taken by itself alone without considering how the national income is spent and how basic amenities and services are reaching the people at large, the Physical Quality of Life Index (PQLI) was developed to throw light essentially on the Quality of Life enjoyed by the working class. The value of this index is the simple average of three statistics, viz. basic literacy rate, infant mortality rate and life expectancy at age one, each on a scale from 0 to 100. It was developed by the Overseas Development Council in the mid-1970s as one of a number of measures created due to the dissatisfaction with the use of GNP as an indication of development. PQLI may be regarded as an improvement, but shares the general problem of measuring Quality of Life on a scale that facilitates comparison. This index does not take account of income per capita directly. The UN Human Development Index that takes care of education, health and income in a somewhat different way captures some more ingredients that influence Quality of Life.

Steps in computing PQLI are

Find the percentage of the entire population who are literate.

Find the infant mortality rate (out of 1000 births) indexed as

$$\text{Infant Mortality Rate} = (166 - \text{infant mortality}) \times 0.625$$

Find the Life Expectancy, indexed as

$$\text{Life Expectancy} = (\text{Life expectancy} - 42) \times 2.7$$

Take the simple arithmetic mean of these three indices to get PQLI.
This index suffers from several limitations. Some of these are

1. There is an overlap between infant mortality rate and life expectancy at age 1.
2. Literacy rate for the entire population is affected by the high rate among the young, and this will overshadow possibly high rate of illiteracy among the adults.
3. PQLI leads to weird comparisons among countries that are very different, but has 'compensating' values for life expectancy and literacy.
4. Far better than literacy are measures of social and cultural development, viz.

National per capita expenditure on primary education;
National per capita spending on employment training;
Average age of mother at first childbirth (tells us more about maternal mortality and permanent injury during childbirth);
Proportion of population living below the poverty line; and
Serious crime rate in crimes per thousand population during the reference period.

13.6.3 Human Development Index

Some of these criticisms are taken care of by the Human Development Index (HDI) which, however, is not defect-free. The HDI is based on only a few parameters captured in terms of three component indices, and HDI is just an unweighted (or equal weighted) average of these three. These parameters have changed over the years, and currently, the parameters taken into account are the following: (1) life expectancy at birth e_0^0 which can be simply understood as the average age at death of a newborn baby; (2) combined gross enrolment ratio which covers all levels of formal education, viz. primary, secondary and tertiary and takes into consideration all persons irrespective of their ages enroled in some level of education; (3) adult literacy with an appropriate definition of literacy, e.g. ability to read and write a simple sentence with understanding and (4) adjusted per capita Gross Domestic Product in purchasing power parity dollar. The adjustment to per capita income is to take note of the fact that beyond a certain level of income, increase in income does make a proportional contribution to increase in quality of living. Even the formula used for adjustment has changed from the somewhat involved Atkinson's formula to the simple natural logarithm currently.

Each component index is based on an average of some parameter and ignores dispersion in that parameter. HDI assumes a linear growth of development with respect to any parameter from the most unfavourable to the most favourable situation. In some cases, these extremes correspond to the current lowest and highest values of the parameter across nations, while in some other cases it goes by the conceivable minimum and the conceivable/attainable maximum values. To care of economic conditions, HDI goes by the concept of international dollars or purchasing power parity dollars, which is somewhat difficult to calculate operationally and discounts that income figure using some monotonic transformation.

Among quite a few limitations of HDI, starting from its deficiency in capturing some important dimensions of human development and, particularly, Quality of Life experienced by an individual in terms of freedom of profession, faith and belief, etc. along with gainful employment matching the individual's potential and expectation, etc., are the following

(a) HDI assumes a linear growth from a state of minimum development (or, better, no development) to the highest attainable state.
(b) HDI is based only on averages and does not consider other features of the distributions of length of life, or per capita income or consumption expenditure in a family or of level of education ever completed among children, etc.
(c) HDI assigns equal weights to all the three dimensions of human development and the corresponding component indices.

In the determinant approach, health of a population would be reflected in healthcare facilities available like number of health centres or number of hospital beds available relative to the size of the population. For education, we would have considered number of schools compared to the population size, the teacher–student ratio, the average distance to be negotiated by a child to reach the nearest primary school and the like. For living standard, percentage of people with monthly per capita consumption expenditure above a minimum threshold would be an important determinant.

13.6.4 Happy Planet Index

The New Economics Foundation in London has come up with a new set of measures that produces surprising results. The somewhat strangely termed Happy Planet Index (HPI) measures for the first time, coherently, both human life and happiness and the impact of an economy on the sustainability of planet earth that provides most of the resources needed to support economy, hence the justification of the name. Objectively, measurable aspects of the HPI include incomes, longevity, infant mortality and levels of education, plus the use of renewable resources, the objective quality of the environment, including air, water and soil and the use of non-renewable resources—thus a country's ecological footprint. Of course, computation of this ecological footprint takes into account nature and extent of industrialization and urbanization and involves a number of assumptions which may not be warranted in all situations.

Happiness is quantified through a set of subjective questions about how people experience the quality of their lives. It is not difficult to expect that people with different attitudes towards life, coming from different socio-economic backgrounds, possessing different levels of education and work experience will have different expectations from the economy, the society and the polity and, hence, will feel different degrees of happiness with the present set-up. Recent discussions on minimalist approach to living with the basic requirements avoiding unnecessary duplication are relevant in this context. HPI includes questions related to how participative people think their democracy is. And this depends on the level of awareness of the people as also their expectations in regard to their duties and responsibilities on the one hand and their rights and privileges on the other.

HPI results in a ranking of countries by human happiness and ecological sustainability. Not very surprising, the richest countries in GDP terms come pretty low on the HPI—and not only because their developed economies create a large ecological footprint. It may be incidentally mentioned that in such economically developed countries, national income is concentrated in the hands of a few and inequality is alarmingly high. Their people experience a relatively low level of life satisfaction. USA occupies the 150th position and Britain comes 108th. The Pacific island of Vanuatu comes at the top, with pretty long lives and a highly enjoyable lifestyle embedded in flourishing local economies, with a low environmental impact. Second to Vanuatu comes New Zealand. Factors like peoples' feeling of well-being, optimism and connectedness, assumed not to be measurable and excluded by economics, are quite important to develop an indicator of Quality of Life.

13.6.5 A Holistic View

Chatterjee (2008) in his treatise entitled 'Human Development and Its Quantification' has provided a very comprehensive and integrated analysis of human development and, that way, of Quality of Life. Of course, the two concepts, one of human development and the other of Quality of Life, are not all the same. The difficulties involved in analysing Human Development have been outlined by Chatterjee as '…it would be seen that the ideal of human development, and in the case of the former (individual development), on outer and inner growth, and when inner growth has been taken into account, different meanings have been attributed to it. Leaders of thought in different countries and periods have looked at life and development from different angles'.

Chatterjee speaks of well conditioning to delineate outer growth and outlines six traits that describe inner growth. In fact, Quality of life, as a subject for discussion, should combine both collective individual and individual aspects, which—in some sense—are related to well conditioning and inner growth. Chatterjee takes a holistic approach to quantifying both these aspects.

For the well-conditioning index, he considers

(1) Percentage of people, the percentage of people above the poverty line represented by international $1 (at 1985 prices);
(2) Net enrolment rate at the primary level;
(3) Net enrolment rate at the secondary level;
(4) Adult literacy rate;
(5) Number of survivors at age 65 in the stationary life table population;
(6) Number of people with age 65+ in the stationary population; and
(7) Size of the stationary population.

As can be easily made out, the first variable takes care of the economic component in a much better way that the currently considered GDP (adjusted) per capita though the Human Poverty Index of UNDP accounts for some more factors linked up with human deprivation. The next three variables in Chatterjee's list are somewhat more comprehensive than the current component of educational attainment, and similarly, the last three variables provide a better picture of the health component in HDI.

Chatterjee's delineation of inner growth bears some semblance with the concept of Happy Planet Index or the Human Happiness Index. Maybe, we can claim that the HPI goes a little beyond Inner Growth Index. The six traits considered by Chatterjee to quantify inner growth and to compute an index for that are

(1) Spirit of one-ness;
(2) Ambience of freedom;
(3) Spirit of equality;
(4) Absence of gender bias;
(5) Creative activity; and
(6) Breadth of awareness.

Each of these composite traits has been conceptualized concretely in terms of several observable variables. Thus, Chatterjee provides a comprehensive framework for assessing inner growth of individuals collectively within a social–economic–political set-up.

Chatterjee delves into a comprehensive analysis of data called from many different sources at the international level to come up with the two indices and, subsequently, with a weighted combination of the two. It is just expected that ranks of different countries on these two indices or the combined one will differ from the HDI ranks.

13.6.6 Other Indices

There have been several other indices to capture different aspects of Quality of Life and, indirectly, of human development. Some of these measures provide ideas about some particular dimension of Quality of Life only, rather than a measure of Quality of Life as a comprehensive multidimensional concept that encompasses all the following. That way, the following indices may be looked upon as components of a composite index of Quality of Life. At the same time, any such index provides a deeper insight into one particular aspect that can be provided by an overall index.

Education index, taking account of literacy, mean years of schooling, etc.

Democracy index, based on people's participation in government policies and actions.

Freedom house, considering freedom of speech, expression, religion and faith.

Legatum prosperity index to comprehend the drivers of and the constraints on creation of a prosperous society.

Gini coefficient to reveal the extent of inequality in incomes.

Gender parity index, gender-related development index and gender empowerment measure to reflect discrimination based on gender in access to opportunities for development and in achievements.

Child development index to consider child health, child education and child labour.

13.7 Science and Technology to Improve Quality of Life

What can be the role of Science and Technology in understanding and assessing Quality of Life? On the one hand, benign applications of S&T tend to offer more and more goods and services of better and better quality for human consumption. On the other hand, such applications enhance human expectations and the comparison between the two becomes complicated. However, science can provide a better framework for analysing Quality of Life, identifying its determinants and correlates, and suggesting avenues for improving it. Proponents of Science, at least some of them, can argue that a better understanding of science and its applications can go happily along a sober and saner way of life that can dissociate Quality of Life from the pure materialistic concept of consumption. Lives of great men do substantiate this point of view. One contribution of S&T is undeniably true that it opens the door of the human mind to know more about him, about others, about the unseen world and thereby enhances Quality of Life. This contention treats knowledge and information as greater needs for humanity, compared to material conveniences and comforts. And this is where, as said earlier, Science and Technology play a big role. Similarly, the need for communication among people, far and near has become an important determinant of Quality of Life, being an essential element of services required to ensure a good level of living. As the human family grows and begets many complicacies, movements and expression of individuals as well as their behaviours and actions become more and more circumscribed. These restrictions run counter to a rich Quality of Life and show the absence of any necessary linkage between economic development and Quality of Life. Freedom in various walks of life is a very significant need and can be enjoyed even in a so-called primitive society or in an undeveloped economy.

In the above context, one feels tempted to relate the mental make-up of an individual to Quality of Life as is perceived by him/her. And this make-up may be partly due to lineage and upbringing but is predominantly acquired through practices that enrich the human mind and enlighten the soul within. This encourages us to listen sincerely to the precepts of wise men and women who enjoyed, according to them, the highest Quality of Life, but might have intentionally denied the existence of many earthly conveniences. One wonders if Quality of Life is a purely personal assessment, is totally free from considerations of material entities—their production and distribution—and is entirely a function of the state of realization of one's self. In such an eventually, should we proceed any further to analyse Quality of Life? Let the wise men answer.

Of course, attempts to produce more goods and services of better and more diverse quality and to ensure access by many more people to such goods and services through more even and effective distribution mechanisms will continue and, in the process, more energy will be consumed and greater depletion of non-renewable resources will take place, our common environment will get more polluted posing greater problems to our health and eventually leading to a deterioration in our Quality of Life. Is the problem really that cyclic in nature? Let us ponder over the matter and evolve a commonly acceptable solution.

Certain unwelcome impacts of an inefficient administration lead to violation of some basic human rights. The primary right, viz. right to life, is sometimes violated or taken away by the state in terms of capital punishment inflicted on criminals. But the more serious violation takes place through murders and even attempts to murder. One can think of deaths due to avoidable accidents caused by poor and unsafe maintenance of facilities and equipments by the state machinery.

One might argue on the role of Science and Technology in arresting human rights violation. However, deaths due to avoidable accidents can definitely be reduced by taking due advantage of technological developments. And a large-scale awareness coupled with a scientific temper can contribute to better Quality of Life through creation of social and cultural harmony. In fact, Science and Technology can contribute significantly in building adequate social infrastructure especially for the disadvantaged through construction of low-cost houses, cheap solar power devices, affordable supply of potable water, etc. And social infrastructure is as important as physical economic infrastructure in the context of national development.

13.8 Concluding Remarks

Quality has generally been spoken of—if not quantified—in relation to certain norms. Thus, in respect of various strands of life like food, clothing, housing, environment, health, education, personal and social conduct and overall development of creative and constructive potentials, quality norms are sometimes set out of reference to the existing circumstances or outside the feasibility of their improvement. Such norms are bound to remain ineffectual. In fact, some thinkers argue—and quite rightly—that quantification of Quality of Life diverts our attention more and more to the small sections of human activity than the totality of human fulfilment. And, excess is in some sense an impediment towards human fulfilment. According to Subramanyan (2003), Quality of Life emerges out of the ability of human beings to ask the right questions (not to be stifled in that), solve the right problems and transcend the existing world, and to follow, revivify or remodel it. In this sense, Quality of Life cannot be confined to local or country-specific considerations.

Quality of Life and human development, as these are currently comprehended somewhat narrowly, are not synonymous unless we speak about the entire human race or the quality of life experienced by humanity in its entirety.

References

Chatterjee, S. K. (2008). *Quantification of human development—A holistic approach*. Sankhya, 70, Series B, Part two.

Dasgupta, P. (2001). *Human well-being and natural environment*. New Delhi: Oxford University Press.

Dreze, J., & Sen, A. (1995). *India: Economic development and social opportunity*. New Delhi: Oxford University Press, New York: Human Development Reports, UNDP.

Lawler, E. E. (1982). Strategies for improving quality of work life. *American Psychologist, 37* (2005), 486–493.

Marvia, P. H., & Lawler, E. E. (1984). Accounting for the quality of work life. *Journal of Occupational Behaviour, 5*, 197–212.

Mukherjee, S. P. (2003). Quality of life. In B. Das & G. Bhattacharya (Eds.), *Quality of life* (pp. 29–31). Howrah: Ashirvad Agency.

Nussbaum, M. (1993). Non-relative virtues: An Aristotelian approach. In M. Nussbaum & A. Sen (Eds.), *The quality of life*. New Delhi: Oxford University Press.

Sen, A. (1992). Capability and well-being. In M. Nussbaum & A. Sen (Eds.), *The quality of life*. New Delhi: Oxford University Press.

Subramanyan, K. G. (2003). Some thoughts on quality of life. In B. Das & G. Bhattacharya (Eds.), *Quality of life* (pp. 24–28).

Taylor, J. C. (1978). An empirical examination of the dimensions of quality of working life. *Omega, 6*(2), 153–160.

Watson, I., Buchanan, J., Campbell, I., & Briggs, C. (2003). *Fragmented futures: New challenges in working life*. Sydney: Federation Press.

www.cmriindia.nic.in.

Chapter 14
Quality—Information Technology Interface

14.1 Introduction

The word 'industry' has, of late, acquired a comprehensive generic meaning, viz. a human enterprise that carries out the basic functions of procuring, processing and providing. Each of these functions has three discernible facets, viz. planning, executing (as planned) and verifying (to check conformity of execution with plan). In recent times, the ambit of quality has embraced this broad definition of industry and the scope of Quality Management has expanded significantly to stimulate and absorb many new approaches, tools and techniques.

Recent advances in production technologies and in use of sensors and even of robots with embedded softwares have brought in a big change in data capture and Management now faces the need to take decisions and to initiate actions almost instantaneously with the advent of processed data providing some information about processes and systems.

In this context, applications of IT in terms of software usage for monitoring and controlling various processes and in facilitating the adoption of computation-intensive methods for quality improvement exercises have become an indispensable support on the one hand and a welcome augmentation on the other. Expert systems are installed as decision support to corrective, preventive and improvement actions. As the usage of softwares increased rapidly by skilled professionals or mathematicians or engineers or statisticians and by less skilled supervisors and even operators, the demand for softwares with better quality at cheaper cost was heightened. As one consequence of this scenario, IT products and related processes have now been brought under the arena of formal Quality and Reliability analysis. Software quality assurance, software reliability prediction, Capability Maturity Models for software development processes, etc., point to the rapidly increasing role of Quality Management concepts, methods and practices in the IT sector.

On-line monitoring and control of complicated processes involving a multiplicity of quality characteristics and taking into account possible time dependence

© Springer Nature Singapore Pte Ltd. 2019
S. P. Mukherjee, *Quality*, India Studies in Business and Economics,
https://doi.org/10.1007/978-981-13-1271-7_14

of each, using recent techniques for process monitoring, change point detection, and process adjustment would not have been possible without support from IT, both by way of handling complex computations in no (real) time and in terms of automatic actions on the concerned processes. In fact, many of the currently used initiatives like Six-Sigma are oriented towards effective and imaginative uses of statistical techniques and IT facilitates the adoption and practice of these techniques on shop floors and office environments. No longer can we avoid sophisticated and efficient tools for quantitative analysis like nonlinear regression, MANOVA, OA designs and response surface methodology, data-dependent or sequential plans, neural networks, set-covering algorithms and similar other developments. At the same time, we cannot afford the luxury of a leisurely traditional approach to the applications of these tools. IT shows the way.

Another important role being played by Information Technology is in the area of Management Information Systems and Decision Support systems to help management in taking effective actionable decisions regarding Quality Management in its broadest sense. In fact, without an appropriate and comprehensive information base about both internal and external issues connected with Quality Management, which is updated continuously, augmented whenever found necessary and analysed to derive proper inputs to decision-making, right when a problem arises, Quality Management will suffer from delays and deficiencies.

In India, some forward-looking industries who have been known for their quality products and services as also some others who have been compelled to pull up their quality and to turn around have been making good use of IT in controlling and improving their processes. Software industry in India has also risen up to the occasion, and several units in this sector can boast of having developed and implemented Quality Management Systems that meet the requirements of Level 5 of the CMM standard.

In this chapter, we first attempt a somewhat sketchy review of applications of Information Technology by way of softwares and expert systems which have facilitated implementation and maintenance of Total Quality Management in manufacturing and service organizations, followed by a short discussion on the use of softwares for application of specific control, assurance and improvement methods and associated techniques. Then we come to consider different aspects of software quality, focusing on the software development process, passing through software quality testing to software reliability.

14.2 IT and TQM

The importance of IT in TQM has been discussed widely in the literature, and Ang et al. (2001) refer to many such articles published in a variety of journals. Even though there exist substantial examples and anecdotal evidence to illustrate the critical role of IT in the success of TQM, quantitative, empirical evidence to confirm such a claim is still lacking. Ang et al. carried out a study to investigate the extent to which IT has been used in Malaysian public agencies. Nine dimensions of TQM were considered, viz. leadership, strategic planning process, output quality assurance, supplier quality assurance, important innovations, information and analysis, human resource utilization, customer satisfaction and quality results. On the whole, the use of IT in support of TQM exceeded the moderate level, i.e. above 4 in a 7-point scale. Over 85% of the organizations reported that they have used a moderate to maximum feasible amount of IT to support 'important innovations' and 'information and analysis'. For both 'customer satisfaction' and 'strategic planning process', 70% of the organizations reported that the use of IT was 4 or more. The findings of this study are somewhat different from the findings in the Australian banking industry, who found quality customer services and product issues most strongly influence IT planning and not IT applications that support innovations.

A whole lot of softwares have been developed over the last two–three decades to facilitate Quality Management activities and are being used on a large scale in the context of both manufacturing as also software development processes. In fact, softwares provide support to implementation, monitoring and evaluation of Total Quality Management function in quite a few organizations. Some of the available softwares are specific to some particular activities within the broad ambit of Quality Management, some others help organizations in maintaining compliance with standards like ISO 9001 Standard or the Malcolm Baldridge National Quality Award requirements, acting more or less as consultants. Of course, expert systems provide a stronger support to Quality Management.

SQC pack developed by PQ systems—as the very name indicates—is quite helpful in operating control charts for different sub-group quality measures. Similarly, GAGE Trak developed by GAGE Trak as well as Gage Manager by Altegra support measurement system analysis including calibration activities. Process Street developed by Goodwinds takes care of process control in a broad sense. Compliance Quest developed by Compliance Quest, qms Wrapper by PM & QMS Software, My Easy ISO by Effivity Technologies and Gensuite by Gensuite focus on procedures and documents needed to achieve compliance to some standards. In fact, the first one considers both ISO 9001 QMS as also ISO 14001 Environmental Management System CAPA Manager by Adaptive Business Management Systems has been designed to help implementation of TQM. IIQMS developed by Harrington Group International and Minitab 17 by Minitab are also providing support to a wide range of quality-related activities. Isolocity is a cloud-based software developed more recently to facilitate implementation and maintenance of a Quality Management System.

It is somewhat evident that we have to take recourse to IT for the purpose of collection and instantaneous analysis of information about different steps and elements of the entire production process in order to decide on as also to trigger appropriate corrective and improvement actions and enhance quality and productivity. Beyond the production arena, it can also be used with profit to develop and maintain a comprehensive database covering customers and changes in their requirements over time, suppliers and other business partners about their changing offers and demands, employees with their skills possessed and needed for augmentation besides data pertaining to various projects completed and being run currently along with the existing and the potential product and service profiles. And all this information is required for a successful implementation of TQM.

14.3   IT-Enabled Quantitative Analysis

While simple statistical tools, applied imaginatively, have yielded useful results in terms of improvement of quality in products, processes and services, many of the classical tools used in the context of Quality Management have undergone significant ramifications and sophistications to reflect real-life situations and to ensure enhanced efficiency. In most cases, this has resulted in the development of quantitative (if not strictly statistical) methods and techniques which can be applied in real-time analysis, control and modification of processes only through computerization and automation. Such applications have been greatly facilitated by the development of appropriate softwares. To illustrate this point, we consider some recent contributions in two selected areas only, viz. Monitoring and Control of Multiple Quality Characteristics and application of Fault Tree Analysis for evaluation of reliability for complex and highly reliable systems on which failure data are rarely available.

14.3.1   Multivariate Statistical Process Control

The availability of cheaper and more robust sensor technology has resulted in industrial processes becoming extensively instrumented with measurements being routinely recorded for a wide variety of characteristics on data acquisition systems. A consequence of this is that the scope of applications for univariate SPC tools has been limited. Tools for analysis of multivariate data pertaining to multiple characteristics of a product or an in-process unit have to be used, with necessary modifications wherever necessary, to deal with multiple quality characteristics expressed in different units of measurement, correlated among themselves to different extents and having different impacts on overall quality. The assumption of multivariate normality for the joint distribution is not warranted in many situations. Problems may arise if all or some of the characteristics are binary. Hotelling () proposed the multivariate extension of the X-bar chart, using the T-square statistic and developing a control ellipsoid, to detect a change in the process mean vector. There have been several alternative procedures suggested, including some relatively sophisticated ones.

An important problem is to decide on the particular characteristics which were not within their prescribed limits when a point falls outside the control ellipsoid and that circumscribes the use of this chart for the purpose of process control actions.

A later approach was to use principal component analysis (PCA) to reduce the dimensionality of the data either on process or on product quality variables. Similarly, canonical correlation analysis or partial least squares can be used as a technique that reduces the dimension for process and product variables simultaneously. In the dynamic bi-plot (Sparks et al. 1997), dimension reduction is carried out by using the singular value decomposition. However, in such dimension reduction-based approaches, deviations along the minor axes are difficult to detect. The dynamic bi-plot or the Gabriel bi-plot has certain advantages over usual multivariate process control techniques.

A completely different approach has been to transform the multiple characteristics into dimensionless entities and to subsequently combine them into a scalar which can be dealt with by usual univariate process control schemes. Proposed by Harrington () and used by Mukherjee (), this approach also suffers from the difficulty to identify the misbehaving characteristics in case of an out-of-control point. In all the three approaches, we badly need the use of appropriate softwares to carry out the computations fast and a fast computation is a must to monitor the process and to initiate any action, if needed, as soon after a cause of lack-of-control as possible.

Multidimensional scaling (MDS) can be used to graphically display a p-dimensional vector for a unit or a batch for which p characteristics have been noted on a two- or, at most, three-dimensional plane and the distance of these points from a reference point corresponding to the desired state can be easily tracked sequentially as points arrive.

Cox (2001) explains the use of multi-way PCA (MPCA) in case we have several, say k, batch runs of a process, and we get three-way data, viz. batch run x variable observed x observation number within a run. Assuming n observations on each of p variables for a batch, we can use PCA to the matrix of order $k \times np$. Cox also refers to an alternative approach viz. use of canonical decomposition (CANDECOMP).

MDS and SPC can be harmonized in several ways. The first is to add to the MDS configuration as each new observation is made. As new points are added sequentially, out-of-control situations will hopefully be highlighted by outlying additional points. CANDECOMP model can be taken as MDS which nowadays finds interesting applications in complex data-rich processes.

All these approaches and many others in the context of multivariate process control are mostly computer-based, and their applications are essentially software-enabled. In fact, without the help of Information Technology it would have been quite a difficult task to capture data on multiple quality characteristics of the sampled units or items and to plot the multivariate data on a control chart with some dimension reduction through the use of dimensionless transforms for the different characteristics which could be combined in terms of the geometric mean or by plotting the first two principal components on a bivariate control chart with a rectangular control region or by developing a control ellipsoid and plotting the sample points. Use of appropriate software enables one to get results fast.

14.3.2   Use of Fault Tree Analysis

Recent times have seen the development and use of many complex systems with huge numbers of components arranged in non-standard complicated configurations. And most of such systems demand extremely high reliability, since any failure of such a system can lead to disastrous consequences. Usual attempts to enumerate critical path sets or critical cut sets—as the case may be—do not help us much to evaluate reliability of such a system in terms of component reliabilities. It goes without saying that components assembled within the system are highly reliable.

Fault Tree Analysis or generally Event Tree Analysis is a method using a logic diagram with Boolean gates, linking components at the bottom layer in terms of events, viz. failures or successful operations through different layers of sub-systems to the top event, viz. system failure or success. The analysis of large fault trees is generally tedious and obtaining cut sets on a personal computer may be slow. Fortunately, almost all fault trees can be reduced by using techniques like (1) modularization, (2) binary decision diagrams, (3) bit manipulation, (4) truncation and approximation, (5) partitioning cut sets, and (6) object-oriented programming. These methods are difficult to apply and are mostly approximations.

Xie et al. (1998) have provided some simple fault tree reduction principles and a simplified MOCUS algorithm using reduced fault tree to derive various importance measures involved in system reliability analysis. Use of fault trees and their reduction as also application of MOCUS algorithm and similar techniques require use of relevant softwares.

14.3.3   Other Applications

Multi-attribute decision-making methods, viz. Technique for Order Preference by Similarity to Ideal Solutions (TOPSIS) and a nonparametric procedure called Operational Competitiveness rating (OCRA) to ranking the technical measures generated from customer needs, instead of using the conventional and simplistic Simple Additive Weighting (SAW) in the context of a Quality Function Deployment exercise are known to be quite useful. While it is not impossible to apply TOPSIS or OCRA without taking help of softwares, use of the latter makes the application easier and quicker.

14.4   On-Line Control of Processes

14.4.1   Use in Process Control

There exists a good volume of literature on detection of change point while monitoring a process continuously in terms of a statistic based on inspection results contaminated by noises. Most of the detection algorithms work off-line in the sense that they test the cumulative evidence for any significant change and if such a change has been suggested, the unknown change point is estimated. Better than this post-mortem analysis will be on-line detection of the change point followed by a process stoppage and correction. A comprehensive account of change point problem analysis can be found in Basseville and Nikiforov (1993). In their paper entitled 'A Bayesian On-Line Change Detection Algorithm with Process Monitoring Applications', Sarkar and Meeker (1998) develop a Fortran program using NAG routines and obtain graphical output through S-plus.

Computers are increasingly being used to monitor the performance of complex systems using various alternative monitoring schemes involving essentially some form or the other of cumulative sums os a suitable statistic. Box et al. (2003) describe the waterfall chart to evaluate the performance of such monitors. This chart is computer-based on a specific detection threshold and assumes that the process standard deviation is known.

14.4.2   Use of Expert Systems

Western Electric Company developed a handbook in early 1940s that supplemented the use of control charts in those early days, by noting the pattern of points on a control chart for sample mean associated with a specific cause for assignable variation in the process mean. Subsequently, a list of possible such patterns along with the underlying causes of assignable variations could be used by operatives whenever an unusual pattern of points was revealed on a control chart. A modern-day extension of such an augmentation that takes due advantage of computers and command softwares can be christened as an expert system

Expert systems can provide valuable expert knowledge to process engineers who must make important process control decisions quickly. This has three important components, two in-built, viz. the knowledge base and the inference engine while the third is developed by itself from relevant external inputs. The crude augmentation done by Western Electric can be construed as an analogue of the first two elements of an expert system. The expert system for process quality control captures the test data, analyses it, informs the production engineer of further information needed to diagnose the cause(s) of the unusual variation in the data, and upon entry of the needed information, provides the engineer with a course of action to remedy the situation.

Pfau and Zack (1986) define ES as computer systems encoded with human knowledge and expertise that solve problems at an expert level of performance in a specific problem area or domain. Problems, in this context, go much beyond identification of causes of assignable variation and do include complex problems of analysis of the input data.

SPC-Pro is built using EXSYS, an expert system shell that has the ability to interact with external programs. SPC-Pro links to Lotus 1–2–3 for graphing and to SYSTAT for statistical analysis. SPC-Pro implements the set of Western Electric rules for analysing and SPC chart plot.

In building the expert system, the knowledge base and inference engine need to be customized for the type of process that we are trying to control and improve. This is due to differences among production processes in terms of the sources of variation and in terms of the actions recommended to reduce or eliminate assignable causes of variation.

The analysis process proceeds through a series of levels. The first level of analysis performed by the expert system is to identify a general assignable cause category. The second level of analysis then examines probable root causes of the general assignable cause category. The root cause analysis might proceed, in a specific application, to additional levels of analysis until the user of the expert system has discovered the basic root cause appropriate corrective or preventive actions for specific root causes can also be stored in the knowledge base.

A Gap Analysis Expert System was developed by Khan and Hafiz (1999) for ISO 9000 using Crystal 4 is a PC-based expert system shell developed by Intelligent Environments; it is a structured rule-based shell and provides a very user-friendly environment, being menu-driven. This expert system has four sections, viz. Introduction, Recommendation, Gap Analysis and Troubleshooting. The most important section is the gap analysis one, designed to provide computerized pre-audit information about prerequisites arranged in some hierarchy with 20 high-level issues associated with ISO 9000 Standard Implementation and the related low level issues. The knowledge base consists of approximately 500 rules.

14.4.3   Use in TQM

Executive information systems, expert systems and decision support systems can be utilized in supporting strategic planning of Total Quality Management, human resource management and supporting top management leadership. Franz and Foster (1992) have developed a decision support and expert system called the Total Quality Management System (TQMS) to assist management in the design of an integrated TQM programme and in developing a TQM implementation strategy. Needless to say, all these management support systems must be built on the foundations of a string information system for other functions including quality databases to continuously update quality-related information and computer-controlled quality measurement systems to compute various quality measurement routines. (Gupta and Sagar 1993).

These are just two illustrative uses of IT in Quality Management, focusing on on-line control and improvement of processes and products, which has become almost mandatory in a global situation where delayed off-line decisions and actions will be too late to deliver the goods. However, there exist many non-core activities which significantly influence business prospects and results. For example to carry out a vendor selection exercise or to develop some measure of customer satisfaction through properly designed sample surveys, we have to collect and subsequently handle a large mass of data. Without adequate use of IT, it may be pretty time-consuming and even difficult to collect the variegated data required for the purpose. And a reasonably quick and efficient analysis has to depend on IT.

14.5 Off-Line Quality Activities with IT

While on-line control of a production process is a must to meet product quality requirements, improvement activities take place after results of on-line monitoring and control are available and before the next run of production starts. During such off-line improvement activities (which, of course, may have to be preceded by some control activity that cannot be worked out without some analysis is carried out at the back end), we may need many more items of information than just the process parameters which are controlled on-line. In fact, we have to eventually examine the dependence of the response or output of a process on various relevant parameters of materials, machines, operators, ambient conditions and other process parameters besides the most important design parameters.

Such a dependence analysis requires data to be generated through properly designed experiments which have to be carried out not for routine production purposes but to examine the impact of each of the controllable factors associated with the design and the process. Industry has to use the minimum possible number of runs of such a multi-factor experiment which will allow exploration of the response surface followed by location of the optimum combination of factor levels. We may have to start with a large number of factors which are likely to affect the response variable and carry out some screening experiments to identify the relevant factors which should be included in the main experiment and kept at several possible levels.

We have a good volume of literature on the subject, including multi-response experiments. And Taguchi methods have been widely used for both designing such experiments and analysing the data obtained therefrom using signal-to-noise ratio as the parameter to be optimized. Orthogonal arrays and linear graphs were advocated by Taguchi in this context. He spoke of a three-level design process, viz. system design, parameter design and tolerance design and of robust designs by explicitly taking account of uncontrolled 'noise' factors likely to impact the response.

It is true that designs for industrial experiments had been developed and were being used several decades back, when computing facilities were meagre. However, analysis of data arising from alternative designs even with multiple responses has been greatly facilitated by Information Technology. This is particularly so when data are censored or some data are missing or the number of data points is less than the number parameters to be estimated and we tend to use partial least squares approach or when maximum likelihood estimation of parameters like regression coefficients involves transcendental equations, and we are compelled to adopt the Expectation–Maximization algorithm or to use Monte Carlo Markov chain approach and in similar other complicated situations. In fact, the use of algorithms and softwares has become almost routine if we have to keep pace with relevant features of data emerging from various types of experimental designs and the recent developments in statistical tools that are appropriate to deal with such features.

14.6   Information Base for Quality Management

While the use of computational softwares depends on the extent to which Quality Management is oriented towards quantitative analysis, the use of information bases pertaining to internal and external issues or of a single consolidated information base comprising all items of information relating to the market, the supply chain, the production apparatus, the transportation–storage–distribution network, the customer feedback responses, the interactions with regulatory bodies, the findings of audits and reviews and the like has become almost a mandate for a manufacturing or service organization that seriously contemplates over its survival and growth in a fiercely competitive environment. And this calls for an effective database management system in place.

Recent advances in analysing large data sets and making inferences from 'big data' may encourage developing a huge database that captures and preserves the whole host of information, obtained through surveys or arising out of transactions including enquiries and quotations or by accessing websites of different linked organizations. The latter could be existing and potential vendors, existing and potential customers, standards organisations, regulatory bodies, accredited laboratories to carry out calibration of test and measuring equipments, social activist organizations reacting on discharges of effluents to neighbouring water bodies and soil systems as also air over the region, and many others directly or remotely connected with the existence and operation of an organization.

Data on profiles of existing and potential vendors and customers could be structured or unstructured, quantitative or categorical or even nominal, quite voluminous, could contain credible as well as doubtful information, could be flowing in from different sources with different velocities and different items of information based on such data may differ in their intrinsic worth or values or weights. Thus, all the five characteristics of 'big data', viz. volume, variety, velocity, veracity and value are exhibited by such a data set. Since business is all for customers, their needs and expectations, their purchasing power and habits, their level of satisfaction with the current products and services offered, their dealings with the supplier in case of a complaint or a dispute and similar other features have to be accessed and analysed to initiate launch of a new product or change in the existing mechanism to address customer complaints. In fact, for decisions regarding introduction of a new and costly product and service, a clustering of customers based on the purchasing powers and habits may be taken up first Looking back at the reaction of customers in the past whenever a price rise was effected, likely reaction to the launch of a new, better and costlier product can be examined for each segment of the customer base.

Equally important will be a supplier database, complete with data relating to each purchase order placed with a supplier in terms of the quality and delivery against the order, time taken to replace any defective units or parts delivered and the like, besides a standing part of the data based on the initial vendor assessment exercise reflecting the technical and financial capability of the vendor.

Documentation of the Quality Management System (supposedly on the lines of ISO9001: 2015 Standard) and control of documents and records can and is being widely done in terms of computer-generated soft copies. It is quite easy to incorporate modifications in documents and not-so-difficult to protect records from tampering. At the same time, referral to a document which has been withdrawn from circulation poses no problem with a computerized QMS. And most organizations do have such systems these days.

Effective communication has been stressed in ISO 9001 Standards, and communication can be easily made faster, cheaper and error-free through Information and Communication Technology. Similarly, data generation from different sources —remote as also near—is greatly facilitated by ICT devices. That way, responses in a customer satisfaction survey covering a customer group spread across several urban and rural areas can be captured on hand-held devices with in-built softwares and transmitted in real time for consolidation and tabulation purposes at a central place.

14.7 RFID Technology in Quality Management

Radio-frequency identification (RFID) technology has changed the supply chain management scenario. With many more areas of operation in manufacturing and service enterprises, this technology has a great potential to improve productivity by ensuring just-in-time delivery, reducing levels of inventory, making product identification post-sales easier, cutting down time to locate items in vendor-managed inventories, doing away with human errors in identifying inspection and test status of consignments or supplies received and required in the production area, in capturing relevant data from items or units in process or about individuals and groups within and outside the organization, etc. In fact, one speaks of silent commerce these days and silent commerce is the use of powerful, inexpensive and tiny microprocessors and tags combined with continuous Internet connectivity and sensors to make everyday objects intelligent and interactive, creating new and real-time information and value streams. These developments offer a standardized and scalable approach which can be deployed across an extended enterprise to suppliers, manufacturers, distributors and business partners to provide a reliable and cost-effective visibility at the levels of items, bins and transport boxes.

RFID tags allow automatic data capture, multiple data reads simultaneously, reading beyond sight lines and storage and communication of a lot of information even at the item level.

In the context of a Quality Management System, an efficient supply chain from the vendors to the manufacturer and from the manufacturer to the retailers or distributors is essential. It is also imperative to handle product recall and replacement efficiently to the satisfaction of the customer. Adoption of RFID technology will help locate items with some deficiencies in design or manufacturing or in product release which have already reached the customers or are being held for distribution at some location. And only these items need be recalled for any rectification, if necessary, instead of the costly recall of a whole batch or production run.

14.8 Software Quality Considerations

To define and measure software quality, we must bear in mind that software, unlike engineering products, are malleable: we can modify the product itself—as opposed to its design—rather easily. For softwares, manufacturing is a trivial process of duplication. Creation of softwares requires more of human capabilities than engineering exercises. A software when it is released is almost certain to contain some defects, unlike manufactured items which enter the market usually with no defects. Length of life of a software can be infinitely large. However, most softwares are modified all too often. These differences apart, softwares also have quality characteristics that apply to manufactured items.

The concept of a process plays an important role in today's software development. And this process consists of several different sub-processes, for example gathering information about the requirements of a system, constructing a functional specification of a system, designing a system, testing a system, making it release-worthy and maintaining it during use, besides augmenting it for extended functional requirements. These tasks or activities or sub-processes are different one from the other in their nature.

To improve a defined process, organizations need to evaluate its capabilities and limitations. For this purpose, the Capability Maturity Model (CMM) developed by the Software Engineering Institute at Carnegie Mellon University which supports incremental process improvement and the Testing Maturity Model (TMM) for the function of Testing and the Test Process Improvement Model (TPI) can be followed. In fact, many points made in ISI 9001 Standard have been incorporated in software development process.

There are many desirable software qualities. Some of these apply to both the software and the software development process. The user wants the software to be reliable, efficient and user-friendly. The software developer wants it to be verifiable, maintainable, portable and extensible. The project manager wants the development process to be visible and easy to control. These are generic requirements and different IT products have some additional quality requirements. Thus, information systems are characterized by data security, data integrity, data availability and transaction performance. Real-time systems are characterized by how well they satisfy the response time requirements and have very strict reliability requirements. For Distributed Systems, the development environment must support software development on multiple computers where users are compiling and perhaps linking.

Software quality characteristics are sometimes classified as internal and external. External quality features are visible to the users of the system and are the features in which users are interested. However, only through internal quality features like verifiability is important to meet the external quality of reliability.

To meet these requirements and to test that these have been met, software quality testing followed by debugging has become almost a routine activity in the IT industry. Testing in this context to identify and locate bugs as well as to fix them is not the same as testing of a manufactured item. Attempts have been made to develop optimal testing plans. Predicting the residual number of bugs and esti- mation of software reliability using different models and algorithms have also gone a long way and produced a rich literature.

ISO and IEC have come up with several international standards relating to software both the product and the development process, and many IT industries have installed Quality Systems based on these and other standards. In fact, the Software Engineering Institute (SEI) and the University of Carnegie Mellon have developed the well-known Capability Maturity Model (CMM) to grade the quality-related activities in an IT industry. There are five distinct levels, viz. initial (ad hoc), repeatable, defined, managed and optimized. Recently, an extension of this model called CMM-P has emerged and being tried out in some IT units. There are relevant certification mechanisms also. It is interesting to note that out of the not-so-many industries across the globe certified at the LEVEL 5 status, quite a few are in India and, again, several of them are in the Tata Consultancy Services Group.

14.8.1 Software Quality

Any discussion on software quality has to refer to the study by McCall, Richards and Walters (1977) of quality factors and quality criteria. A quality factor is an external or behavioural characteristic or attribute of a software system. Examples of high-level quality factors are correctness or accuracy, reliability, efficiency, testa- bility, portability and reusability. Such factors have different importance to cus- tomers or users, software developers and quality assurance engineers. Thus, users are more interested in reliability and efficiency than in portability. Developers want to meet customer needs by making their systems efficient and reliable, and at the same time making the product portable and reusable to reduce the cost of software development. The software quality testing team is more interested in the testability of the system so that desirable features like correctness, reliability and efficiency can be easily verified through testing. Some customers may even carry out acceptance tests before taking product delivery.

A quality criterion (Naik and Tripathy 1998) is an attribute of a quality factor that is related to software development. Thus, modularity is an attribute of the software architecture. A highly modular software allows designers to put cohesive components in one module, thereby enhancing the maintainability of the system. Similarly, traceability of a user requirement allows developers to map the requirement properly to a subset of the modules. This way, correctness of the system will increase. Training as a quality criterion relates to development and quality assurance personnel.

Quality factors and quality criteria as propounded by McCall and others are related in a somewhat involved manner. As noted above, quality criteria are features of the software development process, while quality factors are characteristics of the software as a product. Further, quality factors are not all completely independent. If one factor is made to improve, another may be degraded. Some quality factors positively impact some others. One quality criterion may lead to an enhancement in more than one quality factors. In most cases, several quality criteria are related to a single quality factor.

Eleven quality factors defined by McCall et al. can be categorized into three broad groups, viz.

Product Operation, Product Revision and Product Transition. Correctness, reliability, efficiency, integrity and usability can be placed in group one relating to product operation; maintainability, testability and flexibility can easily be clubbed in the second group pertaining to product revision and portability, reusability and interoperability relate to product transition. Following are commonly used definitions of these eleven quality factors.

Correctness is the ability to meet functional requirements specifications. This definition calls for an explicitly stated specification and an unambiguous assessment of meeting the specified functions. For most software systems, such specifications in natural language do not exist, nor is it quite possible to determine unambiguously if these specifications are met correctness also implies ability of the system to meet implicitly expected requirements such as stability, performance and scalability. A correct software system may be correct and yet unacceptable to a customer if it fails to meet the unstated expectations. Correctness is a mathematical property that establishes the equivalence between the software and its specification. Correctness can be enhanced by using tools like high-level language or standard algorithms or libraries of standard modules and adequate testing.

Reliability is formally defined as the probability that the product will perform as expected over a specified time interval when used in a specified environment. Unlike correctness which is an absolute quality feature, reliability is relative and is a customer realization. Customers may accept software failure once in a while. Thus, an incorrect software may be accepted as reliable if the failure rate is very small and the same does not adversely affect the mission achievement. To a commoner, a software is reliable if one can depend on it. Many large software systems are likely to be incorrect when a new function cannot be carried out in all execution scenarios. However, most of them are reliable. Thus, correctness as a concept implies reliability, though the converse may not be true.

Efficiency is one important aspect of performance and is assessed by the extent to which a software system utilizes computing resources like computing power, memory, disc space, communication bandwidth and energy to carry out its intended functions. A software system is efficient if it can use computing resources economically. Thus, by utilizing communication bandwidth a base station in a cellular telephone network can support more users. However, the concept of resource expensiveness has been changing quite fast since some such resources are becoming less expensive by the day.

Integrity relates to robustness or ability of the system to withstand attacks on its security. Thus, it refers to control of access to software or data by unauthorized persons or programmes. This is a crucial requirement with multi-user systems and in network-based applications.

Usability (also referred to as *User-friendliness*) depends on the consistency of the system's user and operator interfaces. In the case of embedded systems with no human interface, usability refers to the ease with which the system can be configured and adapted to the hardware environment. User-friendliness alone is not enough, nor even a bad necessity in all cases. A user-friendly system that fails too often or yields results quite slowly will not be accepted.

McCall et al. have listed 23 quality criteria and defined as follows (McCall et al. 1977). These appear in the International Standard ISO 9126 and are reproduced in Table 14.1.

14.8.2 Source: ISO Standard 9126

This standard defines six broad, independent categories of quality characteristics, viz. Functionality—a set of attributes that bear on the existence of a set of functions and their special properties. The functions are those that satisfy stated or implied needs.

Reliability—a set of attributes that bear on the capability of software to maintain its performance level under stated conditions for a stated period of time (clock time or use time not explicitly mentioned).

Usability—a set of attributes that bear on the effort needed for use and on the individual assessment of such use by a stated or implied set of users.

Efficiency—a set of attributes that bear on the relationship between the software's performance and the amount of resource used under stated conditions.

Maintainability—a set of attributes that bear on the effort needed to make specified modifications (which may include corrections, improvements or adaptations of software to environmental changes or changes in the requirements and functional specifications.)

Table 14.1 McCall's quality criteria

Quality criterion	Definition
Access audit	Ease with which software and data can be checked for compliance with standards or other requirements
Access control	Provisions for control and protection of the software and data
Accuracy	Precision of computations and output
Communication commonality	Degree to which standard protocols and interfaces is used
Completeness	Degree to which a full implementation of the required functionalities has been achieved
Communicativeness	Ease with which inputs and outputs can be assimilated
Conciseness	Compactness of the source code, in terms of lines of code
Consistency	Use of uniform design and implementation techniques and notation throughout a project
Data commonality	Use of standard data representations
Error tolerance	Degree to which continuity of operation is ensured under adverse conditions
Execution efficiency	Run-time efficiency of the software
Expandability	Degree to which storage requirements or software functions can be expanded
Generality	Breadth of the potential application of software components
Hardware independence	Degree to which the software is dependent on the underlying hardware
Instrumentation	Degree to which the software provides for measurement of its use or identification of errors
Modularity	Provision of highly independent modules
Operability	Ease of operation of the software
Self-documentation	Provision of in-line documentation that explains implementation of components
Simplicity	Ease with which the software can be understood
Software system independence	Degree to which the software is independent of its software environment—non-standard language constructs, operating system, libraries, database management system, etc.
Software efficiency	Run-time storage requirements of the software
Traceability	Ability to link software components to requirements
Training	Ease with which new users can use the system

Portability—a set of attributes that bear on the ability of software to be transferred from one environment to another (this includes the organizational, hardware or software environment).

The standard provides a sample quality model that decomposes the six broad features into 20 quality sub-characteristics, not all of which will be involved in dealing with every software system. Thus, the characteristic 'functionality' is broken down into four sub-characteristics, viz. suitability, accuracy, interoperability and security. Similarly, 'reliability' is decomposed into maturity, fault tolerance ad recoverability. Understandability, learnability and operability make up for 'usability'. Time behaviour and resource behaviour are the components of 'efficiency'. Maintainability implies analysability, changeability, stability and testability. Finally, adaptability, installability, conformance (coexistence) and replacability are recognized as quality sub-characteristics under 'Portability'. Each of these 20 sub-characteristics has been defined as an ability/capability.

Each organization must define their own quality characteristics and sub-characteristics that are relevant to their needs. They should identify the level of each relevant feature that should be currently maintained as also the next higher level which they should try to achieve.

14.9   Concluding Remarks

The interface between Quality and IT is so strong and comprehensive that a complete discussion cannot be easily attempted, and the present article is just an indicator of some selected aspects only. Quality and IT support each other and even today the two are inseparable. IT provides the right means to bridge the gap between theoretical researches in the field of quality (with special reference to statistics) and their meaningful applications in industry. Principles and practices of modern Quality Management are being appropriately absorbed in software development processes to ensure better quality of softwares at lesser cost. Development of new and more efficient techniques for data analysis and the growing demand for business analytics and the emergence of big and dynamic data sets has been stimulating development and use of more and better algorithms to be incorporated in offering new and novel softwares. Software reliability occupies a sizeable area in the field of reliability analysis and has given a big boost to Quality Management taken in a broad sense.

References

Adams, T., Ferguson, G. T., & Tobolski, J. (2001). Getting value from silent commerce to-day. Outlook.

Ang, C. L., et al. (2001). An empirical study of the use of information technology to support total quality management. *Total Quality Management, 12,* 145–158.

Basseville, M., & Nikiforov, I. K. V. (1993). *Detection of abrupt changes, theory and applications.* New Jersey: Prentice Hall.

Box, G. E. P., et al. (2003). Performance evaluation of dynamic monitoring systems: The waterfall chart. *Quality Engineering, 16*(2), 183–191.

Chan, L. K., & Wu, M. L. (1998). Prioritising the technical measures in quality function deployment. *Quality Engineering, 10,* 467–480.

Chang, L. L. et al. (1993). An expert system for ISO 9000 quality management system. In *IEEE TENCON.* Beijing: Hong Kong Productivity Council.

Cox, T. F. (2001). Multidimensional scaling used in multivariate statistical process control. *Journal of Applied Sttistics, 28,* 365–374.

Franz, E. S., & Foster, S. T., Jr. (1992). Using a knowledge-based decision support system as a total quality management consultant. *International Journal of Production Research, 30*(9), 2159–21272.

Garvin, D. A. (1984). What does "Product Quality" mean? *Sloan Management Review,* Fall, 25–43.

Ghezzi, C., Jazayeri, M., & Mandrioli, D. (1991). *Fundamentals of software engineering.* New Delhi: Prentice Hall of India.

Gupta, V. K., & Sagar, R. (1993). Total quality control using PC's in an engineering company. *International Journal of Production Research, 31*(1), 161–172.

Hotelling, H. (1947). In Eisenhart et. al. (Ed.), *Multivariate quality control in techniques for statistical analysis* (pp. 111–184). New York: McGraw Hill.

Hwang, C. L., & Yoon, K. (1992). Multiple attribute decision-making: Methods and applications. Berlin: Springer Verlag.

ISO Standard 9126.

Kassicieh, C., et al. (1995). SPC-Pro: An expert system approach for variables control charts. *Quality Engineering, 7,* 89–104.

Khan, M. K., & Hafiz, N. (1999). Development of an expert system for implementation of ISO 9000 quality systems. *Total Quality Management, 10*(1), 47–59.

McCall, J. A., Richards, P. K., & Walters, G. F. (1977). *Factors in software quality* (Vol. 1). Springfield, VA: ADA 049014, National Technical Information Service.

Mutafelija, B., & Stromberg, H. (2003). *Systematic process improvement using ISO 9001: 2000 and CMMI.* Boston, MA: Artech House.

Naik, K., & Tripathy, P. (1998). *Software testing and quality assurance.* Hoboken: Wiley.

Pfau, D. R., & Zack, B. A. (1986). Understanding expert system shells. *Computerworld Focus, 20* (7A), 23–24.

Sarkar and Meeker. (1998). A Bayesian On-Line Change detection algorithm with process monitoring applications. *Quality Engineering, 10*(3), 539–549.

Sparks, R., Adolphson, A., & Phatak, A. (1997). A multivariate process monitoring using the dynamic biplot. *International Statistical Review, 65*(3), 325–350.

Weghell, M., et al. (2001). The statistical monitoring of a complex manufacturing process. *Journal of Applied Statistics, 28,* 409–426.

Xie, M., Tan, K. C., & Goh, H. K. (1998). Fault tree reduction for reliability analysis and improvement. In Basu, Basu, & Mukhopadhyay (Eds.), *Frontiers in reliability.* Singapore: World Scientific.

Chapter 15
Quality Management in Indian Industry

15.1 Introduction

Indian industry reflects a rapidly expanding entity with fast-changing product (and service) profiles, management styles and investment portfolios. At any point of time, a lot of diversity in emphasis on quality and cost among different industry types as also among different units within the same industry category comes to notice. Some industries—in both manufacturing and service sectors—have realized the all-pervasive need for quality improvement as an integral component of strategic management, have put in place quite a few initiatives in that direction and are striving to achieve organizational excellence. A second discernible category is a bit conservative on quality, with necessary (but not sufficient) attention to comply with regulatory requirements or to boost corporate image, without sincerely linking quality with business. The third category with still a sizeable number includes organizations—large, medium and small engaged in manufacturing as also in providing services—can at the best be branded as 'tool-pushers'. Some of them do have a Quality System, sometimes certified for conformity with ISO standards, but they lack a commitment to quality. However, their products and services are not always failing to meet customer requirements, if not expectations.

In a world marked by abrupt and sometimes disruptive developments in business and industry operating in a 'globalized' economy with some possible skewness, Indian industry reveals a rapidly changing scenario in respect of its products and services entering the domestic as well as foreign markets. And this makes it almost impossible to portray a picture of Quality Management in Indian industry that can claim credibility. Isolated studies by some individual investigators as also by some organizations like the Confederation of Indian Industry or the Federation of Indian Chambers of Commerce and Industry do provide some information which are not always comparable in terms of sample size, mode of data collection, analysis of data etc. and do not present similar pictures.

© Springer Nature Singapore Pte Ltd. 2019   307
S. P. Mukherjee, *Quality*, India Studies in Business and Economics,
https://doi.org/10.1007/978-981-13-1271-7_15

The present chapter provides just a sketchy and may be unconsciously biased delineation of Quality Management in Indian industry. This includes findings of a few studies which reveal strengths and weaknesses of Quality Management in Indian industries. A study undertaken more recently by the author in collaboration with a faculty member of a foreign university, using a carefully crafted questionnaire has also been reported. Not many responses could be secured and the author feels that the questionnaire developed for the purpose may be of some interest to quality management professionals and the same has been appended to this Chapter.

15.2 A Peep into the Past

Before the 1920s, most manufactured products were imported, not much industrial activity could be seen and little-organized effort to ensure quality in Indian goods was evident. Some engineering industries started their operations beyond the 1920s, and some of them earned reputation for the quality of their products. It must be added that even during the British rule, some engineering industries including quite a few in the medium- and small-scale sectors were known for the quality of their products, essentially reflecting the sincerity and competence of skilled manpower. After Independence, the country focused on import substitution and there was a great thrust on producing more of consumer goods. Attention to quality suffered a temporary setback.

During the mid-forties, Professor P. C. Mahalanobis foresaw the need for introducing statistical quality control techniques for improving quality of goods produced by Indian industries. A special course was organized by the Indian Statistical Institute in 1945–46 which was attended by 12 persons. There was encouragement from a few persons like C. Tattersall of the Ordnance Testing Laboratory who fully realized the importance of using QC techniques in Indian manufacturing industry. However, Government departments were initially apathetic. Influenced by the pioneering work of Walter A. Shewart of the Bell Telephone Laboratories in Statistical Quality Control, Mahalanobis invited Shewart to visit ISI. Shewart arrived on 22 December 1947 and took a lead in conducting a one-week conference on Standardization in Industrial Statistics at the Presidency College, Calcutta, during 8–14 February 1948, under the auspices of the ISI and the just-formed Indian Standards Institution. This event had 190 participants.

In February 1948, Mahalanobis sought the help of Shewart and some forward-looking industrialists to establish the Indian Society for Quality Control (ISQC) with himself as the Secretary and a Patna-based industrialist B. K. Rohatgi as the President to involve industrialists and management people in the applications of SQC. Regular training programmes to expose people from various industries to methods and techniques of SQC were being organized under the aegis of Chapters of ISQC besides the SQC and OR units of the ISI. The ISQC Bulletin was one among the earliest journals in the field of Quality. In early seventies, two organizations, viz. Indian Association for Productivity, Quality and Reliability in Calcutta and Indian Association for Quality and Reliability in Bangalore, took over the role of ISQC. The National Institution for Quality and Reliability and the National Centre for Quality Management started various promotional activities and a nation-wide quality movement came to occupy the scene. In fact, IAPQR Transactions—a bi-annual publication of IAPQR—started in 1976 is an Indian journal exclusively devoted to quantitative methods for improvement in productivity through quality and reliability.

Meanwhile, the Indian Standards Institution was set-up by the Government of India on the advice and initiative of Professor Mahalanobis. This national body became later known as the Bureau of Indian Standards and played a significant role in promoting quality of Indian goods and services by developing and implementing a whole lot of relevant Indian Standards bearing on different facets of quality and different concepts, methods and techniques that promote the cause of Quality.

These activities in turn led to the starting of the first SQC unit by ISI in Bombay in 1953, followed by two units in Bangalore and Calcutta in 1954. Gradually, SQC units of the ISI came up in several other industrial cities in India. The main objective of the staff of these units was to visit industries and act as consultants in SQC activities of the industries to improve the quality of their products. Today the division of the ISI known as SQC and or division is engaged in both teaching and research besides consultation and promotional activities.

Dr. A. V. Sukhatme who did his Ph.D. on Sampling Inspection under the guidance of E. S. Pearson and was leading the Statistics Division of the Tata Iron and Steel Company contributed a lot to the QC movement in India. Dr. A. K. Gayen who also did doctoral work in England and Professor P. K. Bose of the Calcutta University were among many others who got seriously involved in ISQC activities and lead the QC movement to a greater height in collaboration with quite a few engineers and industrialists.

In 1989, the first Asian Congress on Quality and Reliability was organized jointly by the ISI, the Bureau of Indian Standards and IAPQR. Following up a recommendation from this Congress and appreciating the need for guiding and coordinating the quality movement in the country, the Quality Council of India was set-up.

During the 1980s a strong emphasis on building up a quality movement was laid by the Confederation of Indian Industry which organized programmes involving well-known exponents of quality from abroad to provide necessary technical support to the quality movement. In the early 1990s, a survey by World Competitiveness Report looked at products and services from 41 countries and ranked them. Based on different parameters, India's rank was 39. In 1992, India started the financial reform process by opening up the economy, heralding a new era of competition. Post-2000, quality initiatives have been running parallel to the liberalization movement. In fact, an ISO survey in the last decade of the past century put India in the top ten countries for ISO certificates with 24,660 nationally. It was also the country with the fourth largest number of certified companies. It should be added, however, that this extent of certification to ISO 9000 series of Quality Management System Standards, by itself, does not mean much about Quality Management in India. Of course, quality is highly regarded by the general public in India. The Delhi Metro Rail project has been hailed by many as a 'quality' project. There is a great sense of pride and confidence among Indian workers, particularly in the manufacturing industries. They believe that they can achieve great results even without application of sophisticated tools and high-sounding approaches. The Bombay dabbawallahs demonstrate an ability level to control variability in delivery time around the target that can be easily appreciated as a Six Sigma venture.

15.3 The Current Scenario

In the sunrise sector, comprising—among others—IT and ITES units, confidence in technological advance and in knowledge workers have blinded the industry leaders to the extent that they do not feel the need for any conscious and distinct exercise to ensure quality in their products and services. In fact, creative workers—many of whom are in the IT and ITES sectors—do not like the idea of falling in line with standards like ISO 9001:2000 or CMM.

In a recent study on the state of quality initiatives in Indian companies, opinions were sought on the current attitude of management on the use of Quality Management Practice (QMP) after the liberalization of the Indian economy. Ninety per cent of the respondents felt that the current attitude towards QMP was supportive, the rest feeling not much of a change.

Quality Management Practices adopted by Indian manufacturing industries have taken due advantage from quality circles and small group activities, design of experiments and Taguchi methods, good manufacturing practice, good laboratory practice, Six Sigma projects, failure mode and effects analysis, quality function deployment and similar approaches besides, of course, statistical process control and sampling inspection plans. Some forward-looking companies also prepare themselves for the Rajiv Gandhi National Quality Award or the JRD-QV award, both of which are modelled on the lines of Malcolm Baldrige National Quality Award. A few excellence-seeking companies also participate in the Deming Prize competition.

While various quality initiatives have been taken up by Indian manufacturing industries with varying degrees of success, the Indian Software Industry has done excellently well in regard to quality. In fact, some other service sectors in India are also proving their capabilities to maintain and deliver services of high quality to domestic and overseas customers.

Total Quality Management (TQM) is a frequently used jargon, quite well known to industry executives, practised in a large number of industries incorporated in several quality-related education programmes. However, hard and soft aspects of TQM are separately followed in most industries, some eDOEmphasizing the soft HR-related aspects only while some focus on quantitative analysis, , QFD, FMECA, and many similar techniques which are understood by only a few workers in an industry. Not many top managers are involved in the quality exercise, though many play a supportive role.

People specializing in industrial engineering or operational research and management services are not really convinced about the role of TQM (as distinct from SQC) to improve organizational excellence or to enhance productivity. Not much is done to record costs of poor quality and hence the lack of a concrete evidence to motivate corrective and preventive actions.

15.4 Implementation of Six Sigma

Some of the hurdles Indian industries are facing in implementing Six Sigma approach are common to other countries also and may be cited as under

1. Lack of constancy of purpose.

As customer needs and expectations are changing very fast and organizations labour hard to comprehend and meet these, management priorities and objectives change frequently and decisions and strategies fail to cascade down the different levels of people effectively.

2. Lack of focus.

Some organizations take up several initiatives simultaneously like Kaizen, Quality Circles, ISO-9000, TQM, QS-9000, Six Sigma. People within the organization are confused in the absence of a proper integration of the initiatives, and cannot devote the time, attention and resources to the successful implementation of any one initiative.

3. Failure to appreciate implications and imperatives of any approach.

Successful implementation of the Six Sigma approach requires knowledge and skill about the respective business processes, in-depth knowledge of different tools and techniques like QFD, descriptive as well as inferential statistics, DOE and response surface methodology, besides ability to convince and lead people. However, many consultants and even MBB's do not have the above-mentioned qualities. This leads to the usual quality-circle type (cause-and-effect diagram solution) of solving a problem that does not help achieve a quantum jump in profits or customer satisfaction.

4. Improper project selection.

Six Sigma projects must be selected in line with the organization's goals and objectives. Some organizations fail to have SMART (specific, measurable, achievable, relevant and time-bound) goals and objectives while some others are not able to link the projects to their goals and objectives. In some cases, the project scope is too large to be completed within the stipulated time frame, the project team members do not have any authority and responsibility or everything is forcibly considered as projects.

5. Lack of resources.

Many organizations these days work with minimal workforce burdened with many responsibilities and cannot spare their people for comprehensive training. They look out for shortcuts and some even look upon such training as unwarranted, in view of the fact that they recruit educated and skilled people.

6. Lack of coordination among functions.

This leads to improper selection of critical-to-quality characteristics, incorrect or inadequate data leading to improper analysis and inappropriate solutions.

7. Concentration on trivial many rather than the vital few.

Sometimes, the CTQ's are easy to attack and not much return on the project objective is selected for the quiche closure of the project.

8. Foreclosure of projects.

Sometimes projects are discontinued due to organizational restructuring. In a few cases, achievement of goals of the team may change due to changes in statutory/ regulatory requirements, taxation policies, etc. Improvements are made by god or government, not by team members.

9. Non-availability of data.

Not too unoften relevant and adequate data do not exist, may be time-consuming and expensive to collect and people may avoid data collection out of apprehensions.

10. Impatience to get results.

Some organizations are impatient to get quick results and, failing to derive that, lose faith in the efficacy of the initiative. This slows down implementation and eventually leads to its withdrawal.

11. Ineffective change management.

Any major quality initiative will suggest changes in some of the business processes—the way these are carried out, the people involved, the resources deployed and the results expected—and changes will always invite some resistance which has to be reasoned out. This requires a good blend of technical and managerial skills that is not easy to come across.

15.5   TQM in India: A Survey

A questionnaire survey on the status of TQM in Indian manufacturing companies was conducted by Tata Management Training Centre, Pune, India. The questionnaire covered a total of 310 questions grouped into a few categories that encompass various aspects of TQM practices in these companies. The copies of the questionnaire were sent to thirty-one Indian companies where TQM-related activities were initiated in some form or the other. The survey was conducted during the period June 1998–February 1999. Responses were received from twenty-two companies and fifteen of these were finally selected on the basis of the following criteria: (i) authors' acquaintance with the companies, (ii) practice of TQM in some form or the other for at least last two years, (iii) one of the major players in the concerned field of business is a member of the Confederation of Indian Industry (CII) and (iv) readiness to share detailed information.

The management practices in these companies evolved round the critical processes that represented combined value-added organizational activities. The processes were divided into two broad categories: primary and secondary. The primary processes related directly to the production of goods that aimed at achieving TQM goals as immediate objectives. The secondary processes attempted to assist the effective execution of the primary processes. The responses, after these were appropriately grouped, were sought for the following major questions:

- What is the nature and extent of the involvement of top management in promoting and monitoring TQM activities in your company?
- What are the main TQM focus areas in your company?
- What has prompted your company to introduce TQM? What are the main driving parameters for its implementation?
- How does your company decide whether TQM activities are being carried out in the desired direction?
- What is the organizational set-up for your company's TQM implementation process?
- How does your company maintain the enthusiasm of employees for change through TQM and keep them motivated?
- What are the critical processes in your company that contribute to the success of TQM?
- What are the guiding factors for the identification of critical processes in your company? Does your company have specific measures in quantifiable terms to monitor these processes?
- What processes are benchmarked in your company? Where has the company introduced business process reengineering (BPR)? What have been the major impacts of BPR initiatives?

15.5.1 Major Findings

- As expected, the CEO was found to drive the change process through TQM in as high as 80% of the companies. Although these companies indicated personal involvement of the CEO in promoting TQM implementation process, only 40% stated that the nature of involvement was in the form of a leader/guide/mentor/counsellor. In fact, the survey data showed that the CEO had assigned the responsibility of TQM implementation process to the TQM coordinator in 60% of the companies. Further, only 27% of the companies indicated that the CEO spent over 25% of his time in TQM-related activities.

- The agreement among the companies was not quite close for TQM focus areas where the questions were kept open-ended. Although the focus areas recorded by the respondents covered both primary and secondary processes, the primary business-related direct goals seemed to figure more prominently in their responses. The degree of agreement was rather low in respect of customer satisfaction (46.7%), continuous improvement (40.0%), teamwork (40.0%) and process improvement (33.3%). TQM focus areas like training, vendor development featured very rarely in the responses.

- The companies cited global competition (60.0%) as the most important factor for introducing TQM. The other factors were: decline in profits (40.0%), Improvement of employees' morale (40.0%) and loss of market share (33.3%). Most of the respondents indicated process improvement (80.0%), problem-solving (80.0%), customer satisfaction (73.3%), employee involvement (73.3%) and teamwork (73.3%) as the main driving parameters for the implementation of TQM. Gaps seemed to exist between TQM focus areas and driving parameters with respect to customer satisfaction, teamwork and process improvement.

- Only 40% of the companies indicated a structured mechanism like periodic audit or use of a well-known TQM assessment model to as certain the progress of TQM. The others relied on business results (26.7%), customer satisfaction surveys (13.3%), etc. As regards mid-course corrective actions to monitor the progress of TQM, no dominant practice seemed to emerge in the study. The practice of reviews/deliberations in Quality Councils was the most common one with a response rate of 40%. Some of the desirable practices like re-examination of critical business processes, 'plan-do-check-act (PDCA)' cycle and target-based review of TQM goals were rarely mentioned by the respondents.

- 73.3% of the companies had quality improvement teams in the organizational set-up for the TQM implementation process. Two-thirds of these had either a TQM coordinator or a steering committee. The committees were seen to exist at various levels of hierarchy only in a few companies, 60% of the companies reported the presence of both quality improvement teams and small group activity teams/quality circles. The study, however, showed that the implementation process did not reach the operative level in most of these companies. The awareness of TQM and the importance of its implementation, by and large, remained confined to the managerial level.

- Non-monetary forms of motivation (40.0%) and training (33.3%) seemed to be the two most common forms of maintaining the enthusiasm of employees for change through TQM. The other forms of motivation were: employee involvement (20.0%), highlighting achievements in newsletters and house magazines (20.0%), highlighting achievements in newsletters and house magazines (20.0%) and help from external consultants/experts (13.3%). The study also revealed that the mindset of middle and junior level managers was the most difficult thing to change in the context of TQM (33.3%). This is quite understandable in view of the fact that the need for TQM which stresses customer satisfaction is felt in a competitive market where customer awareness plays a major role in the market share of a company's business. For decades, Indian companies were, however, operating in a fairly closed market where the level of customer awareness was quite low.

- Primary processes were identified as critical processes in most of the companies. These included manufacturing (93.3%), order realization (73.3%), handling customer complaints (66.7%). Very few companies mentioned secondary supporting processes like goal deployment, training, recruitment, appraisal, product development, market research as their critical processes. This observation was in conformity with one of the earlier findings where a marked absence of goals related to secondary processes could be noticed.

- Business goals (86.7%), customer feedback (73.3%) and problematic processes (60.0%) were seen to be the major guiding factors for the identification of critical processes. TQM assessments, 'strength–weakness–opportunity–threat (SWOT)' analysis, 'quality improvement team (QIT)' review, etc., were not mentioned in this category by most of the companies. A conscious and systematic attempt to identify the critical processes was not made by any of the companies. The main focus was on the primary processes which were probably considered to be most essential for attaining business goals. Two-thirds of the companies had quantifiable targets for monitoring the critical processes. These were also reviewed by the management as a part of 'management information system (MIS)'.

- The major processes benchmarked in the company were vendor approval and selection (27.7%), handling customer complaints (20.0%), manufacturing (20.0%), new product development (20.0%) and throughput/cycle time (20.0%). Although more than half of the companies mentioned some of the secondary critical processes like training, empowerment, networking for benchmarking, the survey revealed that these were not dovetailed with the primary processes. As a result, the outcome of such efforts did not contribute directly to the achievement of TQM goals. BPR concept was seen to be practiced in eleven companies in the sample. Process redesign (66.7%) and organizational redesign

(60.0%) were the major reengineered areas. While the implementation of BPR enabled all the companies to reduce cycle time, only ten of these also succeeded in various kinds of cost reduction.

15.6   Conclusion

The article attempts to provide a broad overview of TQM and its evolution through different stages. Several approaches to the concept of quality and various dimensions of product and service quality have been discussed to facilitate a company's endeavour to introduce TQM irrespective of the business in which it is involved.

The findings of the survey on the status of TQM in India, although based on a small sample of companies where TQM was initiated in some form or the other, reveal that these companies are mostly concerned with the mechanics of introducing TQM rather than with its basic philosophy and spirit. The author is of the view that significant changes in the approach to TQM are needed for ensuring the alignment of business practices with TQM goals.

Top management, particularly the CEO, has to lead the process of TQM implementation. He needs to realize that TQM is a powerful supporting process to reach business and other organizational goals. He must not delegate the task of driving TQM to the TQM coordinator.

A marked absence of the recognition of supporting or secondary processes is observed in the survey. The main thrust seems to be on the primary processes alone. Cause-and-effect diagram may be used to establish the linkage between them. The need to develop information sources like SWOT analysis, QIT review, TQM assessment results for arriving at the critical processes is quite important in companies contemplating the introduction of TQM.

The survey also reveals that the assessment and mid-course corrective action in the TQM implementation process are not quite adequate. These should be carried out through internally developed or well-known TQM assessment models, target-based review and the application of PDCA. Though the organization of TQM implementation is mostly in place, it does not seem to percolate down the operative levels in many companies.

In most of the companies studied in the survey, the compensation package of an employee is seen to be primarily linked with his ability to meet the organizational business goals. His contribution to the achievement of TQM goals as a source of performance appraisal is rarely stated explicitly by a large majority of the respondents.

Although it is heartening to note that several companies sampled in the survey have initiated BPR, the efforts are not quite systematic and these do not often establish linkages with the critical processes.

Globally, there have been conflicting reports about the success rate of companies implementing TQM. Research findings indicate that the gaps in the TQM implementation process rather than the principles of TQM are primarily responsible for many reported failures. TQM cannot succeed if it is looked upon as an experiment in isolation. Its success depends on the degree of congruence between business practices and TQM goals. By taking care of the lacunae in the TQM implementation process, Indian companies should be in a position to make significant progress and eventually compete with the world-class ones.

15.7 A Study by the Author

A couple of years back the author collaborated with a senior faculty member of the University of Dublin, School of Business, to conduct an opinion survey among top management representatives of some Indian manufacturing and service industries. It must be stated that in terms of previous experiences of earlier investigators in securing responses from business executives, no statistical sampling procedure was adopted to identify the organizations to be visited. The questionnaire was canvassed by a retired senior executive with a lot of experience about quality in services. Though the number of respondents was rather few, the research questions were framed with a lot of care. A copy of the questionnaire used is given in the Appendix. There were four distinct parts in the questionnaire, viz. impact of quality on practice, attitude of top management towards quality, top management concern for quality and open question on quality. In all, 26 questions with categorical answer codes mostly, except for the last part, were included.

Most of the responses broadly reflected a positive role of quality and a positive appreciation of standards and tools for quality improvement. More than half of the respondents felt that a formal focus on quality in terms of a comprehensive Quality Management System provided a great help to industry in achieving its desired goals (by the way of growth and excellence), while the rest opined that such a focus had done some good in this direction. However, only one of the respondents felt the role of quantitative tools and techniques was essential. Nearly, a half of them acknowledged great help from these tools and techniques applied with discretion. Most of the respondents felt that the role of leadership was essential or of great value. In the perception of all the respondents, top management views quality of services as a necessary adjunct to product quality, either definitely or probably.

Out of the not so many respondents, two did not directly respond to the questions in the survey instrument. They dropped some large hints on Quality Management and drew attention to some issues linked up with the consumers and the market, according to rather low priorities to quality of goods and services.

Considering responses to impact of quality on business practice (possibly implying a focus on quality as is reflected in relevant practices) one finds some concordance among responding industries. And some industries do attach a great importance to quality in its relevant practices, while some others do not appreciate the role of Quality that much.

Of the eight respondents, seven opined that a formal Quality Management System has helped their industries in achieving growth and excellence. Strangely enough, one respondent remarked that a formal focus on quality in terms of a comprehensive QMS was harmful in achieving the organization's goal. One possible explanation could be the perception shared by some executives that a formal focus eventually leads to emphasis on procedures being complied with rather than proper actions being taken to perform well.

The importance of quantitative tools and techniques has a mixed appreciation, five respondents feeling that these are of great help and even essential. Two others opine that some good has been done by these tools and techniques, while observes a negative value for these aids.

On the role of leadership provided by top management responses vary greatly. Three respondents state this role to be essential and two others feel that the role is of great help. To two others, this is only of some value and one respondent—who has a negative focus on quality—associates a negative value to this role. It must be noted that the respondents differ in terms of their professional level and the perceived value of their interactions with top management.

Half of the respondents generally feel that top management does not view quality of services as a necessary adjunct of product quality, the remaining respondents report either 'probably yes' or 'definitely'.

Executives agree that ISO 9001:2008 have definitely helped improvement in quality. A similar finding is about the importance of quality cost Analysis as an exercise to improve quality. This is also true equally about the contribution of Innovation. In fact, in none of these issues, none of the respondents had a negative view.

Coming to quantity–quality relation, responses show remarkable variations, three reporting that quantity expansion does not compromise quality, while two others hold that there is a definite compromise. One is not sure and two others apprehend a compromise.

Cost/price reduction compromising quality, two definitely hold this compromise as a reality, one does not agree at all, two are not sure, two others are apprehensive about such a compromise while two others are not sure about the relation.

On analysing the attitude of top management towards quality, five of the respondents that top management believes in the need for conformity to quality standards, two remain silent while the last one disagrees with the statement that top management has this belief.

Similarly, five respondents agree that management believes in maintaining the optimum level of quality, taking into account organization-specific factors two others disagree while one remains neutral.

As expected, most respondents (7 of the 8) do not feel that quality takes care of itself and needs no special effort; the other executive strongly believes that quality needs no special effort. It may be noted that a product and its quality are inseparable entities in the opinion of some quality managers.

In consonance is the opinion expressed by seven of the respondents that quality and quantity can go together. Only one respondent agrees that these cannot be achieved simultaneously.

On the quality–productivity linkage, most respondents agree that efforts to build quality also lead to enhanced productivity. Only one does not share this view.

On the vexing question whether quality and specially reliability will adversely affect demand and hence production, and hence profit, six executives do not have this apprehension, and two are neutral.

Five of the responding executives agree that quality and reliability add to the value of a product and value-based pricing will more than offset costs of investment in quality. Three were undecided.

Regarding top management's appreciation of quality efforts and offers of necessary support, most respondents have a positive reply, one cannot comment while one differs strongly.

Beyond appreciation, top management is involved in quality efforts and provides necessary leadership in the views expressed by three of the eight respondents. Three remain undecided while two others disagree with this view.

Half of the responding executives feel that quality gets attention only when problems or occasions demand, but not all the time. Three hold a different view and the other person is non-committal.

Agreeing with the above situation, five believe that quality gets continuous and regular care, two are undecided and one does not agree with the view of the first group.

The overall situation is possibly nothing to spring surprise, and there are concomitant factors which beget variations in responses. Of course, there are genuine variations across organizations.

N.B. These eight responses clubbed with the ones already analysed may portray a picture that differs from the one presented above.

15.8   Two Different Cases Not to Be Generalized

As is well known, limited personal experiences by themselves alone cannot be used as inputs to any inference, though these need not be indicative of typically good or bad situations. Similarly, when case studies are conducted on Quality Management practices in a whole host of industries, it is just natural to find a lot of diversity in focus, in practice and in results. In fact, the SQC and OR Division of the ISI have brought out rich volumes of case studies. The material that follows includes an experience of the author that may not reflect even the current situation prevailing in the industrial unit to which the case relates and a second reported by the ISI as representing a level of excellence in Quality Management. Thus, the material reflects some extent two contrasting situations and should be taken with its limitations and no attempt should be made to generalize these situations as characterising similar Indian industries.

15.8.1 Case One

A public sector enterprise manufacturing conveyor belts for the mining and allied Industry as also smaller belts for automobiles and other manufactured products wanted to develop a Quality System on the lines of ISO 9001:1987 version in early 1992. The enterprise was supplying steel-cord reinforced belts for heavy minerals to some foreign countries besides meeting demands from domestic users. It had a quality system of its own and had established itself as a supplier of quality industrial belts. However, the enterprise was failing to generate enough operating surplus beyond what was needed to service debts incurred to augment its manufacturing and specially its testing facilities in the recent past, for a variety of reasons.

The enterprise realized that it would be better to adopt a training-cum-advisory approach rather than a consultancy approach for developing and maintaining a Quality System and eventually getting it certified for compliance with requirements in the international standard. Management accepted the fact that it would take quite some time to come up to the stage of certification and that this had to be a serious engagement for the entire workforce. In fact, the chairman-cum-managing director himself participated in the half-day programme to convey to senior management personnel the spirit behind the standard and its implementation implications.

A prolonged series of training programmes for different levels of people engaged in different processes followed, and the Executive Director would not merely attend most of these programmes but also clarified some doubts and concerns of the participating people. One such programme was also organized for the frontline workers to allay their apprehensions about apprehended additional work to be done by them once the standard was in place. That this implementation would reduce their worries and problems at workplace was highlighted during that programme. Special emphasis was laid on identifying interfaces between several different functions or processes to focus on the concept of a system and provisions of the standard relating to the interfacing operations or functions were duly explained to the participating executives. The concept of business processes leaving aside the core processes at the factory level was not readily accepted and several attempts were made to bring the field staff dealing with customer complaints and their resolution to board.

It came out that, in the words of Crosby, this enterprise could be recognized as a troubled company with a relatively large focus on marketing and support service network. And the 'quality vaccine' was worked out and administered to the benefit of the enterprise. Due attention was paid to requirements of the standard relating to contract review, communication, training, use of statistical techniques, inspection and test status and the like, since these aspects of quality management had not been earlier recognized explicitly. In fact, this recognition led to some re-allocation of duties and responsibilities among the managerial staff members. A meeting with the vendors of raw materials particularly for the compound to be mixed with the fabric was organized so that problems about quality of incoming materials could be sorted out. It was found that in some cases quality requirements were not clearly and completely communicated to the vendors. And their payments were sometimes delayed a lot, causing some resentment at the vendors' end.

The enterprise took two years and a half to put up a Quality system that could easily be certified and continued to maintain its quality activities in full steam thereafter. With some amount of confidence in the benefits of a Quality System, the enterprise now looked at some of the support processes and to its overall business activities. With some extra efforts, a marked improvement in business results followed. But this effort could not be sustained within a work culture that was prevailing those days in the public sector.

Quality improvement alone could not sustain business and corporate management could not rise up to the mark. The enterprise was taken over by a foreign company. Some changes were effected in the existing procedures and practices to pull up organizational efficiency. The new company was accepting more and more rigid specifications for products and components and was making sincere efforts to meet those specifications. The enterprise took up a comprehensive failure mode, effect and criticality analysis of its designs and products and the findings are still acted upon. Later, the enterprise was advised by its principal to learn and adopt the 8D Methodology for Quality Improvement. As expected, business is now stable and satisfactory.

15.8.2   Case Two

The company produces plastic-body ballpoint pens and plastic stationery items.

Products have a good domestic market.

The company markets its products in Europe and USA through foreign agents.

No big complaints received from markets.

Improvements in product quality are achieved through isolated efforts of some individuals engaged in procurement and manufacturing activities, occasionally motivated by adverse results of inspection.

Management appreciates the need to improve quality and to add more functionalities for enhanced product value.

Management would like to have a certificate for compliance with its 'Quality System' to ISO 9001 (2008) Standard.

It approaches a consultancy house for the purpose, who points out the absence of a 'System' currently and the need to develop a system, prior to certification.

This means a lot of internal consultations followed by documentation of processes and procedures in conformity with ISO clauses.

Management feels—and not totally wrong—that a time-consuming documentation, except when done by the consultant, will mean a wastage of resources without a commensurate improvement in quality or productivity.

However, a certificate of compliance with an international standard for Quality Management Systems will boost the image of the company.

A second consultant is finally engaged to ensure a quick certification, at a higher price, with only some cosmetic changes in the existing arrangements.

Management looks upon the certified system as an effective means to boost its image, and not so much as a systems approach towards management of performance in all business processes.

Management has engaged a noted film personality as its brand ambassador in its product promotion ventures.

Management is aware of the fact that the importance to process and product quality as is being currently accorded is sufficient to ensure continuance of certification by a certifying body that can compromise quality of audit with some extra payment.

Inspection and quality control are subordinate to production, procurement and marketing are directly controlled by top management.

New product development is the only function where all senior people are involved and production takes the lead.

Process plans do not exist and internal quality audits are mere rituals.

Such an approach to managing quality-related activities is not rare and not confined to this company only.

There are many ISO-certified organizations—in both manufacturing and service sectors—who really do not follow the standard in spirit and do not fulfil the requirements in their totality, e.g. not much is done on the lines suggested in the last clause 'measurement, analysis and improvement'.

Management of such companies looks upon a Quality System as an arrangement that involves essentially the QC or inspection department and participation of senior executives from other departments like procurement, production, maintenance, instrumentation and marketing only when there is a crisis.

15.9 Concluding Remarks

It must be emphasized that Quality Management in Indian Industries does not reveal a homogeneous situation, characterized by extremely laudable policies, practices and outcomes in some and visibly poor attention paid to quality in some others. A very large number occupies the positions in between the extremes. There are marked variations across regions (with their typical work cultures), product and services profiles, size of workforce, engagements with foreign markets, management styles, investor behaviour and a whole lot of other influences.

In fine, the material reported in this chapter is to be treated as a 'narrow view of a wide diversity'.

Survey on Improving Quality Management

Research contact: Professor S. P. Mukherjee

Introduction to Questionnaire

The questions below are part of a project to help improve Quality Management and Systems. We wish to discover your opinions as an expert in this field.

When you are answering you could do so by relating your answers either to Quality Management in Indian industry in general or to your own Industry.

The questions are in two kinds. The first kind seeks you to tick a box in a semantic differential between two extremes. The second kind seeks a one-line response from you. These answers will not be attributed to you. They will be combined to identify the general view of the entire community.

A.   Impact of Quality on Practice

Q 1. To what extent has a formal focus on QUALITY in terms of a comprehensive Quality Management System helped Indian (or your) Industry in achieving its desired goals (by way of growth and excellence)?

Was Harmful Help	Wasted Time	Irrelevant	Some Good	Great
☐	☐	☐	☐	☐

Q 2. In this context how important has been the role of Quantitative Tools and Techniques?

Negative Value	Irrelevant	Some Good	Great Help	Essential
☐	☐	☐	☐	☐

Q 3. In this context how important has been the role of Leadership provided by Top Management?

Negative Value	Irrelevant	Some Good	Great Help	Essential
☐	☐	☐	☐	☐

Q 4. Does Top Management view Quality of Services as a necessary adjunct of Product Quality?

**Definitely Not   Probably Not   Irrelevant   Probably Yes
Definitely**
   ❑   ❑   ❑   ❑   ❑

Q 5. To what extent have Standards such as ISO 9001:2008 along with ISO 9004 or ISO 17025 helped Improvement in Quality?

**Did Not Help At All Probably Not   Not Sure   May Have Definitely
Helped**
   ❑   ❑   ❑   ❑   ❑

Q 6. Is Quality Cost Analysis an essential exercise to appreciate the role of Quality Improvement?

**Not Needed At All   May Not Help   Not Sure   May Help   Is
Essential**
   ❑   ❑   ❑   ❑   ❑

Q 7. To what extent does Innovation in Design contribute to Quality Improvement?

**Does Not Contribute   May Not   Not Sure   May Contribute   Definitely
Does**
   ❑   ❑   ❑   ❑   ❑

Q 8. Does Quantity Expansion compromise Quality?

**Does Not Compromise May Not   Not Sure   May Compromise   Definitely
Does**
   ❑   ❑   ❑   ❑   ❑

Q 9. Does Cost / Price Reduction compromise Quality?

Does Not Compromise May Not   Not Sure May Compromise Definitely Does

☐   ☐   ☐   ☐   ☐

B   Attitude of Top Management towards Quality

Q 10. Do Management believe that Quality has to be paid for, insisted upon, and that conformity to Quality Standards should be made legally binding?

Agree Strongly   Agree   Neutral   Disagree   Disagree Strongly

☐   ☐   ☐   ☐   ☐

Q 11. Do Management believe that Quality should always be maintained at the optimum level (taking into account factors specific to the organization)?

Agree Strongly   Agree   Neutral   Disagree   Disagree Strongly

☐   ☐   ☐   ☐   ☐

Q 12. Do Management believe that Quality takes care of itself and needs no special effort or attention?

Agree Strongly   Agree   Neutral   Disagree   Disagree Strongly

☐   ☐   ☐   ☐   ☐

Q 13. Do Management believe that Quality and Quantity can never go together?

Agree Strongly   Agree   Neutral   Disagree   Disagree Strongly

☐   ☐   ☐   ☐   ☐

Q 14. Do Management believe that Efforts to build Optimum Quality also lead to enhanced productivity?

Agree Strongly   **Agree**   **Neutral**   **Disagree**   **Disagree**
Strongly
         ☐   ☐   ☐   ☐   ☐

Q 15. Do Management believe that Quality and especially Reliability (long life) will adversely affect demand, and hence production and hence profit?

Agree Strongly   **Agree**   **Neutral**   **Disagree**   **Disagree**
Strongly
         ☐   ☐   ☐   ☐   ☐

Q 16. Do Management believe that Quality and Reliability add more to the value of a product/service and value-based pricing more than any possible costs of investments in Quality?

Agree Strongly   **Agree**   **Neutral**   **Disagree**   **Disagree**
Strongly
         ☐   ☐   ☐   ☐   ☐

C   Top Management Concern for Quality

Q 17. Do Top Management appreciate Quality Effort and provide some / all of the necessary support?

Agree Strongly   **Agree**   **Neutral**   **Disagree**   **Disagree**
Strongly
         ☐   ☐   ☐   ☐   ☐

Q 18. Are Top Management *involved* in Quality Efforts and do they provide the necessary *leadership*?

Agree Strongly   **Agree**   **Neutral**   **Disagree**   **Disagree**
Strongly
      □   □   □   □   □

Q 19. Does Quality get attention only when problems arise or occasions demand?

Agree Strongly   **Agree**   **Neutral**   **Disagree**   **Disagree**
Strongly
      □   □   □   □   □

Q 20. Does Quality receive continual and regular care?

Agree Strongly   **Agree**   **Neutral**   **Disagree**   **Disagree**
Strongly
      □   □   □   □   □

D   **Open Questions about Quality**

Please write one short answer in the box provided

Q 21. What should be the priority to improve quality in Indian (or your) industry?

Response…

```




```

Q 22. In your opinion what is the most significant factor hindering improving quality in Indian (or your) industry?

Response…

```
┌─────────────────────────────────────────────────────────┐
│                                                         │
│                                                         │
│                                                         │
│                                                         │
│                                                         │
└─────────────────────────────────────────────────────────┘
```

Q 23. What is preventing us as an industry from dealing with our quality problems?

Response…

```
┌─────────────────────────────────────────────────────────┐
│                                                         │
│                                                         │
│                                                         │
│                                                         │
└─────────────────────────────────────────────────────────┘
```

Q 24. What should be done to increase understanding of how to improve quality in industry?

Response…

```
┌─────────────────────────────────────────────────────────┐
│                                                         │
│                                                         │
│                                                         │
│                                                         │
└─────────────────────────────────────────────────────────┘
```

Q 25. What is holding us back from working together to improve quality?

Response...

```



```

Q 26. Would you suggest any particular management or systems change that would contribute greatly to improving quality?

Response...

```



```

Response...

References

Agarwal, S. (1993). ISO 9000 implementation in Indian industry proceedings of the 8th ISME conference in mechanical engineering, New Delhi (pp. 638–644).

Banerjee, K., Gundersen, D. E., & Behera, R. S. (2005). Quality management practices in indian service firms. *Total Quality Management & Business Excellence, 16,* 321–330.

Bhadury, B., & Mandal, P. (1998). Adoption of quality management concepts amongst Indian manufactureres. *Productivity, 39*(3), 443–451.

Mathew, T., et al. (2002). Performance improvement through transfusion of TQM and TPM in Indian manufacturing industry. *Industrial Enginnering Journal, 31*(7), 4–10.

Seth, D., & Tripathi, D. (2006). A critical study of TQM and TPM approaches on business performance of Indian manufacturing Industry. *Total Quality Managemet and Business Excellence, 17*(7), 811–824.

Chapter 16
Reliability—An Extension of Quality

16.1 Introduction

Reliability can be looked upon as an extension of quality beyond production, shipment, storage and distribution to the use or deployment phase in the life cycle of a product or service. As has been pointed out in Chap. 1 and spelt out in relevant standards and books, reliability is 'quality of performance" and is the composite effect of 'quality of design' and 'quality of conformance'. It is justifiably understood as 'dependability' as well—and noting that the question of a product being dependable or not does arise mostly, if not exclusively, with the user or customer at the time the product is being put to use or during the period the product is meant to function. And this way, reliability is very much an aspect of overall quality as defined by Deming.

Reliability is linked up with functional characteristics of a product or service as are captured in terms of 'performance parameters" as distinct from 'quality parameters' related to properties and features of materials or processes or checks, etc. In fact, quality of conformance is assessed in terms of specified values or ranges for these quality parameters compared with the values realized during production. Among many performance parameters which can be brought into visualize the mission behind designing and creating a product or service, some can be identified as the critical ones and reliability would be usually defined and measured by considering these critical parameters only. In case of a product or service meant for use during a certain period of time (and not just a point of time), life of the product or service during which it does perform as expected becomes the most important parameter in defining reliability.

Reliability Analysis on the basis of strength built into the product (in terms of quality of design and quality of conformance) and stress encountered by the product during use—developed internally and incidentally as the product functions as also imposed by the external environment in which the product functions—does fall in line with the idea of reliability being an extension of quality.

© Springer Nature Singapore Pte Ltd. 2019
S. P. Mukherjee, *Quality*, India Studies in Business and Economics,
https://doi.org/10.1007/978-981-13-1271-7_16

Going back to history, quality was a concern of the producer, who—in the pre-Industrial Revolution era—would design, manufacture or assemble and check a few products that would be usually appreciated by the customer who would use the product, either for final consumption or for preservation as a utility item or a decorative item. When mass production began with the Industrial Revolution and large numbers of varied products rolled out of the respective machines, quality implying both utility and beauty could not be consistently ensured in each and every product unit. Such a situation created the need for training of producers under more skilled supervisors, inspection of incoming material, of work-in-process and of finished goods to sort out defective units to be either scrapped or reworked for removal of the defect(s) and even replacement of goods sold and found to fail during operation or use. Eventually grew concepts, methods and techniques for Quality control, quality assurance and finally quality management.

Coming to electrical and then electronics products, quality of performance and related issues of safety came to attract a lot of attention from designers and persons engaged in maintenance of plant and equipment, complex products and their accessories as well as of devices used for defence purposes. When such devices and systems failed to function all of a sudden, the consequences were quite severe. This led to the study of reliability of complex products, usually systems. More recently, software reliability came under the scanner and a big volume of literature has thrown up many models, methods and tools for reliability evaluation of software systems.

Failures during use or operation to carry out the functions specified in the mission plan constitute the core of reliability analysis. And, as is true for any analysis exercise, reliability analysis identifies different types and causes of failures, derives estimates of reliability from data which may be available in some simple or overt form, and suggest procedures for reliability improvement. While most of the analysis is generic, some parts are specific to the nature of products under consideration.

Product failures bear a strong analogy with human mortality and even morbidity. Hence reliability analysis has a large ground common to survival analysis and many concepts, measures and methods are common between the two. There exist some subtle differences, of course.

The present Chapter presents a bird's eye view of the several important dimensions of reliability analysis viz. (a) probabilistic: dealing with random variations in the basic performance parameters and throwing up different definitions of reliability (b) engineering: dealing with configurations of different types of systems, importance of components within a system and related issues (c) statistical: dealing estimation of reliability and tests of reliability involved in reliability demonstration exercises and (d) managerial: dealing with optimum redundancy provision, optimum replacement policies and similar other issues involving optimization under appropriate constraints. The focus, of course, is on a comprehensive understanding of reliability, considering different types of products, using different approaches and for different configurations of systems. Though reliability analysis involves a lot of probabilistic and statistical analysis, the general tenor observed in this book has been maintained in this Chapter also.

16.2 Usual Definition of Reliability

More often than not, Performance has been understood narrowly as time up to which the product functions. It is presumed that if the product functions, it functions satisfactorily. Once any of the performance parameters deviates from its specified range or level, the product is taken to have failed. Thus, time up to which a product functions or survival time is the only parameter directly considered in evaluating the reliability of a product. In fact, to be reliable, performance parameters of the product have to satisfy the respective requirements for all times up to the 'mission time'.

Under the simplifying assumption that 'functioning' implies 'satisfactory functioning', Reliability has been understood as Probability of Survival or functioning up to the specified 'mission time.' In fact, the commonly used definition of Reliability, viz. probability that the product/equipment functions satisfactorily for a specified period of time, when used in the manner and for the purpose intended. Dispensing with the implication of 'satisfactorily', we are still left with two more conditions to be taken care of. If a product fails to function for the specified period when used in a manner or under an environment not recommended for the product or when used for a purpose or function different from the one intended, the failure will not be taken as 'unreliability'.

Reliability, according to this definition, revolves round the time up to which the product survives or the time at which fails compared to the specified time indicated in the mission and reflected in the design. It is also recognized that the time at which a product fails will depend on the design and process parameters. Further, different units produced by the same process following the same design are found to fail at different times, the time for a particular unit being unpredictable, being affected by a host of uncontrolled 'chance' factors over and above the assignable design and process parameters. Thus, time-to-failure, T say, or indirectly survival time is a random variable. Hence $[T > t_0]$ is a random event and its probability P – Prob $[T > t_0]$ is taken as reliability. This definition relates to an infinite population of product units which are identically designed, manufactured and tested. Reliability of a single product unit may be possible in a different way.

This definition of reliability starts with a directly observable time-to-failure, which is the same as length of life or simply life in the case of non-repairable (and particularly continuous-duty) products. And we have tacitly assumed that all the critical performance parameters lie within the specified intervals as long as the product functions. It is quite possible, however, that any such performance parameter Y changes with time (or age of the product) and usually changes randomly, the value or level realized at time t being $Y(t)$, which will generate a sample path. The first time that path enters the absorbing state defined by a value or level outside the specified interval, we say that the product has failed. Thus, failure time should be defined as the minimum of first passage (to the absorbing state) times for the different processes $Y_i(t)$, $i = 1, 2,\ldots$.

Nowadays, people recognize differences between Life and Morbidity-Adjusted life. In a similar vein, one can introduce several states of functioning for a product as distinct from the functioning-failed binary classification and can even think of associating different weights or utility measures to durations of stay of a product/equipment in the different states (including states where the product is partly functional, not satisfying all the requirements about critical performance parameters). In such a situation, one can consider the weighted total time of stay in different states of functioning (excluding the absorbing state of failure) to define the Quality-Adjusted Life and derive the survival different states (including states where the product is partly functional, not satisfying all the requirements about critical performance parameters) probability based on the distribution of Quality-Adjusted Life. Random variables corresponding to the durations of stay in the different states may be assumed to be independently and identically distributed for the sake of simplicity. However, it is quite rational to assume non-identical distributions, even if having the same form, and the assumption of a joint distribution may be more justifiable.

16.3   Types of Product

To define reliability, we need a classification of products based on the pattern of use and the possibility or otherwise of repair in case of partial or total failure.

Firstly, we note that certain products are meant to function only at one point of time, not necessarily immediately after production or delivery. The manufactured product may have to be stored for some time and required to function when actuated at some point of time. Most equipments in defence as also in other situations may belong to this category. Usually, such products are not repairable. In such cases, survival up to a specified time is not a concern and probability that the product functions at the point when activated defines probability. Such items are called one-shot or instantaneous-duty products and are generally non-repairable. In case of a repairable one-shot product, reliability is a probability that relates to a population of occasions on which the product was activated.

Many products are meant to function during some interval of time, to remain switched off subsequently and then again activated to function for some more time. This cycle of on and off situations continues. The time intervals during which it is required to function are not the same as the 'mission time', rather the mission time is stated in terms of the number of successful operation cycles and reliability is the probability that the number of successful operation cycles N equals or exceeds a specified number n_0. Thus, reliability becomes Prob $[N > n_0]$, and this probability may relate to a population of copies of the product. These are intermittent-duty products or equipments or repeated-shot ones.

Such items are more often than not repairable. However, in case a product is repairable, we define reliability in a different manner taken up later in this book.

A large majority of products or equipments are meant to function continuously after being activated or turned on. Examples are a power generating system or a display board or monitoring system. For such products, when these are non-repairable, we really have a 'mission time' specified and the product is required to function continuously till this time in order that the product is reliable. In case of repairable continuous-duty products or equipments, we have a different measure of reliability.

Thus, the commonly accepted definition of reliability is really applicable only to non-repairable continuous-duty products or equipments. In all other cases, we need to have different reliability measures.

In the broad category of repairable products, time-to-first failure is not the only variable of interest. Of equal importance in assessing the reliability of the product are times between successive failure and repair times. These, in turn, make for the uptime and downtime for the product during a finite interval or in the limit.

16.4 Availability and Maintainability

One-shot or instantaneous-duty equipments/devices have to operate satisfactorily only at the intended point of time. In such cases, we are interested in what may be called 'operational readiness'. The intended point deployment time may be unpredictable, and the equipment/device has to be ready for use right up to that point of time. Sometimes, we may require the product to function during a period beginning at some intended point of time. The measure of reliability for such a one-shot device or equipment is 'Availability' defined by the following indicator function

$A(t) = 1$ if the item is in the up-state at time t

$= 0$ if the item is down at time t going by the classical two-state analysis.

The availability over a period $(0, T)$ is the integral of $A(t)$ over $(0, T)$. While $A(t)$ is called point-wise availability, this integral is called period availability. The limit of this period availability as T goes to infinity is called limiting availability. This is just the reliability of a continuous-duty non-repairable item.

Availability will be quantified as a probability that the product is in the up-state when required. Point-wise availability of a non-repairable item required to be up at many different points of time can be estimated as the proportion of times the item was really in the up-state. The implicit mechanism is such that being up at a point of time does not ensure an up-state at a different point and the item being down at any point does not rule out the possibility of the item being in the up-state at a subsequent point. This is explained by the fact that properties built into the item vary over time randomly, depending on many environmental and incidental factors.

Most products or devices are meant for intermittent duty or continuous duty and many of these products are repairable. One such a product fails, a diagnosis of failure followed by a corrective action can put the product to its initial or pre-failure state and the product becomes available for use (intended function). Thus, comes in the repair function. Of course, efforts to enhance availability of a product at all times during its useful life go much beyond repair.

When we consider an item that starts in the up-state, remains there for a random duration and fails thereafter, waits and then goes for repair for random times, and then comes back to the up-state, we define availability as

$$\text{Availability} = \frac{\text{Mean time to failure}}{\text{Mean time to failure} + \text{Mean time to repair}}$$

We take for granted that on failure the item goes immediately for repair, else we include the mean waiting time for repair within mean time to repair. Thus, availability is estimated as the long-term proportion of time the item is in the up-state. Similarly, period availability can be interpreted as the proportion of time the item remains up or operational during a given period of time. The item could be down, waiting for or undergoing repair, during the remaining portion of the given period.

Continued availability through necessary maintenance corresponds to maintainability. However, maintainability to ensure both point-wise and interval availability is primarily determined by design features that take due care of ease and effectiveness of inspection and repair or replacement actions. In fact, we focus on

(1) monitoring/inspection at appropriate times and opening the product up, if needed
(2) repairing the components that are going to fail, by replenishing the properties required for their functioning and
(3) replacing the components that have already failed by new ones.

These three activities constitute the core of maintenance. Like any other activity, maintenance as a broad activity as also each of its component activities has to be planned, monitored and evaluated. In many situations, maintenance of a service may imply replacement of some devices or equipments either on failure or in anticipation of failure. While replacements on failure have to be necessarily carried out, preventive replacement can be carried out at planned points of time to reduce chances of failures in between two points of planned replacements. An optimal replacement policy can be worked out to minimize the expected total cost of replacement (considering costs of both failure replacements as ell as preventive replacements) per unit time.

16.5   Types of Failure

Failures take place at random points of time during the life of a product. For subsequent analysis, it is important to distinguish among three types of failure, viz. early or instantaneous failures which occur as soon as the product is put to use or immediately thereafter, chance failures which occur during the useful life of a product and wear-out or degradation failures which, as the name suggests, are attributed to ageing and associated degradation or wear-out of the product. Early failures occur primarily because of faulty use or faulty design. A faulty design is one that could not envisage the stress to be encountered by the product during use and did not provide adequate safety margins against possible stresses. A faulty use could imply some inappropriate use environment for which the product was not intended. Chance failures when at some unpredictable point of time the built-in strength as provided by the design and a compliant manufacturing process fell short of the stress encountered. A wear-out failure is caused by a prolonged use or cumulative exposure to stress leading to loss of properties required for functioning.

Time to early failure has a highly positively skewed distribution with the mode located at time 0, exponentially decreasing to almost touch the time axis around a small value.

Time to chance failure is the most important distribution n reliability analysis and usually follows an exponential r Weibull or some other distribution.

Wear-out failure times follow a negatively skewed distribution rising sharply from an initial value.

Early failures are generally weeded out either by conducting a debugging of the components before these are assembled into a product for an initial time by which early failures take place with a very high probability followed by use of only those components which survive for being assembled into systems. This practice is called 'debugging of components'. While this practice is expected to rule out early failures of systems, we should remember that even when individual components have survived up to a certain time, when assembled into a system mechanical sticking of the components may take place and the system may not necessarily survive up to that time. This is why, a second procedure followed to take care of early failures is to form systems first and to test the systems for an initial period and subsequently to consider only the surviving systems for further use. The latter practice is usually referred to as 'burning in'.

16.6   Stress–Strength Reliability

Another way to look at reliability is to recognize the fact that a product can function (satisfactorily) in terms of the strength built into it by the design and the production. Another way to look at reliability is to recognize the fact that a product can function (satisfactorily) in terms of the strength built into it by the design and the production process. And this strength has to exceed the stress to which the product is subjected during performance. While strength (could be breaking strength, compressive strength, tensile strength, or could be just some similar enabling property) is the outcome of design parameters as also of process parameters, stress is either environmental or incidental. Mostly, stress is imposed by the use environment, may be, by way of, roughness, resistance, interference, excessive heat, etc. Some stress is developed incidentally as the product functions. Many electrical or electronic devices generate heat as these function (particularly for a long duration at a stretch), and this heat is detrimental to their further functioning.

A solid propellant rocket engine is successfully fired provided the chamber pressure X generated by ignition stays below the bursting pressure Y of the rocket chamber. The latter is determined by design and manufacturing processes, while the latter is external and depends on a host of factors, some being uncontrollable. A torsion type stress is the most critical stress for a rotating steel shaft on a computer. Both instantaneous stress as also cumulative stress can lead to failures.

A product fails to function or simply fails when stress to be overcome exceeds the strength built in the product. Both stress and strength are random variables, each being affected by a multiplicity of factors including some which are uncontrollable or uncontrolled and are simply branded as 'chance factors'. Strength may vary from one product unit to another even when design and process parameters are kept at the same levels. For the same product unit, strength may vary over time of use. In fact, in such cases strength is quite likely to decrease monotonically with time or age. Again, such a decrease can also be random and not deterministic. Stress depends on the environment and can vary across operation cycles carried out under different environments. And in the same environment, different units can suffer from different stress magnitudes.

This brings us to a definition of reliability as $R = \text{Prob } [Y > X]$ where X and Y stand, respectively, for stress and strength. Usually, X and Y will be treated as independently and identically distributed random variables. However, cases can arise where these become correlated, e.g. when units likely to be stronger are put to use in harsher or more stressful environments and the apparently weaker ones are used in less stressful situations. The practice of putting stronger copies in more adverse environments motivates the use of a joint distribution of X and Y. If we denote by $f(x,y)$ the joint probability density function, then reliability works out as

$$\int_0 \int_x f(x,y) \, dy \, dx$$

Assuming stress $X(t)$ and strength $Y(t)$ to vary over time, we get two continuous-time stochastic processes $\{X(t), t > 0\}$ and $\{Y(t), t > 0\}$ to define reliability for a mission time t_0 as $R_1(t_0) = \text{Prob } [\inf Y(t) > \sup X(t)]$

$$t \leq t_0 \quad t \leq t_0$$

We can have a moderate requirement and define reliability as

$$R_2(t_0) = \text{Prob}\left[Y(t_0) > \sup_{s \le t_0} X(s)\right]$$

One can consider a Brownian motion or some other processes for $X(t)$ and $Y(t)$ to derive expressions for reliability.

The problem of measuring strength without a destructive test may not be possible in many cases, and we may measure some covariates to work on some relations between Y and some covariate Z. Some would even argue that X and Y are subject to random errors in measurement or that these are damaged (not fully captured) or that these are fuzzy random variables. In the last case, a data-based membership function may be advocated.

16.7 Software Reliability

Softwares, unlike manufactured products, have infinite life generally, can be conveniently augmented or modified otherwise and that way, are malleable and takes a non-trivial time to test before being released in the market. Like manufactured products, these also go through a life cycle, often represented by the waterfall model which starts with *requirements analysis and specification,* and goes through *design and specification, coding and module testing, integration and system testing* to end in *delivery and maintenance.* And, as in the case of a manufactured product, a feedback from the last stage goes back to modification of the first stage.

Important quality characteristics of a software include (1) correctness, reliability and robustness (2) performance (3) user-friendliness (4) verifiability (5) maintainability (6) reusability (7) portability (8) understandability (9) interoperability and (10) productivity. While this is not an exclusive list of quality characteristics of a software, the list is not universally agreed upon. The list indicates some differences between product reliability and software reliability. However, if one closely examines the definition of software reliability as given below, one can easily identify some basic similarity in the two.

Informally, a software is reliable if the user can 'depend' on it. Formally, it is the probability that the software will operate as expected over a specified time interval in a specified environment. This interval may not correspond to a clock time and may well imply a certain large number of operation cycles to be defined appropriately in each case.

Engineering products are expected to be reliable. Unreliable products generally disappear from the market quite early. On the other hand, softwares are commonly released along with a list of 'known bugs'. Users of softwares take it for granted that release 1 of a product is always 'buggy'.

Thus, software reliability is linked up with bugs that may render the software performance to be not acceptable, Bugs may hinder some applications, may render it unusable and may even damage a platform on which it was being used.

Measures of software reliability include: mean time between failures, mean time-to-failure, number of bugs, failure rate and probability of failure-free operation up to the intended time. Software reliability could be estimated in terms of the residual bugs likely to remain in the software that has been released, or in terms of the time at which its application fails or in terms of the time taken to detect and fix or eliminate bugs. Software reliability has grown into a full-fledged discipline with many concepts, models and measures of its own.

16.8 Service Reliability

It has to be noted that a service produced or generated and then delivered to a customer is not necessarily free from some manufactured product(s). In fact, services correspond to a whole spectrum from a completely product-free service like an advice or a counsel or even a training or teaching which does not involve any gadget or infrastructure as is delivered to a client by a lawyer or a counsellor or a teacher to a service heavily loaded with some product(s) like food served in an eatery or repair of some customer–supplied product using some components or devices.

Reliability of a service has been generally comprehended in terms of the following features, viz.

1. Accessibility—Service is available as and when required
2. Adequacy—service provided covers all the faults or failures during use prior to request for service.
3. Uninterruptedness—service should be continuously produced to be completed in the shortest time and should not suffer from intermittent breaks by the service provider.
4. Performance—service meets the customer's needs and expectations so that the serviced product or system functions fully satisfactorily.

Some other features are sometimes added, like

1. Value—negotiate a contract with customer that places a value for a certain level of service requirement.
2. Needs customer participation or input from a wide perspective.
3. Convenience and care in billing operations.

Metrics for service reliability are not unique and may differ from one type of service to another. Thus, the metric for assessing accessibility could be the ratio of success to attempts, where time period, number of customers and successes could be factored in. Metrics like delay time during transaction, delay during set-up, delay after set-up and deviation in service rendered from the customer requirement can be used to assess performance. The number of transactions successfully completed to the number successfully initiated may be a reasonable metric for some purposes. The number of breaks in service production causing delay in service completion and inconvenience to customers in some situations is another important metric.

For certain type of service like repair of a product, reliability can be judged in the usual way in terms of the probability that the repaired product performs its intended function for a specified mission time or for a specified warranty time. The concept of time-to-failure could be introduced, and this probability determined therefrom. In the case of a product-free service like an advice or a counsel, reliability refers to effectiveness as revealed by the outcome in following the advice or counsel in the concerned problem or situation. The focus here is on performance. This is some-what analogous to the reliability of an instantaneous—duty product. That way, accessibility may be taken as an analogue of point-wise availability or operational readiness.

16.9 Life-Testing and Reliability Data

Testing is an integral component in design and development of a new product and in performance assessment of an existing product. The objective of testing is to ensure that the product will satisfy performance criteria, prior to their actual use or deployment in the field. Two basic options for a test plan are (a) component testing and (b) system testing. In the first option, components which make up the system are tested, and based on the results of these component tests, one draws inferences about system reliability. In the second option, the entire system is first assembled and then tested to draw inferences on its reliability. Combinations of these two options may also be tried out.

Going by the stress–strength approach to reliability, there could also exist two options, viz. (a) testing under the usual stress likely to be encountered by the product in use most of the time and (b) testing under higher stress levels. The justification for the second option is to hasten failure and record and analyse failure times even by testing for a relatively short period of time. The second option for life testing is usually referred to as accelerated life tests. The theoretical basis for using such tests is some relation that links stress level with mean life or some other summary measure of the life distribution.

Life and performance tests can be carried out on individual components before being assembled into systems or on systems themselves. System-level tests are usually recommended in situations where functioning of different components within a system is affected by interdependence. The phenomenon of mechanical sticking of components illustrates such a situation. Thus, system-level tests provide a better confidence about system reliability, even though these are relatively expensive. Here we burn-in some copies of the system for a relatively small period during which early failures usually show up. Failed system copies are opened up to identify weak components of undesired interconnections that cause early failures. After necessary corrective and preventive actions only, the system is released to the market.

On the other hand, component testing is less expensive (in some sense) and is almost a necessity to examine the behaviour of individual components. Here components are debugged before being assembled into systems. During the pre-specified debugging period, weak components fail. Only components which survive the initial debugging period are then assembled into systems for release. In fact, results of such tests during the functioning of a system to monitor its performance are essential for the purpose of maintenance. Component tests provide data on component lives which can be subsequently incorporated in an appropriate model to work out the lifetime or failure-time distribution of the system. Even when a system has completely failed, some of its components could still be functioning and component-level tests have to be conducted to identify components which are still surviving. These components can then be used in other systems. Thus, both system and component testing are important in the context of reliability assessment and management.

Autopsy data arise when the system is tested till its failure, after which the same is examined to find out which component(s) have failed. This requires opening up the system and inspecting the state of each component—failed or capable of functioning. Sometimes, components are monitored according to a given observation plan.

The essential idea behind accelerated stress is to compress time as seen by the physical system under study so that the designer can get a preview of the reliability the product. Usually stress acceleration pattern is kept relatively simple, e.g. constant stress or linearly increasing stress, so that simple acceleration models would allow comparison of survival time under accelerated stress with survival time under normal operating conditions. With simple stress patterns, acceleration models may become complicated. Introducing a bit more complex stress pattern may allow better identification of the complex acceleration models. Two new procedures are playing an increasingly important role.

1. Environmental Stress Screening (ESS) to identify manufacturing defects in early production at the system level. This procedure is often not economical.
2. Stress-Life Tests (STRIFE) to identify design defects leading to potential reliability problems in the prototype stage.

Assuming a standard failure-time distribution for reliability estimation, the failure-time distribution under accelerated stress is scaled in relation to the distribution under normal condition by the simple relationship $F_s(t) = F_0(t)$ where is the acceleration factor taking under stress s to normal operating stress 0 and $F_s(t)$ and $F_0(t)$ are the two failure-time distributions.

Since any life test will use energy and destroy, wholly or partly, the units put on test, we have note the cost involved and design the test to keep the cost under control. In fact, there is also another cost, though not direct or visible, associated with the waiting time for the test results to be processed in order that a reasonable estimate of reliability (up to a mission time) can be obtained. In this context, we can have broadly speaking two types of tests, viz. testing until all items fail and testing till a pre-specified test time or testing till a pre-specified number of units fail. The latter tests give rise to censored samples. In the former type of tests, we record failure times of all the units and the test time becomes the last (largest) failure time, which is random. Obviously, this type of test involves the highest number of items used up or destroyed, the highest cost and a random waiting time to reach a decision about reliability. The merit is a complete sample providing more information about the failure-time distribution and convenience in making inference about reliability. In case the test is terminated at a pre-specified time (at which the test is truncated), the number of failures observed is random, but the test time is fixed. The number of items completely used up is random. The sample is generally referred to as Time-Censored or Type I censored. In the other Failure-Censored or Type II censored sample, we have a fixed number of observed failure times and hence a fixed number of items completely used up, while the test time remains random.

   Reliability Testing may involve time testing or failure testing. In the former, we record failure times, while in the second, we simply count how many units failed by a pre-specified test time. Evidently, the second type of testing is cheaper and more convenient.

16.10   Life Distributions

Probability provides a quantification of uncertainty associated with any phenomenon affected by a multiplicity of factors or causes, including some which are beyond our comprehension and hence beyond our control. The latter arises from a host of sources and gives rise to unpredictability about the outcome of any experiment or observation on the phenomenon. Such factors or causes are collectively recognized as 'chance factors or causes' and cannot be individually identified. These chances cause which along with other factors within our knowledge and/or our control, give rise to random variations in outcomes of any repetitive experiment. Thus, in a life-testing experiment observed time-to-failure of the product units put on test reveals random variations which often exhibit some regularity or pattern if a sufficiently large number of units are tested. We assume the units to come out of the same production process following the same design and subject to the same checks and corrections. Any attempt to capture the regularity behind the random variations in times to failure will have to be presented in terms of a probability distribution that gives the probabilities of the failure time lying in small intervals or exceeding or falling short of some specified value(s).

   The concept of probability as a measure of uncertainty involves a repetitive experiment. It should be noted that repetitions may correspond to similar units of a product or to cycles of operation of a single unit. Accordingly, the statement of a given reliability for a single unit has a different connotation compared to such a statement about a product in terms of a production process or a segment thereof. In a software just released, there would generally remain some bugs and the number as well the times these will surface and cause failures are all uncertain. Once we accept that there could be bugs, we are certain that the software will fail sometime or the other, may be even after the intended period of use. However, we are uncertain about the true situation and hence invoke the concept and measures of probability.

In the case of manufactured items, we involve a whole population or a large group of units which are designed, manufactured and tested identically and still differ among themselves in terms of length of life or any functional parameter randomly owing to many unknown and uncontrolled causes. In such a situation, we are uncertain about which units will perform satisfactorily for the intended mission time and which will not and we associate a probability with the entire population to reflect this uncertainty.

Random variables observed in the context of reliability analysis include time-to-failure, time-to-repair, time waiting for repair, time between two consecutive repairs, time of transition from a fully functional state to a partly functional state and the like. These illustrate continuous variables as against discrete variables like number of failures observed during a specified interval or number of components which fail when a complex system fails. Considering a system or even a component subject to periodic shocks like vibration or high temperature or stress, the number of shocks received during a certain period is a discrete variable, while the intensity of a shock illustrates a continuous random variable. Accordingly, we use discrete or continuous probability distributions to work out reliability or probability of any event of interest.

A common discrete distribution describing variations in the number of failures (X), out of a starting number n of units put on test, by a specified time is given by the Binomial probability distribution with probability mass function

$$b(x; n, p) = \text{Prob } [X = x] = {}^nC_x p^x (1 - p)^{n-x} \quad x = 0, 1, 2, \ldots n$$

where p is the probability of a unit failing by the specified time. In case the specified time coincides with the mission time, this probability reduces to $p = 1 - R$ or unreliability. Otherwise, it can be derived from the underlying distribution of time-to-failure. In case, n is large and p is small so that np (the expected value of X) is moderate, the binomial distribution tends to the Poisson distribution 'of rare events' with a mass function given by

$$P(x; m) = e^{-m} m^x / x! \quad \text{where } m = np.$$

In the class of continuous distributions, the most commonly used is the exponential distribution for time-to-failure T with the probability density function given as

$$E(\lambda) = \lambda e^{-\lambda x} \quad \text{which gives Prob } [T < x] = 1 - e^{-\lambda x}.$$

Thus, reliability or probability of survival up to mission time t_0, say, is

$$R(t_0) = \text{Prob}\,[T > t_0] = \exp(-\lambda t_0)$$

If failures cannot take place earlier than time μ because of debugging or burning-in exercise, we can modify the exponential distribution to include μ as a location parameter and write $E(\lambda, \mu) = \lambda \exp[-\lambda(x - \mu)]$.

Important among properties of life distributions are the ageing properties which behave in characteristic ways for different distributions. The underlying measures can be estimated in terms of sample data on time-to-failure or number of failures. A basic property is the failure rate, also called the hazard rate, defined by the following relation.

$r(t)\, dt$ = prob [unit fails in the time interval $(t, t + dt)$ given that it has survived up to time $t = f(t)\, dt/\,[1 - F(t)]$. Usually, failure rate is quite high when the unit is just put to use (indicating high infant mortality), decreases thereafter as the unit gets hardened on the job, remains more or less constant during the useful life period and again rises because of wear-out or degradation. The pattern is often referred to as a bathtub. The average failure rate over the interval $(0, t)$ is therefore

$$A(t) = 1/t \int r(t)\, dt = -\log[1 - F(t)]/t$$

The function $-\log\,[1 - F(t)] = H(t)$ is often called the hazard transform or the hazard function.

The exponential distribution, characterized by a constant failure rate and relatively convenient estimation of the parameter(s) involved, has been found to fit well to data on component life. However, for assembly or system lives, a generalization of the exponential was found by a Swedish physicist S. D. Weibull to provide a better fit. The two-parameter Weibull distribution has a density function given by

$$W(x; \alpha, \lambda) = \alpha\, \lambda\, x^{\alpha-1} \exp\,(-\lambda x^\alpha) \text{ so that } \text{Prob}[T > x] = \exp(-\lambda x^\alpha)$$

The failure rate function for this distribution is given by $r(t) = \alpha\, \lambda\, t^{\alpha-1}$, and this is monotonically increasing if $\alpha > 1$, monotonically decreasing if $\alpha < 1$ and remains constant at λ if $\alpha = 1$ (reducing to the exponential distribution). If necessary, we can incorporate a location parameter in the Weibull density function as well.

16.11   System Reliability

Most of the products in use like a laptop computer or a mobile telephone or a household padlock or a coffee maker, forgetting about complex and costly items like a Television set or a luxury car or an aircraft are systems in which several components—and in some cases a very large number of these—are assembled into sub-systems finally incorporated into the system. Fortunately, such systems are generally quite reliable. It is evident that reliability of any such system will depend on (1) the system configuration in terms of the arrangement of components within the system determining the interdependence in functioning among the components and (2) reliabilities of individual components assembled within the system, including those of any device(s) inserted for detection of component failures and to switch on some components currently not in the active state.

A general consideration of component or system reliability could take into account multiple states of a component or the system at any point in time, e.g. fully functional, functioning with moderately reduced efficiency, functioning with highly reduced efficiency and not functioning or failed. However, in the simple and widely used treatment of the subject, we assume only two states, viz. functioning and failed. Thus, we associate with each component i, a binary state variable $X_i(t)$ which takes the value 1 if the component is in the functioning state at time t and a value 0 otherwise. Similarly, we have the system state variable $S(t)$ at time t. And, probability of the system or any component functioning at time t will be indicated by Prob. $[X_i(t) = 1]$ = Expected value of $X_i(t)$ and Prob. $[S(t) = 1] = E[S(t)]$ assuming that the components or the system to be non-repairable and taking t to be the mission time, these probabilities or expected values are just the corresponding reliabilities R_i and R, respectively.

Among the more common systems are series, parallel and standby systems, besides their mixtures as well. We do use products which have very complex configurations for which system reliability cannot be easily worked out, given component reliabilities. We consider only some simple systems here and give the expressions for system reliability in terms of component reliabilities. The way $S(t)$ is linked to $X_i(t)$ $i = 1, 2, \ldots n$ in an n-component system is often referred to as the structure function. For the sake of simplicity, we assume the components to be functioning independently of one another.

The simplest system—though not an ideal one—is a series system which functions if only all its components function. Thus, a series system will fail if at least one component fails to function. Here, the structure function becomes $S(t) = \pi$ $X_i(t)$ and system reliability becomes $R_s = \pi R_i$ a continued product of positive proper fractions which should be smaller than or at most equal to the smallest fraction. In other words, a series system can at best be as reliable as the least reliable component assembled within it. Assuming all components to have the same reliability R, series system reliability is $R_s = R^n$ and diminishes monotonically with increase in the number of components. To illustrate, if just eight components having the same high reliability of $R = 0.99$ are assembled to form a series system, system reliability will be only 0.92 and component reliability in this case was 0.95, and system reliability will be a meagre 0.66. It may be just imagined how much close one should component reliabilities be in order that a large series system with many components has to have even a moderate reliability of 0.95.

To surmount this difficulty, one option is to provide for extra or redundant components so that the system can function even if some component(s) within it fails (fail) to function. In such a redundant system, we can think of using the extra or spare components in three alternative ways, depending on the situation. We can allow all the components to remain in the active state from the beginning: the system will function as long as the minimum number of components, say $k < n$, required to make the system function remain in the functioning state. In the second alternative, we put only k components in the active state at the beginning, the remaining $n - k$ components remain completely immune from failure, and as and when a component fails, its task will be taken over by a redundant component currently inactive, and the system will function as long k components function. In the third situation, the redundant components are put in a partly energized state and remain subject to less than usual failure rates. These three types of configurations are sometimes referred to as hot standby, cold standby and warm standby systems. The second type is commonly known as a standby system..

In such k-out-of n systems which function as long as any k of the components function, extent of redundancy in design is $r = (n - k)/k$. The case $k = 1$ is usually referred to as a parallel system, and the case $k = n$ defines a series system.

The reliability of a k-out-of n system with a common reliability R for each component is given by $\sum ({}^nC_k)R^k(1 - R)^{n-k}$. The expression for system reliability with different reliabilities R_i of components is not difficult to obtain. The reliability of a parallel system with one redundant component is $R_P = 1 - (1 - R)^2$. This means a two-component parallel system has a reliability of 0.9999 when each component has a reliability of 0.99. In fact, redundancy is introduced to enhance system reliability.

In usual discussions on reliability, we involve only series, parallel and standby systems. Reliability of a standby system cannot be directly obtained from component reliabilities and has to be derived as survival probability for the system in terms of the distribution of system failure time. If we denote by T_i, $i = 1, 2, \ldots n$ time-to-failure (length of life) to component I and by T_S, T_P and T_B failure times of series, parallel and standby systems, then we have the following simple relations

$$T_S = \min T_i, \quad T_P = \max T_i \quad \text{and} \quad T_B = \sum T_i$$

Thus, knowing the probability distributions of component failure times, we can work out the distribution of system failure time in each case and, from there, the probability of the system surviving till the mission time. Usually, component failure times are assumed to be independently distributed with a common distribution form but with different parameters. In some situations, components can be reasonably taken as independently and identically distributed ($i, i. d,$) random variables. Thus, if we denote by $F_i(t) = \text{Prob} [T_i \leq t]$ the distribution function of component life i and S stands for the survival function, then the distribution survival function of series and parallel systems come out as $S_s(t) = 1 - F_s(t) = \pi S_i(t)$ and $F_P(t) = \pi F_i(t)$.

It can be easily proved that the failure rate of a series system is the sum of its component failure rates, while the expression for parallel system failure rate cannot be straightaway derived from component failure rates. Thus, the series system preserves the monotonicity property of failure rate function, while the parallel system does not.

16.12   Reliability Estimation, Demonstration and Assurance

In traditional analysis of quality, we focus on quality of conformance and check for conformity or otherwise of the quality characteristics or parameters of input items or work items in process or finished items, classifying the items as defective or non-defective or noting the number of defects in each item or just noting the parameter values. Considering the production or manufacturing process as a whole, we consider probability of an item being defective or probability that an item will contain a specified number of defects or the mean or standard deviation of a quality parameter as the parameter to be estimated. Most often, these estimates do not require any knowledge of the underlying probability distribution. For example, sample proportions are taken as estimates of probabilities and sample mean or sample standard deviation as the estimate of the corresponding population parameter. Of course, when we carry out a process capability study, we have to estimate the process capability indices by assuming usually a normal distribution for the quality characteristic. Sometimes, such estimates are derived by assuming some specific non-normal distribution also.

Assuming a sample of n units tested up to the mission time t_0 resulting in r failures and $n - r$ units still surviving (their time-to-failure not noted), we have an unbiased estimate of reliability $R(t_0)$ as $R^\wedge = r/n$. This estimate does not need any failure-time distribution to be assumed and its sampling variance is $R(1 - R)/n$. However, if the test is terminated at time $t_1 < t_0$ usually, and we assume a one-parameter exponential distribution for time-to-failure, then the estimate of $R(t_0)$ based on the number of failures r_1 is $(r_1/n)_0^t/t_1$ and its sampling variance is $R_1(1 - R_1)/n$ where $\log R_1 = (t_0/t_1)\log R$. This estimate is only asymptotically unbiased.

Estimates of reliability based on some assumed failure-time distribution are obtained by estimating parameters in the assumed distribution and putting the estimates in the expression of reliability as a function of the mission time and the parameters. If the sample is censored at time t_1 or after r failures have taken place, estimates of parameters will be derived accordingly. Among different methods of estimation available in the literature, method of maximum likelihood and method of moments are commonly used. However, with the Weibull distribution, maximum likelihood estimates cannot be explicitly obtained. Problems may also arise with censored data from other distributions. Sometimes order statistics and quantiles are also used to obtain simple estimates. Estimates of reliability derived this way are only asymptotically unbiased.

Traditional reliability demonstration plans are more or less similar to acceptance sampling plans use in the context of classical quality management, involving a sample size, a stated level of confidence, a certain level of reliability to be demonstrated and a certain probability model for the underlying variable. The simplest case could be to carry out a test for the specified mission time, noting for each unit put on test, whether it fails by the mission (test) time or not and determining a number n of units to corresponding to a specified reliability R (a binomial probability of success) to be demonstrated with a specified confidence $1 - \alpha$. (e.g. 0.99 or 0.95) This gives the number of units to be tested as $n = \log (1 - \alpha)/\log (1 - R)$. This number of units, usually pretty large, can be minimized by using an accelerated life test or by using an appropriate life distribution. As an illustration, to demonstrate a reliability of 0.99 for 20,000 cycles or operation with a confidence of 90%, we need 230 units. However, if we use a Weibull distribution with shape parameter $\beta = 2$, a zero-failure test that runs for $6.77 \times 20,000$ cycles will provide the required information with a sample size of only 5 units. Increasing the sample size will reduce the length of the test. And an increase in the number of units to be tested (and that way used up) as also in the length of the test (during which the units will have to be usually fed with energy) means cost and one has to work a cost-optimal plan in a detailed manner.

As in the context of quality, Quality Control does not suffice to beget customer satisfaction and we have to proceed to Quality Assurance in which records of corrective and preventive actions taken have to be preserved and some tests have to be repeated and some conformity to selected specifications may have to be demonstrated in an interface with the customer. In the context of reliability management, repeat life tests are usually ruled out and reliability assurance is done on the basis of a suitable reliability model, knowledge about reliabilities of components procured from vendors who are required to provide certificates of specified reliabilities, system configuration, anticipated environment in which the product will be required to function and the stress that would be imposed on the product as also results of other related tests and checks carried out within the supplier organization. A check on the quality of softwares used may also be needed in some situations.

16.13   Reliability Improvement

Improvement in the reliability of a component or of a system can be demonstrated through end-of-line or field tests, which usually take a long time, require large samples and equipment solely dedicated for the purpose. To reduce time, accelerated life tests are carried out in harsher conditions. Failure data come too late for any corrective action during manufacture or assembly.

Reliability (of an intermittent-duty or a continuous-duty equipment) during use is enhanced by the following actions

Redundancy (use of standby redundant components)
De-rating (reduction of load or stress—operational and/or environmental) and
Maintenance (inspection, repair and replacement of components or links).

The first—during Design—adds to cost, and to some other undesired consequences like extra space or weight, etc. the other two—before and during use—
lower the effectiveness of the equipment.

'Building In Reliability' is a proactive approach which places emphasis on

(1) identifying input or process parameters which influence the performance
    parameter(s) during use
(2) setting right targets and narrow tolerances for such parameters during the
    design process and
(2) controlling these parameters stringently during the Manufacturing/Production
    process.

The task is to work out the relation between the performance parameter(s) and
these critical design and process parameters from the results obtained through a
properly designed factorial experiment. Subsequently, we have to find out the
optimal combination of design parameter values or levels to maximize reliability.

One way to identify these parameters is to examine the degradation mechanism
for the product (during use), set up some threshold level for degradation or rate of
degradation that leads to failure, relate the threshold with levels of the design/
process parameters and then specify tolerances for such parameters to delay
failure-time or improve reliability.

Properly designed experiments with levels of factors which influence reliability
of the system/product, coupled with an exploration of the response surface, yields
the optimum design for the best attainable reliability. Responses in such experiments are time-to-failure (may be with a censoring plan) along with some other
functional requirement(s) which may enter by way of some constraint(s).

16.13.1 A Typical Case

Consider the following interesting example.

- Quite often performance parameters $Y_1, Y_2 \ldots Y_p$

Should satisfy some constraints which, without any loss of generality, can be
stated as

$Y_i \geq y_i$ $i = 1, 2, \ldots p$ and the product will be treated as having failed once any
of these constraints is not satisfied.

The life test could be terminated after a certain specified time, say t_0, usually
smaller than the mission time up to which the product is expected to function
satisfactorily.

- In the simplest case, the product either survives till t_0 or fails earlier when the
  only performance parameters drop below the specified value.

- In this case, we have to monitor the performance parameter(s) continuously and record time-to-failure as defined above.
- Alternatively, we simply record the time of functioning of the product till the censoring time t_0. After examining the dependence of survival time on the design parameters, we optimize the choice of parameters by maximizing reliability up to t_0 subject to constraints on the performance parameter(s). In fact, we should find out a suitable distribution of time-to-failure and fit a regression equation of the scale parameter on the design parameter. For this, we have to estimate the scale parameter of the fitted distribution for each design point based on its few replications.
- While the simple one-parameter exponential with the pdf. $f(x) = \lambda \exp(-\lambda x)$ is an obvious choice, it may be better to try out the more general Weibull distribution with pdf.

$$f(x; \alpha, \lambda) = \alpha \lambda x \alpha - 1 \exp(-\lambda x \alpha)$$

- In the latter case, it may not be unreasonable to assume that the shape parameter α remains the same across design points and we consider the dependence of the scale parameter λ only on the design points. We estimate α based on all the design points and values of λ for each point.

Reliability required for a device in water supply = 0.95 for

- Volume of 80 gallons delivered per hour
  (acceptable range 75–85 gallons)
- Survival time for at least 2000 h
  Design and manufacturing variables affecting volume
- Thickness (from 2 to 6 in.)
- Tuning time (from 10 to 50 min)
- Width (from 8 to 18 in.).

A central composite design with two levels (high and low) for each of these three factors was chosen to facilitate the exploration of a quadratic response surface.

Thickness u (5 and 3 in.)
Tuning time v (40 and 20 min) and
Width w (15 and 19 in.) used to note
(1) time-to-failure (t) and (2) gallons/hour (y)
in a test censored at 2000 cycles
Response function for y (gallons/hour) taken as

$$y = \alpha + \beta_1 u + \beta_2 v + \beta_3 w + \beta_{12}uv + \beta_{13}uw + \beta_{23}vw + \beta_{11}u^2 + \beta_{22}v^2 + \beta_{33}w^2 + \varepsilon$$

Apart from the 23 = 8 usual factor-level combinations (vertices of the cube spanned by the eight treatment combinations), six axial points with low and high α values taken as thickness (2.3, 5.7) tuning time (13.2, 46.8) and width (8.3,16.7) were each replicated twice, while six replications were made for the central point (4, 30,12.5).

In all, 34 runs of the experiment were carried out with 15 design points and the two responses noted for each run.

ANOVA table for 34 runs and the model chosen.

1 d.f. for each main effect U, V and W, each first-order interaction UV, UW and VW and each quadratic effect U^2, V^2 and W^2, plus 24 d.f. for residual (including 5 for lack of fit).

Data Set

Sl.	State (F/S)	Time to F/S	Thickness	Tuning time	Width	Volume
1	F	561	3	20	10	56
2	F	125	5	20	10	31
3	F	1406	3	40	10	53
4	F	278	5	40	10	31
5	F	814	3	20	15	75
6	F	383	5	20	15	93
7	S	2000	3	40	15	66
8	F	588	5	40	15	92
9	S	2000	2.32	30	12.5	4
10	F	505	5.68	30	12.5	54
11	F	183	4	13.20	12.5	56
12	F	1036	4	46.82	12.5	49
13	F	587	4	30	8.3	47
14	F	961	4	30	16.7	121
15	F	428	3	20	10	52
16	F	167	5	20	10	33
17	F	1608	3	40	10	51
18	F	319	5	40	10	26
19	F	1158	3	20	15	70
20	F	267	5	20	15	93
21	S	2000	3	40	15	70
22	F	1155	5	40	15	89
23	S	2000	2.32	30	12.5	46
24	F	239	5.68	30	12.5	45
25	F	283	4	13.18	12.5	56
26	F	1479	4	46.82	12.5	46

(continued)

(continued)

Sl.	State (F/S)	Time to F/S	Thickness	Tuning time	Width	Volume
27	F	298	4	30	8.30	50
28	F	650	4	30	16.7	115
29	F	726	4	30	12.5	53
30	F	1327	4	30	12.5	48
31	F	964	4	30	12.5	51
32	F	842	4	30	12.5	53
33	F	1185	4	30	12.5	46
34	F	1555	4	30	12.5	48

Apart from the $2^3 = 8$ usual factor-level combinations (vertices of the cube spanned by the 8 treatment combinations), six axial points with low and high α values taken as thickness (2.3, 5.7) tuning time (13.2, 46.8) and width (8.3, 16.7) were each replicated twice, while six replications were made for the central point (4, 30, 12.5). In all, 34 runs of the experiment were carried out with 15 design points and the two responses noted for each run.

The p values in the ANOVA table show that V, W, UW and W^2 are significant. Hence, the revised regression (with U and these 4) now comes out as

$$y = 469.8 - 55.1\,U - 0.2\,V - 56.7\,W + 4.4\,UW + 1.88W^2 + \varepsilon$$

Examining the response surfaces, the best settings for the three factors come out as

$$\text{Thickness} = 2\,\text{in.}, \text{Tuning time} \approx 39\,\text{min}$$
$$\text{Width} \approx 17\,\text{in.}$$

To improve reliability, we can assume some lifetime distribution and regress its relevant parameter which affects the response (time-to-failure) and is likely to vary from one design point to another on the design parameters, work out the optimum values of the design parameters from this regression and determine the optimized reliability which should be a function of the parameter estimated by the regression. In the next few sub-sections, we try the exponential, the Weibull and the log-normal probability model for time-to-failure and proceed to work out the maximum attainable reliability through this approach.

Exponential Failure Model

Suppose the time-to-failure t is assumed to have a one-parameter exponential distribution with density function (3.1). The mean failure time is given by $1/\lambda$, which is a decreasing function of λ. We therefore, model λ as a function of the three factors $(u,\ v,\ w)$, and the relationship is given by $\lambda = \exp(-\mu)$, where

$$\mu = b_0 + b_1 u + b_2 v + b_3 w + b_{11} u^2 + b_{22} v^2 + b_{33} w^2 + b_{12} uv + b_{13} uw + b_{23} vw.$$

Based on the data in Table 4.1, the log-likelihood function is obtained as

$$\text{Log } L = -\sum_{1\mu i} - \sum_{1 t i} \exp(-\mu_i) - (2000) \sum \exp(-\mu_i),$$

where

Σ_1 summation over all i for which $t_i < 2000$
Σ_2 summation over all i for which $t_i > 2000$.

For all calculations, we use the interior point algorithm in the software MATLAB.

The MLE of the regression of $\ln \lambda$ on the design parameters is obtained as

$$\ln \lambda^\wedge = -[0.8594 - 0.6178u + 3.0795v - 5.5910w - 3.5256u^2 - 0.0251v^2$$
$$+ 0.44021w^2 - 0.1155uv + 2.0817uw - 0.1780vw].$$

Corresponding to the optimum setting of the factors, as obtained from the revised regression of volume (y) on u, v and w, the predicted value of λ is 1.46041×10^{-19}, and the reliability at 2000 h is approximately 1.

The problem of improving reliability would be to maximize the reliability at 2000 h, namely $e^{-2000\lambda}$, with proper choice of the design parameters u, v and w, lying in their respective ranges given by $2 \leq u \leq 6$, $10 \leq v \leq 50$, $8 \leq w \leq 18$, subject to the volume (y) being at least 80 gallons per hour.

The optimum choices of the design parameters come out to be: thickness = 2.72 inches, tuning time = 35.91 min and width = 16.69 in., and the reliability at $T = 2000$ is approximately 1, with the volume 80 gallons/h.

Weibull Failure Model
Here we assume that t follows a Weibull distribution with pdf given by (3.2).

The mean time-to-failure, which is $(1/\lambda)^{1/\alpha} \Gamma(1 + 1/\alpha)$,, is clearly increasing in $1/\lambda$, or decreasing in λ. We, therefore, model λ as a function of the three factors (u, v, w), and the relationship is given by

$$\lambda = \exp(-\mu/\sigma),$$

where

$$\sigma = 1/\alpha,$$

$$\mu = a_0 + a_1 u + a_2 v + a_3 w + a_{11} u^2 + a_{22} v^2 + a_{33} w^2 + a_{12} uv + a_{13} uw + a_{23} vw.$$

The log-failure-time is then

$$\log_e t = \mu + \sigma \varepsilon,$$

where ε is the error term distributed with pdf.

$$g(\varepsilon) = \exp[\varepsilon - \exp(\varepsilon)], \, -\infty < \varepsilon < \infty.$$

Based on the data in Table 4.1, and the assumption that α remains the same across all design points, the log-likelihood function comes out to be

$$\text{Log } L = 34 \ln \alpha + (\alpha - 1) \sum_1 \ln(t_i) - \alpha \sum_1 \mu_i$$
$$- \sum_1 t_i^\alpha \exp(-\alpha \mu_i) - (2000)^\alpha \sum_2 \exp(-\alpha \mu_i)$$

where

Σ_1 summation over all i for which $t_i < 2000$
Σ_2 summation over all i for which $t_i > 2000$.

The MLE of α comes out to be $\hat{\alpha} = 4.525$, and the estimated regression of $\log_e \lambda$ on the covariates is obtained as

$$\ln \hat{\lambda} = -[0.0634 - 2.6743u + 0.9593v + 3.5069w - 0.1403u^2 - 0.0136v^2 + 0.1448w^2$$
$$- 0.0136uv + 0.1131uw + 0.0091vw]$$

The optimum setting of the factors obtained from the revised regression of volume (y) on u, v and w, gives the predicted value of λ as 5.30101×10^{-17}, and the reliability at 2000 h is 0.9552.

Here the problem of improving reliability would be to maximize the reliability at 2000 h, namely $e^{-\lambda(2000)} \alpha \approx 1$, with proper choice of the design parameters u, v and w, lying in their respective ranges given by $2 \le u \le 6$, $10 \le v \le 50$, $8 \le w \le 18$, subject to the volume (y) being at least 80 gallons per h.

The optimal choice of the design parameters are obtained as thickness = 2 in., tuning time = 34.063 min and width = 8.593 in., and the reliability at 2000 h is 1 with volume 80 gallons/h.

Log-normal Failure Model
Suppose the time-to-failure t is assumed to have a log-normal (μ, σ^2) distribution. The expected time-to-failure is then given by $e^{\mu + \sigma^2/2}$, which is an increasing function of the location parameter μ. Hence, assuming μ to depend on the design points while the shape parameter σ remains constant, we model μ as follows:

$$\mu = b_0 + b_1 u + b_2 v + b_3 w + b_{11} u^2 + b_{22} v^2 + b_{33} w^2 + b_{12} uv + b_{13} uw + b_{23} vw.$$

The log-likelihood function then comes out to be

$$\text{Log } L = -(m/2)\ln(2\pi) - (m/2)\ln(\sigma^2) - \sum_1 \ln(t_i) - (1/2\sigma^2)\Sigma_1 (t_i - \mu_i)^2$$
$$+ \sum_2 \ln \Phi\big((\ln(2000) - \mu_i)/\sqrt{\sigma^2}\big),$$

where

Σ_1 summation over all i for which $t_i < 2000$
Σ_2 summation over all i for which $t_i > 2000$
m   number of observations with $t_i < 2000$
$\Phi(\cdot)$ cumulative distribution function of a standard normal variate.

$$\Phi(t) = 1 - \Phi(t).$$

The MLE of the fitted regression of μ on the covariates comes out to be

$$\mu^\wedge = 0.0958 + 0.1818u + 0.2317v + 0.3376w + 0.4179u^2 - 0.0095v^2$$
$$- 0.2337w^2 + 0.4839uv + 0.6302uw - 0.1665vw,$$

and the MLE of σ is 3.4138. The optimum values of the covariates that maximize the reliability at $T = 2000$, subject to the constraint that the volume is at least 80 gallons/h, are estimated to be thickness = 10.92 in., tuning time = 29.9 min and width = 10.49 in. The corresponding estimate of the reliability at $T = 2000$ is approximately 1, with the volume 80 gallons/h.

One thus finds that optimum choices of design parameters vary from one underlying model linking the relevant parameter to the design parameter values to another, all of them varying from the settings yielded by the response surface analysis. Fortunately, however, different models yield optimal choices which all ensure an almost perfect reliability.

16.14   Concluding Remarks

Quality of a product (including service as something produced) that is required to carry out some physical function during its life is a combination of the quality of conformance and the quality of performance. And these two depend on the quality of design. Good quality of conformance is necessary but not sufficient. Features to ensure good quality of conformance as also good performance quality should be properly identified and incorporated in the design.

Reliability analysis draws heavily upon probability models as well as stochastic process models and has grown to be a subject on its own merit, as somewhat distant from quality assurance and management. Even software reliability analysis as one domain of reliability analysis has attracted a lot of attention from research workers. Reliability of structures, multi-state reliability, reliability taking account of multiple causes of failure, reliability estimation from autopsy data and a host of similar other problems continue to engage researchers from various disciplines including metallurgical engineers.

Just as Quality Cost Analysis has been an important concern of Quality Management, Warranty Analysis has also been a big issue with manufacturers and Warranty Analysis including problems of prediction has gained a lot of importance these days.

References

ATT. (1990). *Reliability by design*. Indianapolis, IN: ATT.

Barlow, R. E., & Proschan, F. (1975). *Statistical theory of reliability and life testing—Probability models*. New York: Holt, Rinehart and Winston.

Blischke, W. R., & Murthy, D. N. P. (1996). *Product warranty handbook*. New York: Marcel Dekker.

Condra, L. W. (1993). *Reliability improvement with design of experiments*. New York: Marcel Dekker.

Hahn, G. J., & Shapiro, S. S. (1967). *Statistical models in engineering*. New York: John Wiley.

Henley, E. J., & Kumamoto, H. (1981). *Reliability engineering and risk assessment*. New Jersey: Prentice hall.

Hoyland, A., & Rausand, M. (1994). *System reliability theory: Models and statistical methods*. New York: Wiley.

Kalbfleisch, J. D., & Prentice, R. L. (1980). *The statistical analysis of failure time data*. New York: Wiley.

Meeker, W. Q., & Escobar, L. A. (1998). *Statistical methods for reliability data*. New York: Wiley.

Montgomery, D. C. (2012). *Design and analysis of experiments*. New York: Wiley.

Nelson, W. (1982). *Applied life data analysis*. New York: Wiley.

Prendergast, J. (1996). Building in reliability Implementation and benefits *Journal of Quality and Reliability Management, 13*(5).

Reliasoft. (2014). Reliability engineering resources (155).

Viertl, R. (1988). *Statistical methods in accelerated life testing*. Vandenhoeck & Ruprecht.

Xie, M., Tan, K. C., & Goh, K. H. (1998). Fault tree reduction for reliability analysis and improvement, series on quality, reliability and engineering statistics. *World Scientific, 10*(4), 411–428.

Chapter 17
Statistics for Quality Management

17.1 Introduction

Planning before production, monitoring during production, evaluation at the end of production line and estimation of performance during use or deployment of any product or service delineate the ambit of Quality Management. Quality Planning—which has to be taken up along with Product or Service Planning—is an interdisciplinary activity wherein statistics (both as data and as a scientific method) has to play a crucial role in view of the uncertainties associated with most entities involved. Science, technology and innovation provide the hard inputs into this activity, and Statistics coupled with Information Technology is to enhance the contribution of each input, judged by its role in the overall 'quality' of the output taken in a broad sense. In the context of a concern for sustainability, this broad sense would remind us of the definition offered by Donkelaar a few decades back.

Scientific, technological and managerial aspects of quality are no longer confined to Statistical Quality Control (SQC), dating back to the days of W. A. Shewhart when control charts were introduced to monitor and control variations in product quality. Over the years, the concept and even the quantification of quality have vastly expanded to embrace a wide array of entities from designs and processes, through products and services, to environment and life. Quality control, quality assurance, quality improvement, Quality System and similar other phrases have come to be recognized distinctively in terms of corresponding implications and imperatives. Methods and techniques of quantitative analysis have been integrated with recent developments in Science and Technology, including those in behavioural and management or decision sciences.

Responding to the enormous diversity of industrial situations and recognizing the importance of statistical concepts, methods and techniques in all the function, viz. quality planning, quality assurance and quality improvement, a wide spectrum of statistical methods and techniques has found justifiable and useful applications in the arena of quality—from elementary summarizing measures to sophisticated

© Springer Nature Singapore Pte Ltd. 2019
S. P. Mukherjee, *Quality*, India Studies in Business and Economics,
https://doi.org/10.1007/978-981-13-1271-7_17

multivariate analysis and designs of multi-response experiments. Many such techniques and tools have been developed to meet needs of manufacturing and service industries and have found gainful applications in quality-conscious and forward-looking enterprises.

There have been significant researches in the broad area of Statistics for Quality Management over the last few decades, though essentially in terms of the application of relatively modern and occasionally sophisticated statistical methods and tools.

Quality Management—regarded as the totality of all activities bearing on quality—has witnessed a number of paradigm shifts over the last few decades. The following are some among the major shifts.

Focus shifted from products to processes.

Proactive process adjustments beyond reactive product rectifications recognized as essential.

Control over the entire production process (usually multi-stage) oriented to take care of dependence among stages deemed as a necessity.

Greater emphasis is placed on Planning (design) as well as experimentation than on execution.

Data-based decisions, data-dependent (sequential) plans and procedures are preferred.

Greater use of computers, artificial intelligence, robotics and expert systems is contemplated.

These shifts notwithstanding, statistical theory and methods have continued to remain effective instruments for defining, measuring, analysing, controlling and optimizing quality. Of course, applicable statistical methods and techniques have undergone appropriate ramifications and modifications, which have been influenced by applications of statistics in other branches of knowledge like psychometry, market research, control system analysis and which, in turn, have impacted on the use of statistics in hitherto unknown areas.

A lot of theoretical research has been done over the years to come up with more efficient and robust procedures for monitoring and controlling quality during production as also for improving quality of processes, products and services through on-line and off-line activities making ample use of relevant statistical techniques and thereby providing the desired quantitative support to modern Quality Management philosophy and practice. Some of these researches have incidentally created new knowledge in statistical methods, besides offering new applications of existing knowledge in subjects like probability and statistics, psychology and decision sciences.

In the present chapter, only contributions that have already found wide applications or have potentials for gainful applications in industry have been briefly discussed in terms of a few selected contributions—mostly in the area of on-line process control. Some references to contributions in other areas have been touched upon. Research workers will find the references a useful repository of knowledge and a base to work further.

17.2 Need for a New Paradigm

Professionals/practitioners find that imaginative applications of simple and conventional techniques of quantitative analysis continue to yield effective results in the area of Q&R, e.g. seven tools of SPC (old and new). On the other hand, researchers feel that new and sophisticated techniques become necessary to represent reality, e.g. renewal processes with reward structures (in the context of process control), discounted cash flow analysis (while analysing quality costs), prior distributions of incoming quality (in the study of acceptance sampling), chance constrained programming (to develop optimal response surface designs for experiments with multiple responses), Coxian distributions (to represent service times in the problems of queues, inventories and maintenance).

A simplistic study of SQC and reliability analysis that has characterized the traditional applications of statistics tends to generally overlook a few apparently trivial but really significant points, like:

1. A control chart without a set of rules for process adjustments (based on a process model and the plotted points) is not a tool for process control.
2. Estimates of process parameters derived from control charts should be regarded as 'conditional'.
3. A production process is generally multi-stage with the output quality any stage depending on output qualities at some preceding stage(s).
4. Acceptance sampling plans are mostly curtailed or semi-curtailed.
5. Performances of process and product control plans are better understood (particularly for the purpose of comparison) in terms of time or cost, rather than O.C. and ASN functions.
6. A combined analysis of both variable and attribute information (as is done in CLGS plans or Attri-Vari charts) leads to more effective decisions than an analysis based on either, exclusively.
7. Measurements on quality characteristics are often subject to errors or uncertainties.

Regarding control charts, hardly any distinction is made between charts used to test homogeneity among some initially collected samples or rational sub-groups and those meant for controlling current production. In the first case, the advantages and deficiencies of an (X, R) chart compared to ANOVA preceded by a suitable test for equality of variances are not discussed. The procedures for developing even three-sigma limits for the second case (which is the important purpose at hand) are not considered. Many textbooks refer to limits on an X-chart as $X = +R$ without caring for the fact that when the first sub-group has been collected and its mean and range calculated, there are no X and R values available. While the superiority of moving average and moving range charts and of the exponentially weighted moving average chart over the traditional X–R charts is casually covered in some syllabi. Details about various EWMA charts and their variants are not discussed. Attri-Vari charts do not find many takers.

We tend to forget that what is needed in industry is a process control plan (in terms of a triplet, viz. sample size, sampling frequency and control limits) so determined that the cost of detecting a specified shift in some process parameter(s) or a pattern of such shift per unit time is minimal, taking into account costs of both Type I and Type II errors. While there exist a plethora of papers on this subject, very few curricula include these designs for optimal process control plans. In fact, control charts should be looked upon as devices to detect change points in the underlying item quality distribution. Process capability indices and process potential indices are not properly differentiated; estimation of process variability from sub-groups of item quality measures or from large sample and subsequent estimation of even the simplest process capability index are not usually covered. In the area of industrial experiments needed to find out optimal levels of controllable factors affecting quality, adequate emphasis is not given to a discussion on fractional factorial experiments and also on response surface methodology. In most industrial situations, controllable factors correspond to continuous variables and we have to explore the top of the response surface based on the available responses to only a few treatment combinations by using various search algorithm.

Statistical techniques introduced in the first half of the last century have lost their efficacy and even relevance in the present-day context, when production processes in both manufacturing and service industries derive strengths from advances in technology with the consequent improvement in quality. New and better materials are now treated by advanced methods and subjected to automatic and efficient control mechanisms that have dispensed with human decisions and actions to ensure high levels of conformity with even stricter specifications, and we no longer use per cent defective or number of defects per unit as any metric for quality. We have replaced such terms by measures like parts per million or billion defective and defects per million opportunities, respectively.

All this calls for radical changes in process control techniques. It must be admitted that simplicity is a great virtue, and traditional tools were pretty simple for the operators and supervisors to apply those tools conveniently. In the same breath, we should also remember that opportunity costs for not using more efficient tools should not over-ride the simplicity impact. Coming to control charts introduced by Shewhart for on-line control of a manufacturing process, it is found that some of the statistics plotted traditionally as the sub-group quality measure are no longer relevant. And some of the rules to interpret points outside a control limit should also be modified to make better use of information available through process inspection.

In case of inspection of items by an attribute characteristic resulting in a dichotomy between defective and non-defective items, a control chart for number of defectives d in sub-groups of a constant size n_i, we may recall that the upper control limit (UCL) on the d-chart was $n\pi + 3\sqrt{[n\pi(1 - \pi)]}$ where π is the process fraction defective which can be interpreted as the probability that an item inspected randomly from the process will turn out to be defective. In the present-day context, process quality π is quite small, say 0.001 and a sub-group size of $n = 25$ should be in order. With this, the UCL works out as 0.505 approximately. Even in the case of $n = 50$ and $\pi = 0.005$, UCL comes to be smaller than 2.0. The implication is that once a sub-group contains a single defective item we get an out-of-limit point on the chart in the first case or when we get more than one such defective item, we get a similar point on the chart. The fallacy of using a d-chart in the first case is obvious. In such situations, one of the following three actions could be taken, viz. (1) increase sample size to make the UCL at least 2 so that an out-of-UCL point could be taken as a signal (2) introduce a stricter definition of a defective item so that π could increase and consequently UCL also could increase, e.g. if an item was earlier defined as defective if its breaking strength was less than a specified value v, and we now define the item as defective if its strength is less than $v_0 > v$ and (3) control some other quality characteristic related to the characteristic being currently controlled.

Option 1 is not always available, and we could possibly explore option 2 resembling an accelerated test or option 3. With the existing procedure, the sample size may have to be taken as a million or at least a thousand.

A similar problem will arise with the p chart. It will be better to use a run chart and consider the number of consecutive non-defective items between two defective items. This number will be having a geometric distribution with parameter $1/\pi$ which will ensure an UCL that makes sense.

The existing literature on control charts does not consider the problem of process adjustments based on a point out of a control limit. In fact, designs of economic control charts are based on the assumption that a process adjustment made on the basis of such a point takes the process back to its initial state of control, irrespective of the distance of this point from the central line. It is not improbable that a point out-of-control limit is a false indication, and if we make any adjustment based on its value, we may introduce an assignable variation in a process that has been in its initial state of control.

In the area of sampling inspection plans, we generally cover Dodge-Romig plans— single, double and multiple, using both LTPD as also AOQL criteria, referring to the tables prepared by Dodge and Romig. We also include variable sampling inspection plans—single and double, in cases where the population s.d. is known and where it is estimated by the sample s.d. In some universities, some ideas about continuous sampling plans CSP 1 and its modifications, viz. CSP 2 to CSP 5 and SP-A and SP-B proposed by Wald and Wolfowitz, are given. Rarely, do we refer 5 201 Parts 1 and 2. Our teaching does not conform to practice nor does it motivate students to aspects of such plans which can provide good food for research. Thus, we do not explain the curtailed and semi-curtailed plans, the Bayesian sampling plans, price-adjusting sampling plans and a whole host of other issues connected with design and implementation of sampling inspection plans.

17.3   Some Recent Developments

The ambit of statistical methods has been expanded to cover broadly all methods for quantitative analysis, with or without an established statistical foundation, e.g. FTA (fault tree analysis, FMEA (failure mode and effect analysis) and EMECA (failure effect and criticality analysis). Of course, FTA has been used in applied probability as a tool for estimation of small probabilities (of system failure).

Statistical Methods for Quality and Reliability was initially accepted in terms of their application value. Gradually, extensions, generalizations and refinements were being attempted to meet real-life applications, to become more efficient and even to stimulate further research. Some of these proposed by practitioners like the Mil Std. 105 or 414 found their statistical support later on through Markov chains and transition probabilities to justify the proposed switching rules. So is true of several procedures suggested by G. Taguchi for off-line and on-line Quality Control.

Developments in statistical methods and techniques bearing on quality reported in the literature can be broadly grouped as extensions, generalizations and refinements. As an illustration, it can be pointed out that the traditional control chart has been extended to multivariate set-ups (through Hotelling's T^2 chart or Andrew's Plot or through Harrington's or Mukherjee's use of dimensionless transforms), to sequences of dependent observations (using Markov Chain analysis) to non-normal situations (e.g. through Edgeworth expansions), to multi-stage processes taking care of inter-stage dependences (in terms of Zhang's cause-selecting control charts) and so on. The control chart has also been generalized as a scheme to monitor a production process and to exercise on-line control (through appropriate process adjustments). Thus, the control chart has came to be recognized as a stopping rule. The economic designs of process control plans has been a significant refinement in formulation and, more importantly, in evaluation of process control plans in terms of real-time costs and benefits.

Confining ourselves to developments which make of efficient use of not-so-sophisticated techniques but carry the potential for gainful applications, we will first discuss the cause-selecting control charts initiated by Zhang (1984) where we speak of local and overall quality in different steps of a multi-step production process. Conceptually quite different from multivariate control charts which should be and are applied to deal with several interdependent quality characteristics noted simultaneously at any one step, these charts take care of the sequence of interdependent characteristics along the production line.

Moving average and moving range charts are useful in situations where it takes considerable time to measure a certain quality characteristic and inspection results become available at a rate too allow to form rational sub-groups. Roberts outlined a procedure for generating geometric moving averages and showed that tests based on geometric moving averages compared favourably with multiple run tests and moving averages tests. The geometric moving average chart is particularly suitable when successive inspection results form an auto-regressive series. Ferrell suggested the use of median and mid-range charts using run-size sub-groups for controlling certain processes. The run-size sub-groups for controlling certain processes. The run-size sub-group is defined as a set of observations in one run-up, or run-down including both the end points.

The cumulative sum control charts and their variants have gained extensive popularity in the last two decades. These charts are based on a work done by Page in 1954. Fage asserted that the cumulative sum schemes are much more sensitive than the ordinary Shewhart control chart, specially to moderate deviations from the target. A comprehensive account of these charts was given by Johnson and Leone who showed that these charts could be developed from a pair of revered sequential probability ratio tests.

Recently, Bauer and Hackl have used moving sums of recursive residuals that have been used for sequentially testing the constancy of the process rolling. In the case of unknown variance, the latter is recursively estimated from observations prior to those which are actually used in the numerator of the moving sum. Hackl also discusses the use of non-sequential moving sum statistics. Non-sequential moving sums are superior in power to similarly defined moving sums of squares of residuals in case the processes level does not remain constant. Bauer and Hackl have subsequently demonstrated the use of sequentially applicable test procedures based on moving sums of squares of residuals has also been considered.

In recent years, considerable attention has been paid to economic designs of control charts. Generally, the sample size, the sampling interval and the control limit factors have been optimally determined for several types of control charts. A pioneering paper in the area of cost modelling of quality control systems is that of Girshick and Rubin. The economic criterion used by them is the expected income from the process. The optimal control rules depend on the solution to complex integral equations. Bather, Ross, Savage and White have investigated generalized formulations of the Girshick-Rubin model. Their results also are primarily of theoretical interest as they do not lead to simple process control plans. Most investigations on economic designs of conventional Shewhart control charts can be branded as semi-economic in that either the proposed model did not consider all relevant costs or no formal optimization techniques were applied to the cost functions. For example, Weiler suggested that for $\overline{X}$ chart the optimum sample size should minimize the total amount of inspection required to detect a specified shift. Similar semi-economic analyses have been carried out by Weiler, Aroian and Levine, Cowden, Barish and Hauser, Mukherjee, etc. Taylor suggests that sample size and sample frequency should be determined based on the posterior probability that the process is in an out-of-control state. Dynamic programming methods are utilized extensively in the development. Duncan proposed a fully economic model of a Shewhart $\overline{X}$ charts used to monitor a process in which the mean of the quality characteristic shows a linear trend over time. The case of multiple assignable variations has been considered by Duncan as well as Knappenberger and Grandage. Saniga has developed a model for the joint economic design of $\overline{X}$ and R charts. Several authors have investigated single assignable cause economic models of the fraction defective control charts. Mention may be made of the works done by Ladany, Ladany and Alperovitch, Chiu, Gibra, Montgomery and Heikes, etc.

Economic designs have also been developed for cumulative sum control charts by Taylor, Goel and Wu and Chiu. These relate to a single assignable cause situation. Montgomery and Klatt have developed the economic design of T^2 control chart in case of several quality characteristics. Harris extended this study to the multiple assignable cost case. Saniga and Shirland and Chiu and Wetherill report that very few practitioners have implemented these economic designs. Two reasons for the lack of implementation of this methodology have been cited by Montgomery. Firstly, the mathematical models and their associated optimization schemes are relatively complex—something that will be gradually taken care of by the availability of computer programmes. A second problem is the difficulty in estimating costs and other model parameters.

Multi-characteristic control charts have been discussed by Ghare and Torgerson (1968) for monitoring the central tendency of a number of measurable quality characteristics on one control chart. This chart is particularly effective when two or more characteristics are correlated. Mukherjee (1971) suggested a single control chart for a desirability index in such situations. Most of the work done on $\overline{X}$ charts assume that variance of the quality characteristic is stable and known. Little has been done to study the effects of unstable variability on the design of these charts. In the area of economic designs, the distribution of time the process remains in control is generally assumed to be exponential. Baker shows that the memoryless property of the exponential distribution may be significantly misleading in certain cases. The use of control charts for acceptance purpose is not yet fully developed. The blend of control chart techniques and acceptance sampling procedures is very desirable. The role of the computer for process control purposes is also not fully utilized. Quesenberry (1991) suggested SPC Q charts for start processes and short or long runs in the multivariate set-up. He suggested computing Q statistics defined as

$$Q_r = P^{-1}\{H_{r-2}[\sqrt{(r-1)}/\sqrt{r}(X_r - X_{r-1})/S_{r-1} \quad r = 3, 4, 5 \ldots$$

where P is the probability integral to the left under the standard normal distribution and H_r is the distribution function of the student's t-distribution with r degrees of freedom. The Q statistics are identically and independently distributed as normal with zero mean and unit s.d.

These Q charts give exact control over false alarm rates, when no prior data or estimates are available.

A somewhat related development has been the multivariate diagnostic scale often referred to as Mahalanobis-Taguchi strategy. (See Taguchi and Jugulum 2000, 2002). The use of this strategy to predict abnormal conditions or environments which can cause problems needing corrective action can be described as follows:

1. Select appropriate features and (multivariate) observations that are as uniform as possible yet distinguishable in terms of Mahalanobis D^2—statistic, representing normal behaviour, to form a Mahalanobis space to be regarded as the base or reference point of the scale. [Find out the useful set of variables using orthogonal arrays and signal-to-noise ratios.]
2. Identify conditions outside the Mahalanobis space and compute D^2 for these conditions/observations (from the centre of the Mahalanobis space) and check if they match with the decision-maker's judgement.
3. Calculate signal-to-noise ratios to determine the accuracy of the scale to detect departures from the reference or base.
4. Monitor the conditions using the scale and based on D^2 values take appropriate corrective action on the process.

Control by gauging has been an established practice in industry because of its operational convenience and economy. Numbers T_1 and T_3 of items (in a sub-group), respectively, exceeding and falling short of an upper and a lower gauge limit provide a natural basis for such a control procedure. The question of deciding on symmetrical gauge limits seems to have been first examined by Tippet and then, on his lines, more comprehensively by Stevens. Others contributing to the method include Mace, Bhattacharyya, Mukherjee and Das. In fact, the latter two have made the most detailed investigation into the problem of developing an optimum process control plan by gauging.

Most-if not all—production processes involve more than one step and successive steps through which inputs pass are not generally independent. In fact, if at each step a single quality characteristic is recorded, the value or level of this characteristic at a particular step is often likely to depend on the value or level of the characteristic(s) noted in the earlier step(s). Characteristics noted in different steps are expected to be different one from the other. In such multi-step processes, a Shewhart chart can be maintained at each individual step, but to interpret points on the different charts in terms of the entire process becomes difficult. It must be appreciated that a multivariate control chart like the Hotelling's T^2 chart does not provide a solution in the case of a multi-step single-variable situation. In a somewhat convoluted approach, one can—at least in theory—consider the variable characteristics noted in the different steps as following a joint distribution which can be assumed to be multivariate normal. However, in this case, one complete production run will correspond to one point on the chart and only when several such runs have been completed one can talk about a possible state of control or otherwise. Since any idea about process quality comes only at the end of a run, it becomes too late to detect any undesirable behaviour of the quality characteristic at any individual step. And this is what we need to control the process.

A different approach to this problem was proposed by Zhang (1984, 1992). Basic to Zhang's cause-selecting control charts is the concepts of overall quality and specific quality. Overall quality is that quality which is due to the current sub-process and any previous sub-processes. Specific quality is that quality which is due only to the current sub-process. Standard Shewhart charts are designed to discriminate between chance and assignable causes. The cause-selecting charts of Zhang are designed to further distinguish between controllable assignable causes and uncontrollable assignable causes. Controllable assignable causes are those assignable causes that affect the current sub-process but no previous sub-processes. The uncontrollable assignable causes are those assignable causes affecting previous sub-processes that cannot be controlled at the current sub-process level.

Let X represent the quality measurement of interest for the first step of the process, and let Y represent the quality measurement of interest for the second step. The observations X_i and Y_i are paired, perhaps being measurements on the same item of production. One of the most useful models is the simple linear regression model.

$$Y_i = \beta_0 + \beta_1 X_i + \varepsilon_i$$

where it is assumed that the ε_i's are independent $N(0, \sigma^2)$ random variables. It is usually assumed that the regressor variable is non-random, but that is not reasonable in this case. This problem is handled by treating the regression as conditional on the value of the regressor. The model used to construct the cause-selecting chart need not be linear.

The cause-selecting chart is a Shewhart or other type of control chart for the cause-selecting values, Z_i, where

$$Z_i = Y_i - \hat{Y}_i$$

are the residuals generated by the model used. The initial set of n observations is used both to estimate the relationship between X and Y and to establish control limits for future observations. The centre line for the Shewhart cause-selecting control chart is $\overline{Z}$ where

$$\overline{Z} = \frac{1}{n} \sum_{i=1}^{n} (Y_i - \hat{Y}_i).$$

An advantage of the cause-selecting chart over the Hotelling T^2 chart is that it is easier to determine which process is out of control. The Hotelling T^2 chart may indicate that the entire process is out of control but does not indicate which step of the process is out of control.

Another advantage of the cause-selecting approach is that it does not require a linear relationship between quality measurements on the processes. Use of the Hotelling T^2 chart assumes a multivariate normal distribution, which implies a linear relationship between two dependent quality measurements. Therefore, when the relationship is nonlinear as in the examples of Zhang (1984, 1992), the usual Hotelling T^2 chart would not be appropriate unless the variables could be transformed to achieve multivariate normality.

Wade and Woodall (1993) recommend using prediction interval endpoints as the control limits. Using prediction limits is reasonable since the cause-selecting charts involve using a data set to estimate the relationship between the two quality measurements and then calculating the control limits to use for future observations.

Several procedures have been suggested for monitoring multivariate process variability, starting with the sample covariance matrix as a sample point, e.g. using charts for the generalized variance, or vector-variance chart or charts by using some diagonalization of the covariance matrix. However, these need relatively large sample size to ensure a small probability of raising a false alarm. Djauhari et al. (2016) developed a chart based on the Cholesky decomposition of the covariance matrix that can be used even with a small sample size.

Control of finite length production processes has received increasing interest since many industries that implement the so-called just-in-time (JIT) philosophy try to lower inventories by reducing lot sizes or lengths of production run. Crowder (1992) addressed the problem of quality control in short production runs from an economic perspective. More recently, Weigand (1993) considered the case of optimal control in finite production runs. Castillo and Montgomery (1993) extended Ladany's (1976) process control plan to the case of an X-bar chart and worked out conditions under which Ladany's model gives better designs than Duncan type models (1956) using response surface designs. Ladany fixes the length of the cycle T—which he subsequently optimizes—and minimizes the total cost during a cycle. The objective function (in four variables) becomes more complicated, and a grid search algorithm is generally used to minimize it.

17.4   Statistics for Quality Assurance

The use of sampling inspection for quality appraisal and for reaching decisions like acceptance, rejection, rectification, devaluation, down-grading on lots of items was formalized by Dodge and Romig, nearly seven decades back. Since then major developments have taken place by way of new plans, new methods for determining plan parameters, modifications of existing plans for meeting exigencies of special applications, new approaches for deriving sampling plans, based on simultaneously on both variables and attributes, etc. Economic designs of sampling plans and Bayesian sampling plans have engaged a lot of attention from current research workers. More recently, Mukherjee revisited Dodge-Romig plans as solutions to problems in a game between the provider and the customer and to complex problems of constrained integer programming problems, noting that the sample size 'n' and the acceptance number 'c' are bound to be integral; the objective function, viz the expected number of items to be eventually inspected (in the acceptance-rectification case), is a nonlinear probabilistic step function and the constraint that the customer's risk has a specified value cannot be exactly satisfied.

It is seen that the operating characteristic curve, the prior distribution of process quality and the cost parameters are three important factors in the design of a sampling plan. Among the several schemes, the fully economic approach considers all these three elements. In case parameters in probability distributions are estimated in terms of sample data, we have the economic-statistical approach.

The fundamental problem in sampling inspection is to strike the appropriate balance between quality and costs. From this point of view, plans based on fixed points on the OC-curve should be applied with great care because it involves an arbitrary element and a subjective judgement, and presents more difficulty than one might think, except perhaps for experienced users in familiar situations. The economic incentive schemes are most suitable for consumers such as the Government or big industries who purchase a large quantity from the manufacturer. The schemes may not be very effective to the general buyer. The minimax schemes aim at minimizing costs without assuming the knowledge of a prior distribution. This suffers from the disadvantage that the value of fraction defective p corresponding to the minimax solution may well lie outside the range of values of p which actually occur.

The economic schemes based on Bayesian theory are more precise and scientific, leaving much less to judgement. However, the process curve and the loss functions will have to be specified, and it is sometimes difficult to obtain information about these quantities. Another argument against this approach is that is assumes the process curve to be stationary, in the long-term sense. Fortunately, work to date has indicted that the Bayesian scheme is robust to errors in the assumed process curve and loss functions, except the break-even quality. In many circumstances, a semi-economic design similar to that of Wetherill and Chiu (1974) is desirable because (1) it attains some specified protection, expressed in terms of a point on the OC-curve, (2) it minimizes costs subject to the above restriction, (3) it requires little prior information and (4) it involves the use of only one simple table.

There has not been so much work done on sampling by variables as on inspection by attributes. A method using frequency distributions in incoming inspection is proposed by Shainin (1950). This plan is called the lot plot method. Early schemes based on computed OC-curves and the normality assumption are presented by Bowker and Goode (1952). They give the OC-curves of many known σ and unknown σ plans. Lieberman and Resnikoff (1955) give a comprehensive account of inspection by variables with theory, tables and references. Their tables are reproduced in the MIL-STD-414 (1957). The plans in MIL-STD-414 incorporate a check on the process average, and a shift to tightened or reduced inspection when this is warranted.

Investigations of different aspects of variable acceptance sampling have started. Owen has studied the problem for the unknown sigma case by properties of the non-central t-distribution in a series of papers (see Owen 1967, 1969); Das and Mitra (1964) and Rossow (1972) has investigated the effect of non-normality for one-sided plans. The determination of the acceptance constant k based on sample range and mean range is further studied by Mitra and Subrmanya (1968).

The economic aspect of variable sampling is mainly approached in two directions. Stange (1964, 1966a) presents a minimax scheme. The paper by Grundy et al. (1969) discusses economic solutions for the sampling problem with a normal prior distribution. The batch quality is measured by the average, instead of the fraction defective. Related studies are further presented by Wetherill and Campling (1966) and Dayananda and Evans (1973).

A large number of sampling plans are designed for the 'curative' aspect of curing bad-quality production (when it arises) by preventing bad batches reaching the customer. These plans may provide, perhaps as a secondary function, an incentive to induce the manufacturer to improve his quality; but the effect of incentive is not specifically studied. The incentive aspect is often of great importance to the customer who operates the sampling plan.

An early paper that studies the effect of a sampling plan on the change of the process curve is presented by Whittle (1954). Whittle assumes that for a constant batch size the process mean $\bar{p}$ of the batches received depends on the sample size n by

$$\bar{p} = p_0 + p_a \exp(-cn)$$

Whittle uses this model to discuss the best allocation of a given total of inspection efforts over a number of different products.

The importance of incentive is explained by Hill (1960). In his paper, he discusses the requirements of an incentive scheme and indicates the lines upon which such a theory might be developed. In another paper, Hill (1962) discusses the features of a number of standard tables, with some reference to their incentive function based on a switching device of the inspection levels.

A number of standard attribute sampling inspection tables have been widely applied in industry. Among these are ASF (1944), SRG (1948), MIL-STD-105D (1963) and DEF-131A (1966). These tables index the plans by AQL and incorporate a scheme of changing between normal, tightened and reduced inspection.

Another mechanism of inducing incentive is to vary the unit price paid to the manufacturer for accepted lots by making it dependent on the number of defective items observed in the sample. The rationale leading to these price differential acceptance plans is that the consumer is willing to pay a higher price for a product with a lower fraction defective because of the savings he will realize through reduced losses. The producer, therefore, will find it more profitable to supply reasonably good quality. Durbin (1966), Roeloffs (1967) and Foster (1972) have written on this scheme. Incentive contracts based on price differential policies but not directly involving the applications of acceptance sampling are studied by Flehinger and Miller (1964) and Enzer and Dellinger (1968). Mohapatra and Banerjee as well as Dey extended Foster's idea of price—adjusting sampling plans.

The present theory treats the problems of sampling inspection as static, the only exception being the switching rules. In practice the prior distribution may cause the frequency of outliers to increase and there exists an interaction between the system of sampling inspection used and prior distribution. A dynamic theory with a feedback mechanism taking these factors and information from previous inspection results into account is needed.

Standard theory assumes all samples to be random. In practice, proportionate stratified sampling is often used, resulting in a somewhat steeper OC-curve. No serious attempt to utilize other types of samples (used in sampling theory)—stratified or cluster sampling—has been made as yet.

17.5   Statistics for Quality Improvement

Improvement in process and product quality involves technological, statistical and economic considerations and draws heavily upon human efforts. Among the important dimensions of quality discussed in Chap. 1, quality of design is the fundamental one and improving quality of design implies changes in the levels or values of features of materials and components on the one hand and operations and checks carried out on those on the other hand. Thus, improvement in design quality entails investigations like identifying various features (or factors) which are likely to affect process/product quality as suggested from domain knowledge, checking which of these features have significant effects on quality, working out the quantitative relation connecting the significant factors and quality, and determining the optimal combination(s) of factor levels that corresponds to the best quality (as assessed through some measure of the process or product under improvement). These steps except the first require data on the factors and corresponding quality measures through properly designed experiments.

Industrial experiments and their designs are somewhat different from experiments and experimental designs used in agriculture or in other fields. A sampling unit in an industrial experiment means a run of the experiment with an assigned set of feature levels and the response or quality parameter found at the end of the run. Most often, units are produced sequentially since the runs of the experiment take place in the same plant or the same processing facility. Costs of experimentation are non-negligible and hence the need for designs with minimum possible number of design points which can provide information on the main effects and some technologically important lower-order interactions of interest to the experimenter. This is why full factorial designs are avoided, and we take recourse to fractional factorial designs. Box-Behnken designs or Plackett–Burman designs are sometimes used. Orthogonal arrays along with linear graphs delineate the commonly used Taguchi designs. Taguchi designs are developed to achieve optimality along with robustness of the response. Central composite designs with varying number of replications for design points located at the vertices and in the centre of the factor space also find applications in industrial experiments.

Screening experiments focus on scooping out factors which affect the response significantly from the relatively large set of factors likely to impact the response are used to derive the regression, often taken to be quadratic, of the response variable on levels of the factors and to test the regression coefficients for their significance. Only the significantly affecting main effects and first-order interactions are subsequently used to find out some working knowledge about the optimum design point so that levels of the factors can be appropriately chosen for the full-scale experiment to result in the response surface. Optimum-seeking experiments where detailed information about the response surface is not a mandate can be conducted by using sequential designs to minimize the number of design points and hence the total cost.

It is true that designing an experiment that provides adequate information at the minimum possible cost requires some knowledge of statistics at an advanced level.

Similarly, dealing with multiple responses corresponding to different independent or correlated quality parameters invokes somewhat complicated optimization methods like holding some parameters at some pre-fixed desired levels and optimizing only one or even simultaneously optimizing all the response variables by taking advantage of goal programming approach. However, with easy access to design inventories or banks built up by different research organizations and the development of relevant softwares, applications of such designs have become a common practice in forward-looking industries today.

17.6   Statistics in Six-Sigma Approach

The Six Sigma approach has been adopted by organizations seeking to achieve high levels of customer satisfaction and business growth by solving business problems which in many cases can be related to some quality issues. A Six-Sigma organization is not just a 'tool-pusher'. At the same time, Six-Sigma implies a highly disciplined and quantitatively oriented approach to performance improvement. Quite naturally, several mathematical and statistical tools and techniques are involved in this approach. However, the nature of process analysis and optimization determines the types of such tools and techniques to be used in a particular project. Some project may need only some simple tools for data collection, presentation and interpretation. Some other project may call for the application of sophisticated and even not-so-well-known tools and techniques.

Since most of the variables—categorical or continuous—in process analysis and subsequent optimization are affected by a multiplicity of causes and are consequently unpredictable, they are random and hence the need to apply probability calculus and statistical methods for dealing with such variables. Broadly speaking, we need the following types of analysis and hence the corresponding tools and techniques.

Dependence Analysis—taken care of through Categorical Data Analysis, Correlation and Regression Studies. Depending on the situation, we may have to use nonlinear regression models or logistic regression models. In fact, screening of factors to be eventually included in the experiment can be based on tests of significance applied to the regression coefficients.

Analysis of Factor—Response Relations in terms of ANOVA (Analysis of variance) and ANCOVA (Analysis of Covariance) along with their multivariate generalizations. Subsequently, we have to express yield (observed response or some transform thereof) in terms of a regression with those factor variables which significantly effect yield. This provides us with the response surface which has to be explored to find out the optimal design point as closely as possible.

Optimality Analysis—in terms of Methods to reach the optimum point on the Response Surface or to find the best combination of levels of controllable factors.

Besides, quite a few other supportive analyses and some techniques for data collection and presentation including exploratory data analysis are also generally taken up in a Six-Sigma project.

Since the Six-Sigma approach talks of the defect rate m (called DPU or the number of defects per unit, noting that a defect corresponds to a non-conformity that begets customer dissatisfaction) and yield Y, i.e. proportion of defect-free units, we make use of the Poisson probability model to derive yield in terms of the defect rate as $Y = \exp(-m)$. For a multi-step process, the rolled throughput rate is obtained as the geometric mean of yield rates at the different steps. Of course, the normal probability model is used quite a lot in carrying out most of the analyses mentioned earlier.

Sometimes, data have to be collected from a continuous stream of incoming materials or in-process items or finished goods, and a sampling rate has to be worked out. This rate or fraction may even be allowed to depend on the results obtained sequentially.

Techniques like ANOVA are useful in many contexts, e.g. in establishing measurement system capability. In fact, to determine repeatability and repro-ducibility of measurements or test results, the ANOVA approach is widely adopted. Speaking of designed experiments where ANOVA and ANCOVA have to be used to analyse the resulting data, we need to define the experimental material and the experimental units as also the response variable(s) appropriate to the context. Somewhat sophisticated designs are needed in some cases to minimize effort or maximize efficiency in estimation of effects. Quite often we have to use fractional factorial experiments to minimize the number of design points. For optimum-seeking experiments, sequential experiments may be preferred. In some cases, mixture designs are the appropriate ones.

For locating the optimal combination of factor levels or the optimal design point from a study of the response surface, we need to know and use method of steepest ascent or some other gradient search method. The problem becomes quite com-plicated if we deal with more than one response variable.

Tools used in the context of hypothesis testing and estimation based on some parametric models or on a distribution-free approach have to be applied to evaluate and compare effects, to establish relations and to yield ideas about improvements. Dealing with attribute or categorical data, we need to adopt slightly different pro-cedures. And since normality is assumed in most estimation and testing procedures, it will be wise to test for this normality itself before proceeding further.

Incidentally, for this problem as well as for many others, there are no unique tests or estimation procedures, and one has to make a wise choice of any procedure in a given context. These days, many of these inference procedures take account of any prior information or knowledge about the underlying parameter(s) to make the procedures more effective. The desirability of going for such sophisticated proce-dure involving many choice problems like that of the prior distribution(s) has to carefully examined, since there would always remain the need for an early decision that is at least admissible.

17.7 Growing Need for Research

In most of the areas where research has been going on, one can identify scope for further refinements or greater robustness. Some of these areas are briefly covered in what follows.

The adjustment of a process, necessitated by an alarm signal from a control chart, requires an estimate of the current process mean. In this connection, one should not overlook the point that what are observed in such a case are values of the conditional variable.

$X* = \{X|(X > UCL)U(X < LCL)\}$ and X. Wiklund (1992) obtained the m.l.e. of process mean using a doubly truncated normal distribution and also suggested a few modified maximum likelihood estimates (m.l.e.'s) to avoid too small m.l.e.'s which might induce costs that are higher than the gain in income from a too small adjustments towards the target. ML estimation using other types of truncation or other forms of conditioning on X corresponding to other rules for raising alarms has not yet been studied.

Some recent studies reveal that performance of control charts in detecting deviations in process location or dispersion designed on the basis of estimating values of in-control process parameters based on sample data analysed in what is known as Phase I by potting, and interpreting the appropriate control depends a lot on the Phase I sample. And because of sampling variations associated with such estimates, control charts behave quite unpredictably in Phase II. A recent article by Zwetsloot and Woodall (2017) provides a comprehensive discussion on this issue.

The literature on SQC is literally flooded by papers/articles/notes on economic designs of process control plans, using various control charts, coupled with diverse adjustments along with their impacts on the underlying process, assuming different process models, introducing a variety of cost components and taking recourse to different approximations in optimizing the objective function. One thing runs in common—a constant sampling interval. A recent argument has been that the sampling intervals should depend on sample observations. Based on a particular sample, the next sampling interval should be shorter or longer depending on whether the current observations indicate some change in the process or not. The analysis of a VSI process control plan is obviously complicated, but the optimum VSI plan is expectedly better than the corresponding optimum with a fixed sampling interval. A lot remains to be done here. Secondly, magnitudes of shift in the process mean are generally taken into account through their effects on the process fraction defective and that way on the process loss. Rarely, a probability distribution has been associated with these magnitudes (Wiklund 1993). Economic designs are based on out-of-limit points as the only warnings, an exception being the work of Parkhideh and Parkhideh (1996) where flexible zone runs rule are used in addition.

By reducing lot sizes or lengths of production run, Crowder (1992) addressed the problem of quality control in short production runs from an economic perspective.

More recently, Weigand (1993) considered the case of optimal control in finite production runs. Castillo and Montgomery (M) extended Ladany's (1973) process control plan to the case of an X-chart and worked out conditions under which Ladany's model gives better designs than Duncan type models (1956) using response surface designs. Ladany fixes the length of the cycle T—which he subsequently optimizes—and minimizes the total cost during a cycle. The objective function (in four variables) becomes more complicated, and a grid search algorithm is generally used to minimize it.

To sharpen competitive analysis based on data collected through a Quality Function Deployment (QFD) exercise, most of which are ordinal and quite a few of which are obtained through Likert scaling, there have been attempts to use techniques like Theory of Ordered Preferences based on Similarity to Ideal Situation (TOPSIS) or Operational Competitive rating Analysis (OCRA). More recently, Multiple Moments along with bootstrap sampling has been applied. The issues involved attract many more possibilities, using statistical tools and softwares.

17.8 Concluding Remarks

Nearly ninety years back, elementary statistical tools were first introduced to control quality during manufacture in the Bell Telephone Laboratories. Since then, many developments have taken place in the domain of statistical methods and techniques which found useful applications in many fields of scientific and industrial investigations. The Second World War provided a big boost to the use of such methods and techniques to enhance efficiency in decisions and actions. The importance of scientific management of quality in manufacturing and subsequently in services was appreciated in developed and developing countries. Statistical methods and tools were introduced in many industries and problems of measuring, analysing, controlling and improving quality motivated development of new statistical methods and tools which could be integrated with developments in manufacturing technology and could be conveniently implemented through softwares. In fact, specialized softwares for SQC as well as TQM and related tasks appeared on the scene and greatly facilitated the application of even sophisticated statistical tools.

Researches continue on many intriguing problems in applying statistical methods and tools, for example those around the normality assumption of the underlying probability distribution or the assumption of independence among different quality characteristics or the assumption that every level of each factor in a factorial experiment is compatible with every level of some other factor or the assumption that in a mixture experiment the proportions need not satisfy any specific but natural constraints, etc. Solutions are still awaited in many such problems. The focus has shifted from on-line quality control to off-line quality improvement activities. However, even for on-line control purposes emphasis on multivariate data and economic efficiency of the change point detection mechanism has been growing.

References

Alwan, L. C. (2000). *Statistical process analysis.* New York: McGarw Hill.

Antleman, G. R. (1985). Insensitivity to non-optimal design theory. *Journal of the American Statistical Association, 60,* 584–601.

Aroian, L. A., & Levene, H. (1950). The effectiveness of quality control charts. *Journal of the American Statistical Association, 65*(252), 520–529.

Baker, K. R. (1971). Two process models in the economic design of an X-Chart. *AIIE Transactions, 3*(4), 257–263.

Barnard, G. K. (1959). Control charts and stochastic process. *Journal of the Royal Statistical Society: Series B, 21*(24), 239–271.

Basseville, M., & Nikiforov, I. V. (1993). *Detection of abrupt changes, theory and applications.* New Jersey: Prentice hall.

Bather, J. A. (1963). Control charts and the minimization of costs. *Journal of the Royal Statistical Society: Series B, 25*(1), 49–80.

Bissell, A. F. (1969). Cusum techniques for quality control. *Journal of the Royal Statistical Society, Series C, 18*(1), 1–30.

Bowker, A. H., & Goode, H. P. (1952). *Sampling inspection by variables.* NY: McGraw Hill.

Box, G. E. P., Hunter, W., & Hunter, J. S. (2005). *Statistics for experiments: Design, innovation and discovery.* Hoboken, New Jersey: Wiley.

Burr, I. W. (1969). Control chart for measurements with varying sample sizes. *Journal of Quality Technology, 1*(3), 163–167.

Burr, I. W. (1967). The effect of non-normality on constants for $\overline{X}$ and R Charts. *Industrial Quality Control, 11,* 563–569.

Castillo, E. D., & Aticharkatan, S. (1997). Economic-statistical design of X-bar charts under initially unknown process variance. *Economic Quality Control, 12,* 159–171.

Castillo, E. D., & Montgomery, D. C. (1993). Optimal design of control charts for monitoring short production runs. *Economic Quality Control, 8,* 225–240.

Chiu, W. K., & Wetherill, G. B. (1974). A simplified scheme for the economic design for $\overline{X}$ charts. *Journal of Quality Technology, 6*(2), 63–69.

Chung-How, Y., & Hillier, F. S. (1970). Mean and variance control chart limits based on a small number of subgroups. *Journal of Quality Technology, 2*(1), 9–16.

Collani, E. V. (1989). The economic design of control charts. Teubner Verlag.

Craig, C. C. (1969). The $\overline{X}$ and R chart and its competitors. *Journal of Quality Technology, 1*(2), 102–104.

Crowder, S. V. (1992). An SPC model for short production runs. *Technometrics, 34,* 64–73.

Das, N. G., & Mitra, S. K. (1964). The effect of non-normality on sampling inspection. *Sankhya, A,* 169–176.

Dayananda, R. A., & Evans, I. G. (1973). Bayesian Acceptance-Sampling Schemes for Two-Sided Tests of the Mean of a Normal Distribution of Known Variance. *Journal of the American Statistical Association, 68*(341), 131–136.

Del Castillo, E., & Montgomery, D. C. (1993). Optimal design of control charts for monitoring short production runs. *Economic Quality Control, 8,* 225–240.

Djauhari, M. A., et al. (2016). Monitoring multivariate process variability when sub-group size is small. *Quality Engineering, 28*(4), 429–440.

Does, R. J. M. M., Roes, K. C. B., & Trip, A. (1999). Statistical process control in industry. Netherlands: Kluwer Publishing

Donkelaar, P. V. (1978). Quality—A valid alternative to growth. EOQC Quality, 4.

Duncan, A. J. (1956a). The economic design of X-bar charts used to maintain current control of a process. *Journal of the American Statistical Association, 51,* 228–242.

Duncan, A. J. (1956b). The economic design of—Charts when there in a multiplicity of assignable causes. *Journal of the American Statistical Association, 66*(333), 107–121.

Duncan, A. J. (1956c). The economic design of X-charts used to maintain current control of a process. *Journal of the American Statistical Association, 51*(274), 228–242.

Durbin, E. P. (1966). Pricing Policies Contingent on Observed Product Quality. *Technometrics, 8* (1), 123–134.

Enzer, H., & Dellinger, D. C. (1968). On some economic concepts of multiple incentive contracting. *Naval Research Logistics Quarterly, 15*(4), 477–489.

Ewan, W. D. (1968). When and how to use on-sum charts. *Technometrics, 5*(1), 1–22.

Ferrell, E. B. (1964). A median, midrange chart using run-size subgroups. *Industrial Quality Control, 20*(10), 1–4.

Ferrell, E. B. (1958). Control charts for log-normal universes. *Industrial Quality Control, 15*(2), 4–6.

Foster, J. W. (1972). Price adjusted single sampling with indifference. *Journal of Quality Technology, 4,* 134–144.

Flehinger, B. J., & Miller, J. (1964). Incentive Contracts and Price Differential Acceptance Tests. *Journal of the American Statistical Association, 59*(305), 149–159.

Freund, R. A. (1960). A reconsideration of the variable control chart. *Industrial Quality Control, 16*(11), 35–41.

Freund, R. A. (1957). Acceptance control charts. *Industrial Quality Control, 14*(4), 13–23.

Freund, R. A. (1962). Graphical process control. *Industrial Quality Control, 28*(7), 15–22.

Ghare, P. M., & Torgerson, P. E. (1968). The multi-characteristic control chart. *The Journal of Industrial Engineering,* 269–272.

Ghosh, B. K., Reynolds, M. R., & Van Hiu, Y. (1981). Shewhart X-bar s with estimated variance. *Communications in Statistics, Theory and Methods, 18,* 1797–1822.

Gibra, I. N. (1971). Economically optimal determination of the parameters of X-control chart. *Management Science, 17*(9), 633–646.

Gibra, I. N. (1967). Optimal control of process subject to linear trends. *The Journal of Industrial Engineering,* 35–41.

Girshick, M. A., & Robin, H. (1952). A bayes approach to a quality control model. *Annals of Mathematical Statistics, 23,* 114–125.

Goel, A. L., & Wu, S. M. (1973). Economically optimal design of cusum charts. *Management Science, 19*(11), 1271–1282.

Goel, A. L., Jain, S. C., & Wu, S. M. (1988). An algorithm for the determination of the economic design of X-charts based on Duncan's Model. *Journal of the American Statistical Association, 63*(321), 304–320.

Goldsmith, P. L., & Withfield, H. (1961). Average runs lengths in cumulative chart quality control schemes. *Technometrics, 3,* 11–20.

Grundy, P. N., Healy, M. J. R., & Ross, D. H. (1969). Economic choice of the amount of experimentation. *Journal of the Royal Statistical Society B, 18,* 32–55.

Hald, A. (1981). *Statistical theory of sampling inspection by attributes.* Cambridge: Academic Press.

Hawkins, D. M., & Olwell, D. H. (1998). *Cumulative sum charts and charting for quality improvement.* Berlin: Springer Verlag.

Hill, I. D. (1960). The Economic Incentive Provided by Sampling Inspection. *Applied Statistics, 9*(2), 69.

Hill, I. D. (1962). Sampling Inspection and Defence Specification DEF-131. *Journal of the Royal Statistical Society. Series A (General), 125*(1), 31.

Hillier, F. S. (1969). $\bar{X}$- and R-charts control limits based on a small number of subgroups. *Journal of Quality Technology, 1*(1), 17–26.

Iqbal, Z., Grigg, N. P., & Govindaraju, K. (2017). Performing competitive analysis in QFD studies using state multipole moments and bootstrap sampling. *Quality Engineering, 29*(2), 311–321.

Jackson, J. E. (1956). Quality control methods for two related variables. *Industrial Quality Control, 12*(7), 4–8.

Johns, M. V., Jr., & Miller, R. G., Jr. (1963). Average renewal loss rates. *Annals of Mathematical Statistics, 34,* 396–401.

Johnson, N. L. (1966). Cumulative sum control charts and the Weibull Distribution. *Technometrics, 8*(3), 481–491.

Johnson, N. L., & Leone, F. C. (1962a). Cumulative sum control charts: Mathematical principles applied to construction and use, Part-12. *Industrial Quality Control, 18*(12), 15–21.

Johnson, N. L., & Leone, F. C. (1962b). Cumulative sum control charts: Mathematical principles applied to construction and use, Part II. *Industrial Quality Control, 19*(1), 29–36.

Johnson, N. L., & Leone, F. C. (1962c). Cumulative sum control charts: Mathematical principles applied to construction and use, Part III. *Industrial Quality Control, 19*(2), 22–28.

Kemp, K. W. (1961). The average run length of a cumulative sum chart when a V-Maak is used. *Journal of the Royal Statistical Society, 23,* 149–153.

King, S. P. (1959). The operating characteristics of the control chart for sample means. *Annals of Mathematical Statistics, 23,* 384–395.

Knappenberger, H. A., & Grandage, A. H. (1969). Minimum cost quality control tests. *AIIE Transactions, 1*(1), 24–32.

Kotz, S., & Lovelace, C. (1998). *Process capability Indices in theory and Practice.* London: Arnold Press.

Ladany, S. P., & Bedi, D. N. (1976). Selection of the optimal set-up policy. *Nabal Research Logistics Quarterly, 23,* 219–233.

Ladany, S. P. (1973). Optimal use of control charts for controlling current production. *Management Science, 19*(7), 763–772.

Lave, R. E. (1969). A Markov Model for quality control plan selection. *AIIE Transactions, 1*(2), 139–145.

Lieberman, G. J. (1965). Statistical process control and the impact of automatic process control. *Technometrics, 7*(3), 283–292.

Lieberman, G. J., & Resnikoff, G. J. (1955) Sampling Plans for Inspection by Variables. *Journal of the American Statistical Association, 50*(270), 457.

Mitra, S. K., & Subramanya, M. T. (1968). A robust property of the OC of binomial and Poisson sampling inspection plans. *Sankhya B, 30,* 335–342.

Montgomery, D. C. (2004). *Introduction to statistical quality control.* New Jersey: Wiley.

Montgomery, D. C., & Klatt, P. J. (1972). Economic design of T^2 control charts to maintain current control of a process. *Management Science, 19*(1), 76–89.

Moore, P. G. (1958). Some properties of runs in quality control procedures. *Biometrika, 45,* 89–95.

Mukherjee, S. P. (1971). Control of multiple quality characteristics. *I.S.Q.C. Bulletin, 13,* 11–16.

Mukherjee, S. P. (1976). Effects of process Adjustments based on control chart evidences. *IAPQR Transactions, 2,* 57–65.

Mukherjee, S. P., & Das, B. (1977). A process control plan based on exceedances. *IAPQR Transactions, 2,* 45–54.

Mukherjee, S. P., & Das, B. (1980). Control of process average by gauging I. *IAPQR Transactions, 5,* 9–25.

Nagendra, Y., & Rai, G. (1971). Optimum sample size and sampling interval for controlling the mean of non-normal variables. *Journal of the American Statistical Association,* 637–640.

Owen, D. B. (1969). Summary of recent work on variable acceptance sapling with emphasis on non-normality. *Technometrics, 11,* 631–637.

Owen, D. B. (1967). Variables sampling plans based on the normal distribution. *Technometrics, 9,* 417–423.

Page, E. S. (1962). A modified control chart with warning limits. *Biometrika, 49,* 171–176.

Page, E. S. (1964). Comparison of process inspection schemes. *Industrial Quality Control, 21*(5), 245–249.

Page, E. S. (1954). Continuous inspection schemes. *Biometrika, 41,* 100–115.

Page, E. S. (1963). Controlling the standard deviation by Cusums and Warning Lines. *Technometrics, 6,* 307–316.

Page, E. S. (1961). Cumulative sum charts. *Technometrics, 3*(1), 1–9.

Parkhideh, S., & Parkhideh, B. (1996). The economic design of a flexible zone $X\bar{A}^-$-chart with AT&T rules. *IIE Transactions, 28*(3), 261–266.

Quesenberry, C. P. (1991). SPC Q-charts for start-up processes and short or long runs. *Journal of Quality Technology, 23,* 213–224.

Quesenberry, C. P. (2001). The multivariate short-run snapshot Q chart. *Quality Engineering, 13*(4), 679–683.

Reynolds, J. H. (1971). The run sum control chart procedure. *Journal of Quality Technology, 3*(1), 23–27.

Roberts, S. W. (1966). A comparison of some control chart procedures. *Technometrics, 8*(3), 411–430.

Roberts, S. W. (1959). Control charts based on geometric moving averages. *Technometrics, 1*(3), 239–250.

Roeloffs, R. (1967). Acceptance sampling Plans with price differentials. *Journal of Industrial and Engineering, 18,* 96–100.

Rossow, B. (1972). Is it necessary to assume a Normal distribution in applying Sampling Schemes for variables? *Qualitat Und Zuverlassigkeit, 17,* 143.

Ryan, T. (2000). *Statistical methods for quality improvement.* New Jersey: Wiley.

Sarkar, P., & Meeker, W. Q. (1998). A Bayesian On-Line Change Detection algorithm with process monitoring applications. *Quality Engineering, 10*(3), 539–549.

Scheffe, H. (1949). Operating characteristics of average and range and range charts. *Industrial Quality Control, 5*(6), 13–18.

Schilling, E. G. (1982). Acceptance sampling in quality control. New York: Marcel Dekker

Shainin, D. (1950). The Hamilton Standard lot plot method of acceptance sampling by variables. *Industrial Quality Control, 7,* 15.

Stange, K. (1966). Optimal sequential sampling plans for known costs (but unknown distribution of defectives in the lot), Minimax Solution. *Unternehmensfurschung, 10,* 129–151.

Stange, K. (1964). Calculation of economic sampling plans for inspection by variables. *Metrika, 8,* 48–82.

Taguchi, G., & Jugulum, R. (2000). *The Mahalanobis-Taguchi strategy: A pattern technology.* New York: Wiley.

Taguchi, G., & Jugulum, R. (2002). New trends in multivariate diagnosis. *The Indian Journal of Statistics, Sankhya Series B, Part, 2,* 233–248.

Taylor, H. M. (1968). The economic design of cumulative sum control charts. *Technometrics, 10* (3), 479–488.

Tiago de Oliviera, J., & Littauer, S. B. (1965). Double limit and run control charts. *Revue de Statique Appliques, 13*(2).

Tiago de Oliviera, J., & Littauer, S. B. (1966). Double limit and run control chart techniques for economic use of control charts. *Revue do Statistique Appliques, 14*(3).

Truax, H. M. (1961). Cumulative sum charts and their application to the chemical industry. *Industrial Quality Control, 18*(6), 18–25.

Wade, M. R., & Woodall , W. H. (1993). A review and analysis of cause-selecting control charts. *Journal of Quality Technology, 25,* 161–168.

Wadsworth, H. M, Stephens, K. S, & Godfrey, A. B. (2002). *Modern methods for quality control and improvement.* Berlin: Springer Verlag.

Weigand, C. (1993). On the effect of SPC on production time. *Economic Quality Control, 8,* 23–61.

Wetherill, G. B., & Campling, G. E. G. (1966). The decision theory approach to sampling inspection. *Journal of the Royal Statistical Society. Series B (Methodological) 28,* 381–416.

Wetherill, G. B., & Chiu, W. K. (1974). A simplified attribute sampling scheme. *Applied Statistics 22*

Wheeler, D. J. (1995). *Advanced topics in statistical process control.* Knoxville: SPC Press.

Whittle, P. (1954). Optimum preventive sampling. *Journal of the Operational Research Society, 2,* 197.

Wiklund, S. J. (1992). Estimating the process mean when using control charts. *Economic Control Charts, 7,* 105–120.

Wiklund, S. J. (1993). Adjustment strategies when using Shewhart charts. *Economic Quality Control, 8,* 3–21.

Xie, M., Goh, T. N. & Kuralmani, V. (2002). *Statistical models and control charts for high quality processes.* New York: Kluwer Academic Publishers.

Zhang, G. (1984). Cause-selecting control charts: Theory and practice. Beijing: The People's Posts and Telecommunications Press.

Zwetsloot, I. M., & Woodall, W. H. (2017). A head-to-head comparative study of the conditional performance of control charts based on estimated parameters. *Quality Engineering, 29*(2), 244–253.

Zhang, G. X. (1992). *Cause–selecting control chart and diagnosis, theory and practice.* Denmark.

Index

© Springer Nature Singapore Pte Ltd. 2019

S. P. Mukherjee, *Quality*, India Studies in Business and Economics,

https://doi.org/10.1007/978-981-13-1271-7